CONSERVATION IN THE CONTEXT OF A CHANGING WORLD

Concepts, Strategies, and Evidence

Recent developments in ecological theory point the way to a stewardship approach that promotes biocultural diversity and ecosystem resilience. In addition, the escalating pace of anthropogenic environmental change makes it clear that conservation strategies which incorporate social as well as ecological dimensions are essential. This thoroughly updated version of *Conservation in the Context of a Changing World* covers a broader geographic, historical, and cultural scope that integrates material from the natural sciences, social sciences, and humanities. Contemporary and comprehensive, this book provides essential material for understanding tradeoffs between different options for resolving complex issues, including climate policy, the contrasting interests of different groups, the roles of Indigenous peoples, biopiracy, human–wildlife conflict, and new governance models such as co-management. Sources of evidence about the natural world and the roles of local and traditional people are emphasized. This is a vital resource for making informed decisions about controversial issues in conservation.

BERTIE J. WEDDELL is retired from Washington State University and former Principal and founder of Draba, a natural resource conservation business. She has been teaching, writing, and consulting about conservation for four decades. Dr. Weddell has provided expert testimony on wetland ecology regarding Tribal and federal water rights. She is the author of *Conserving Living Natural Resources in the Context of a Changing World* (Cambridge University Press, 2002).

CONSERVATION IN THE CONTEXT OF A CHANGING WORLD

Concepts, Strategies, and Evidence

BERTIE J. WEDDELL

Washington State University

CAMBRIDGE
UNIVERSITY PRESS

Shaftesbury Road, Cambridge CB2 8EA, United Kingdom

One Liberty Plaza, 20th Floor, New York, NY 10006, USA

477 Williamstown Road, Port Melbourne, VIC 3207, Australia

314–321, 3rd Floor, Plot 3, Splendor Forum, Jasola District Centre, New Delhi – 110025, India

103 Penang Road, #05–06/07, Visioncrest Commercial, Singapore 238467

Cambridge University Press is part of Cambridge University Press & Assessment,
a department of the University of Cambridge.

We share the University's mission to contribute to society through the pursuit of
education, learning and research at the highest international levels of excellence.

www.cambridge.org
Information on this title: www.cambridge.org/9781108986502

DOI: 10.1017/9781108985987

First published 2023

A catalogue record for this publication is available from the British Library

Library of Congress Cataloging-in-Publication Data
Names: Weddell, Bertie J., 1948– author.
Title: Conservation in the context of a changing world : concepts,
strategies, and evidence / Bertie J. Weddell, Washington State University.
Description: Cambridge, United Kingdom ; New York, NY : Cambridge
University Press, 2023. | Includes bibliographical references and index.
Identifiers: LCCN 2023017032 | ISBN 9781108986502 (paperback) |
ISBN 9781108985987 (ebook)
Subjects: LCSH: Ecosystem management. | Biodiversity conservation. |
Conservation of natural resources. | Conservation biology. | Natural
resources – Management.
Classification: LCC QH75 .W4198 2023 | DDC 333.95/16–dc23/eng/20230705
LC record available at https://lccn.loc.gov/2023017032

ISBN 978-1-108-98650-2 Paperback

Additional resources for this publication at www.cambridge.org/weddell.

In memory of Jim

Contents

Preface

Overview

Every society makes rules about who may use plants and animals and when, where, and how such uses are permitted. Even in societies that we think of as being guided by rationality and science, values and assumptions are central to interactions with the natural world. In this book, we will see how shifts in the dominant voices in conservation narratives have led to shifts in dominant strategies, from an emphasis on utilitarian values (Part I) to the protection of nature from people (Part II) to stewardship of resilient ecosystems and diverse cultures (Part III). We will evaluate the evidence for and assumptions behind different concepts that guide the management of living natural resources and ideas about what uses of those resources are considered legitimate. And, we will see how ideas about whether nature tends to be in a state of balance that people inevitably disrupt has profound implications for conservation.

The book is designed to promote an understanding of past and current challenges and opportunities in *conservation* – the formal or informal regulation of the uses of living natural resources – by considering how ecological context influenced the development of different approaches. It begins with a utilitarian approach to regulating harvests of featured species and then proceeds to preservationist strategies for protecting and restoring vulnerable populations and habitats with the goal of maintaining populations and communities of wild organisms. Finally, I end with a stewardship approach that seeks to conserve complex and resilient ecosystems which integrate the conservation of biodiversity with the well-being of diverse human communities.

I started from the premise that it is more useful for students to learn about the historical conditions that gave rise to different approaches to conservation and the strengths and limitations of each than to study a single approach as the best one. The book is organized into three parts, each of which describes the historical context and assumptions, the conceptual framework, the principal strategies, and the challenges and advantages of a particular approach to conservation. I hope that by showing how each strategy operates within a specific world view, has made important contributions, and has definite limitations, this volume will encourage readers to view science and conservation as ongoing processes rather than as static entities.

Part I shows how the academic disciplines of forestry, wildlife management, range management, and soil science developed in response to the unregulated exploitation of living natural resources after Europeans colonized the Western Hemisphere, Africa, parts of Asia, and *Oceania* (the central and southern Pacific Ocean). Professionals in these disciplines seek to regulate the exploitation of economically valuable resources such as timber, game species, forage for *livestock* (cattle, sheep, goats, and pigs), and soil so that supplies will continue to be available. To accomplish their objectives, utilitarian conservationists attempt to enhance populations and habitats that provide economic benefits and to reduce or eliminate processes and species that are deemed detrimental. They often focus on a small number of natural processes, such as density-dependent population growth and the development of stable plant communities. An underlying assumption is that managers can maximize the flow of useful products by controlling or compensating for forces that upset a natural world which tends to be in balance.

Part II covers efforts by preservationist conservationists to preserve natural places and living things regardless of their economic value. Some of the roots of this movement go back to nineteenth-century efforts in the American West to preserve wild spaces for their intrinsic beauty and spiritual value. A century later, awareness of environmental problems and accelerating losses of species led to a different goal for preservationist conservation: the protection of the diversity of life at all levels of organization, which came to be known as *biodiversity*. Whereas utilitarian conservation attempts to conserve natural resources *for* people, preservationist management typically seeks to protect those resources *from* people. Preservationist resource managers apply insights into processes such as extinction and colonization, and the consequences of population size and isolation, to this challenge. This approach assumes that protection from human use sustains species and ecosystems.

Part III investigates an alternative perspective – that nature is in a state of flux to which people are integral. This approach, which I call stewardship conservation, draws on insights from both utilitarian and preservationist conservation, but it suggests other ways of thinking about our place in nature and of managing ecosystems to sustain biological diversity and human well-being. It was fostered by a variety of practical, theoretical, and ethical considerations that highlight the need for an approach to conservation which emphasizes the variability and complexity of nature and encompasses the activities of people – including those who are usually left out of mainstream conversations about conservation – as part of the natural world. The underlying assumption here is that integrating the conservation of biological and cultural diversity with compatible resource uses can be sustainable.

The order in which I discuss these different approaches parallels the order in which I encountered them in my career. When I began studying wildlife biology at Washington State University in the mid-1970s, college courses and texts in conservation dealt primarily with a utilitarian approach, emphasizing regulated harvests of species that were hunted, trapped, or fished. There was also some discussion of the roots of two other themes, preservationist and stewardship approaches to conservation. Preservationist ideas were voiced in early controversies over the protection of natural areas like the park at Yosemite, as well as in some writers' fears of a timber famine. In addition, in the nineteenth century, several European and American scientists published studies of interactions between organisms and their environment, foreshadowing a more integrated approach to stewardship of complex ecosystems, although it was not until the 1950s that ecologists undertook many detailed studies of ecosystem functions. But initially the dominant approach was utilitarian conservation that emphasized maximizing economically valuable yields from the natural world.

About a decade after I started my studies, the Department of Game in my state was renamed the Department of Wildlife, a typical change for state agencies around the USA at that time. This reflected a shift in thinking by scientists and managers from an emphasis on game to recognition of the importance of rare and endangered species and the designation of reserves for their protection. At around the same time, management to maintain fundamental ecosystem processes and to include the activities of people in conservation began to get some traction.

By the turn of the twentieth century, I was writing and teaching about conservation and human rights. Players with a variety of perspectives and agendas joined conversations about what conservation should entail. Many of these players advocated conservation through *sustainable use* that integrates present needs without compromising the needs of future generations. This approach sees people as integral to the natural world, and it treats disturbances as important processes in dynamic ecosystems. Its proponents call for scenarios that protect biodiversity while addressing human well-being and cultural diversity.

The distinctions between the different approaches are not rigid. Many resource managers and conservation professionals hold views that are a combination of the different approaches covered in this book and practice strategies that synthesize elements of the different styles. But even though reality is complicated, and conservationists don't really fit into pigeonholes, understanding the different schools of thought that have influenced the theory and practice of conservation during the past century and a half can help us to put current challenges in context. If we understand the assumptions underlying various policies, we are in a better position to evaluate them. Organizing those ideas into three categories – utilitarian, preservationist, and stewardship – can help us do that.

The history of conservation in the Western world is like a musical score in which different instruments fade in and out over time, only to come to the fore later. Utilitarian conservation favors regulated use of wild plants and animals and the suppression of disturbances which remove those resources (Chapters 1–4). However, preservationist thinking, most visible in the resistance to the damming of the Tuolumne River in Yosemite Valley in 1901 and the founding of the Wilderness Society in 1937, provided a subtext. The preservationist theme grew louder in the latter half of the twentieth century, when many conservationists looked at the causes of extinction and rarity along with ways to address threats to biodiversity in general and to small, fragmented populations in particular (Chapters 5–8).

The utilitarian and preservationist approaches are really two extremes on a continuum. Nevertheless, many books present the philosophical context of conservation as a dichotomy between a human-centered viewpoint (*anthropocentrism*) and a life-centered viewpoint (*biocentrism*). Scholars in several disciplines have recently pointed out, however, that the utilitarian and the preservationist perspectives often share a similar philosophy: the idea that natural systems tend to be in a state of equilibrium that excludes people. At the same time, some scientists suggest that there is an alternative way of thinking about the natural world and our role in it.

Balance and Flux

Apart from the hostile influence of man, the organic and the inorganic world are ... bound together by such mutual relations and adaptations as secure, if not the absolute permanence and equilibrium of both, a long continuance of the established conditions of each at any given time and place, or at least, a very slow and gradual succession of changes in those conditions. But man is everywhere a disturbing agent. Wherever he plants his foot, the harmonies of nature are turned to discords. The proportions and accommodations which insured the stability of existing arrangements are overthrown.

(Marsh, 1874:34)

This statement by George Perkins Marsh – a nineteenth-century American diplomat, conservationist, and writer – expresses a concept that can be traced in Western thought as far back as ancient Greece: the idea that nature in the absence of human intervention is in a state of balance that changes little over long periods of time. In the nineteenth century, this view became a credo for conservationists. (People define "nature" in many ways, but in Western culture most definitions of nature involve the idea that nature is not significantly affected by people.)

The "balance of nature" figures so prominently in discussions about conservation that is worth looking at in more detail. In scientific formulations, balance – or

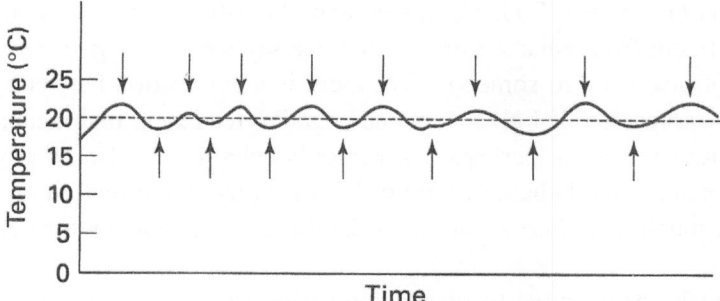

Figure P.1   Changes in the temperature of a hypothetical room regulated by a thermostat with a set point of 20°C (indicated by the dashed line). Arrows pointing down indicate points at which the thermostat turns the heater off. Arrows pointing up indicate points at which the thermostat switches the heater on.

equilibrium – is defined as a state in which there is no net change in a system. Suppose that 55% of a mature forest has not burned for 200 years. From time to time, fires burn some patches of this forest, converting them initially to open fields. At the same time, however, young forest patches age. Eventually they become mature. If the rate at which mature forest is created equals the rate at which other mature patches are destroyed, the amount of mature forest will remain constant, and the forest as a whole might be said to be in a state of equilibrium.

The idea of equilibrium is closely connected with the idea of self-regulation. If a self-regulated system is truly at equilibrium and something happens to cause it to deviate from that equilibrium, then we would expect to see compensatory changes that move the system back to its equilibrium state. A thermostat is a familiar example. If a room's temperature is regulated by a thermostat set at 20°C, then the thermostat should cause the heater to stop producing heat when the room becomes warmer than 20°C. When the heater shuts off, heat loss exceeds heat production and the room's temperature falls, restoring the temperature after a while to 20°C. With the heater off, the room continues to cool, until its temperature drops so far below 20°C that the thermostat causes the heater to turn on again, thereby initiating a compensatory production of heat designed to return the room's temperature to 20°C (Figure P.1). (Thermostats vary in the precision with which they do this. A very sensitive thermostat will turn off the heater when room temperature rises just slightly above 20°C; a less sensitive thermostat will not respond until the temperature rises several degrees. But regardless of whether temperature fluctuates a lot or a little, over time the thermostat should maintain temperature near a set level.)

This type of regulation, in which change in a variable in one direction sets in motion a compensatory change that causes the value of the variable to change in the opposite direction (thereby tending to return it to its original level), is termed *negative feedback* (*see QP.1*). The meaning of "negative feedback" in this context is quite different from what we mean when we say we give negative feedback, in the sense of criticism, to someone. But there is a connection between these two uses of the expression. If I give someone negative feedback, that means I am dissatisfied with something, perhaps that person's behavior. So I probably want the person to change their behavior. I am trying to reverse the direction of something (behavior), much as a thermostat alters the direction of temperature change in a room.

The opposite of negative feedback is *positive feedback*, in which change in a variable in one direction sets off further change in the same direction. Because positive feedback involves causing something that is changing to change even more in the same direction, it can result in sudden, explosive change. An atomic reaction is an example of positive feedback. Positive feedbacks are involved in exponential growth (Section 2.1.1.1) and some processes related to climate change (Section 10.3.6).

It is easy to conceive of a population that is regulated at a set level. This is termed the *carrying capacity* of its environment (Section 2.1.1.3). If such a population increases above that level, then there will be a decrease in reproduction and/or an increase in mortality until the population declines. If it drops below that level, then reproduction will increase or mortality will decrease, or both, allowing the population to grow until it reaches the carrying capacity. If there is not a long time lag between the changes in population size and compensatory adjustments, then this hypothetical population will remain fairly stable.

Understanding feedback is important. The consequences of not recognizing positive feedback can be serious.

For many decades, the prevailing scientific theories about populations and communities hinged on the idea of equilibrium. Marsh's concept of a harmoniously balanced natural world that people perturb permeates much scientific and popular writing. In this view, the nonhuman world is like a pendulum, characterized by a tendency to return predictably to its starting point. Ecologists grounded in equilibrium theories focus their attention on populations that are in equilibrium with their resource base, plant communities that return to an equilibrium or stable climax state after they are disturbed, and species assemblages in which rates of extinction and colonization are in balance. We will see, however, that many scientists now question the view of the natural world as tending toward equilibrium most of the time. The idea that people are outside of a balanced natural world has also been called into question for a variety of philosophical as well as scientific reasons.

The assumption that the natural world is in a balanced state if people are excluded can lead to two quite different strategies for management. Either we can try to protect that balanced state (a preservationist approach) or we can manipulate it

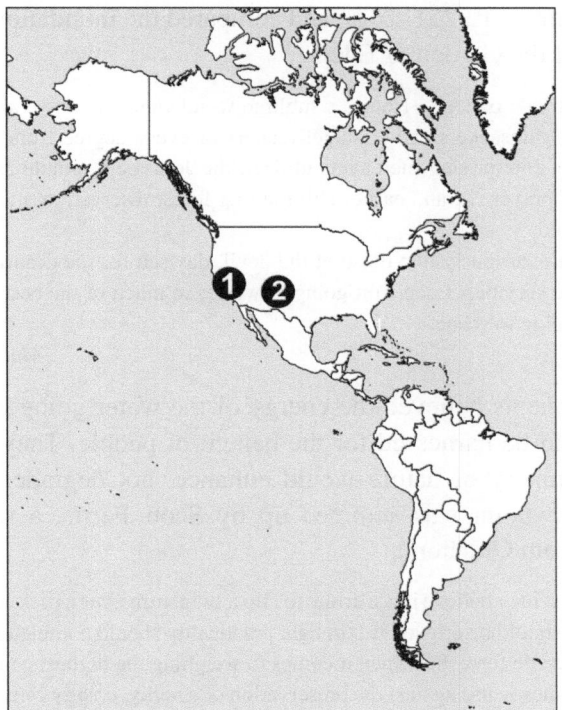

Figure P.2 Locations of: 1, Yosemite National Park, California; 2, Cerro Grande grasslands, New Mexico. Map created by Eva Strand using Esri, DeLorme World Countries Generalized Data & Maps for ArcGIS 2013, with permission.

with a utilitarian approach. In the words of the American ecologist Daniel Botkin, if nature is like a watch, we can "appreciate the beauty of the watch" or we can "attempt to take the watch apart and improve it" (Botkin, 1990:156). The examples below, from North America and Africa, illustrate these two approaches.

In North America, the roots of this controversy go back over 100 years. The debate over the building of Hetch Hetchy dam in the scenic Yosemite Valley (Figure P.2) was one of many conflicts in North America between those who wanted to preserve nature and those who wanted to use it. President Lincoln set aside the Yosemite Valley for public use in 1864. Initially, it was administered by the state of California. In 1901, San Francisco city officials proposed damming the Tuolumne River in the park to provide power and water for its residents. A bitter controversy ensued.

John Muir, a champion of wilderness and later the founder of the Sierra Club, argued that the sublimely beautiful canyon should be left in its natural state for people to appreciate. He described the falls as "harmonious and self-controlled," "without a trace of disorder – air, water and sunlight woven into stuff that spirits might wear" (Muir, 1912:251–252), and compared the inundation of the canyon to the destruction of the Garden of Eden:

Our magnificent National parks ... Nature's sublime wonderlands ... have always been subject to attack by despoiling gainseekers and mischief-makers of every degree from Satan to Senators.... Thus long ago a few enterprising merchants utilized the Jerusalem temple as a place of business instead of a place of prayer ...; and earlier still the first forest reservation, including only one tree, was likewise despoiled.

   Their arguments are curiously like those of the devil, devised for the destruction of the first garden – so much of the very best Eden fruit going to waste; so much of the best Tuolumne water and Tuolumne scenery going to waste.

(Muir, 1912:256,257,260)

The dam's proponents believed the energy of the water going over the falls was wasted and should be harnessed for the benefit of people. They argued that controlling the machinery of nature would enhance, not degrade, the value of the canyon. This viewpoint was summed up by Scott Ferris, a US Congressional Representative from Oklahoma:

These patriotic earnest men believe it is a crime to clip a twig, turn over a rock or in any way interfere with Nature's task. I should be grieved if I thought practicality should completely drive out of me my love of nature in its crude form, but when it comes to weighing the highest conservation, on the one hand, of water for domestic use against the preservation of a rocky, craggy canyon, allowing 200,000 gallons of water daily to run idly to the sea, doing no one any good, there is nothing that will appeal to the thoughtful brain of a commonsense, practical man.

(Quoted in Ise, 1961:92)

The argument that people are entitled to use nature's resources was put even more emphatically by Representative Martin Dies of Texas, who stated "God Almighty has located the resources of this country in such a form as that His children will not use them in disproportion," and implied that to utilize them was to follow "the laws of God Almighty" (quoted in Ise, 1961:92).

   In 1913, Congress passed a bill authorizing the project, and the valley was dammed. These two views of our relationship to nature might seem to be light years apart, but they have a lot in common. They are grounded in the same world view – the idea that people are separate from a natural world that tends toward a stable equilibrium. In one case, people are superior to nature and entitled to manipulate it, dominate it, control it, use it, and improve upon it. In the other, the natural world is pure and good, while people are morally tainted and alienated from nature but long to be reunited with it. In either case, we are outside of that which is in balance without us. In these views, humanity is either better than or worse than the nonhuman world but not part of it.

   The dam's supporters were utilitarian in their approach to resource management; they advocated the utilization of economically valuable natural resources.

The dam's opponents took a preservationist stance; they argued that conservation should involve the protection of natural places from exploitation by people.

It might at first seem odd that those who wished to dominate nature accepted the view that nature is in balance and people are outside of that balanced world. Yet if we return to the writings of George Perkins Marsh, who so clearly articulated the idea that nature is in balance, it becomes evident that this perspective is quite compatible with the idea that people are separate from nature and entitled to manipulate it. In Marsh's view, humanity was above nature, "of more exalted parentage" than "physical nature" and belonging to "a higher order of existence" (Marsh, 1874:34). Consequently,

man [and domesticated animals and plants] … cannot subsist and rise to the full development of their higher properties, unless brute and unconscious nature be effectually combated, and, in a great degree, vanquished by human art. Hence, a certain measure of transformation of terrestrial surface, of suppression of natural, and stimulation of artificially modified productivity becomes necessary.

(Marsh, 1874:37)

Marsh felt that by changing nature, people had "effected … changes which … resemble the exercise of a creative power" (Marsh, 1874:10, 37). Although he believed that civilization had gone too far in transforming the natural world, Marsh saw no contradiction between the idea that nature is harmonious and the idea that people should manipulate nature's harmonies.

If we turn our attention outside North America, we can again find examples of utilitarian and preservationist management plans that are rooted in the balance-of-nature perspective. Kenya's Tsavo National Park was set aside as a preserve by colonial authorities in 1948 (Figure P.3). At the time of its creation, most of the park was densely vegetated with trees, and the premier attraction was its elephant and rhinoceros populations (Sheldrick, 1973). When the park was formed, the Indigenous[1] people who had lived in the area were prevented from using the land for traditional hunting or livestock grazing. Wildlife viewing became the only permitted land use.

The colonial managers took a hands-off, let-nature-take-its-course approach, with the expectation that the park would continue to support trees, elephants, and rhinos. By the late 1950s, however, it had become clear that the elephants were destroying the trees and preventing their regeneration, and widespread elephant

---

[1] The terms *Indigenous peoples* and *Aborigines* refer to descendants of the earliest human occupants of an area. *Native Americans* (also referred to as *Amerindians* or Indians), *Alaska Natives*, *Inuit* (arctic peoples of Alaska, Canada, and Greenland), and *Native Hawaiians* are Indigenous peoples of the United States. The Canadian constitution recognizes three groups of Indigenous peoples: *First Nations* (Indians), *Métis* (some groups of Canadians and Americans with First Nations and European ancestry), and Inuit. Along with Indigenous people of Siberia, the Inuit used to be called Eskimos, but some people now consider that a derogatory term. Indigenous people of Australia and New Zealand are usually referred to as Aborigines or Aboriginals. There are thousands of groups of Indigenous peoples throughout the world.

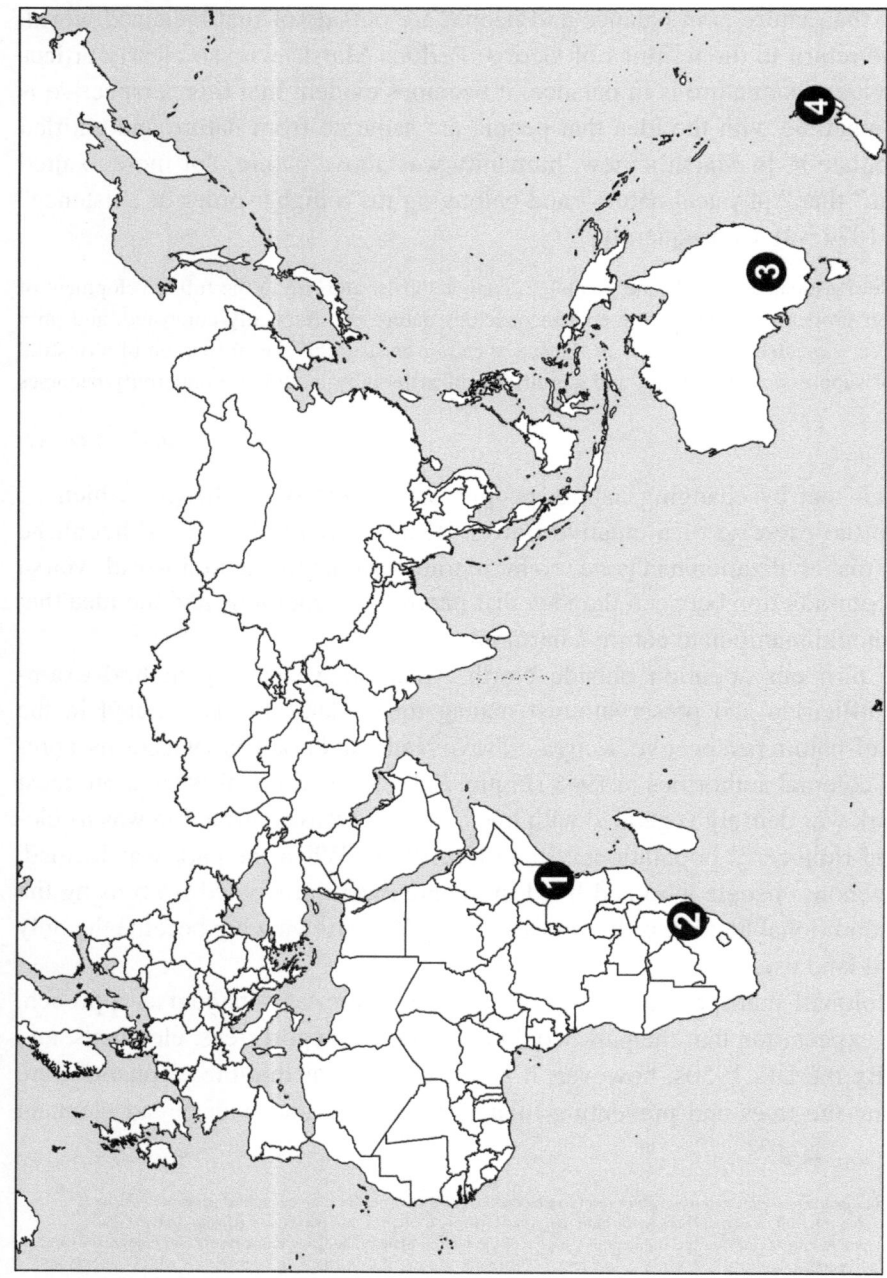

Figure P.3 Locations of: 1,Tsavo National Park, Kenya; 2, Kruger National Park, South Africa; 3, Yathong Nature Reserve, Australia; 4, Waikato, New Zealand. Map created by Eva Strand using Esri, DeLorme World Countries Generalized Data & Maps for ArcGIS 2013, with permission.

mortality seemed imminent because of this habitat degradation. The colonial wild-life researcher Ian Parker reported that "many visitors who saw the ravaged wood-lands were appalled. The vast expanses of dead and battered wood were likened repeatedly to the Somme battlefields of the First World War" (Parker and Amin, 1983:71).

The park's management argued that such die-offs were part of a natural cycle, and so they continued to follow a strategy of minimum intervention. Things got worse instead of better, however. The effects of habitat alteration were com-pounded by severe droughts in the 1960s and early 1970s, and as a result, thou-sands of rhinos and elephants died. After Kenya became independent in 1963, park managers continued the same approach.

By 1973, grassy areas had replaced woody vegetation throughout the park. Elephants, rhinos, and other *browsers* (species that feed on woody plants) had declined, whereas *grazers* (those that feed on grasses and grasslike plants) such as zebras and gazelles had increased markedly. Ironically, the policy of eliminat-ing people and letting nature take its course contributed to a dramatic alteration in the landscape and its wildlife, instead of perpetuating a stable community as managers had envisioned (Botkin, 1990, 2012; Rogers, 1999). (It is possible that other factors which changed at the same time had contributed to the elephants' decline. For instance, the restriction of their movements might have increased the potential for them to destroy their habitat.)

In South Africa's Kruger National Park (Figure P.3), managers pursued a dif-ferent strategy of preserve management. In a decidedly hands-on program, they intervened to control the balance of nature by culling lions, elephants, and *ungu-lates* (hoofed mammals), constructing wells, building dams, burning vegetation, and controlling diseases. Like the managers of Tsavo, the managers of Kruger tried to maintain the habitats and species that were prevalent at the time the park was set aside. A highly manipulative approach to conservation was used to keep a "natural" area in a desired state. The connection between equilibrium thinking and this type of management is less obvious than in the Tsavo example, but it is equally strong. In fact, the balance-of-nature viewpoint was explicitly accepted in the proclamation setting aside the area in 1898 (Rogers, 1999).

Tsavo and Kruger were managed in strikingly different ways, yet both these strategies were grounded in the idea that nature tends toward balance and stability. If both hands-on and hands-off management are rooted in the equilibrium view-point, one might ask if any other alternative is possible. But if we stop assuming that nature tends to be in balance, new possibilities emerge.

As a result of several high-profile controversies about resource manage-ment, most people are aware of the tension between preservationist and utili-tarian approaches to resource management, even if they do not use those terms to describe the situation. Unfortunately, the popular media have presented the debate as if these were the only two alternatives, posing questions like: do we want owls or jobs (Box 9.4)? This is not a helpful dichotomy.

If they had had a stewardship approach that assumes people are part of nature, managers might have been in a position to consider the effects of traditional hunters on the elephant populations. That approach is grounded in a different view of nature, one in which natural systems are often in a state of flux and people are an integral part of that flux.

What Next?

The utilitarian and preservationist approaches to conservation are appropriate under certain circumstances, but each also has limitations. For example, early game managers often set out to create habitats with a lot of edge because some game species make a lot of use of such habitats. This resulted in many small patches of forest interspersed with open country. In terms of the objectives of game managers, this was appropriate. Later, however, it became apparent that some species need forest interiors, and those species do not do well in highly fragmented landscapes. Most of them are not hunted species. Their needs were overlooked by early wildlife biologists, and now some of those species are threatened or endangered partly because of this management. If our objective is to maximize populations of game species, creating edges might be a good idea. But if we seek to preserve enough habitat to maintain viable populations of all species, maximizing the amount of edge is not a good way to do this. Changing objectives often lead to changing priorities and strategies.

Students may well ask "If managers no longer believe that maximizing edge is a good strategy, then why should I bother to learn about it?" There are several reasons why learning about outmoded ideas is important. First, there is no single correct way to conserve resources. The point is that managing for a high proportion of edge habitat has disadvantages but may be appropriate under certain circumstances, such as when the goal is to benefit edge-dependent species. Second, we have no special corner on the truth. To act as if we do is arrogant and only invites our successors to wonder how we could possibly have been so naive. Every generation in every cultural setting focuses on certain things and develops insight about those things. Likewise, every generation and culture has its own blinders and prejudices. Our predecessors did, and so do we. Third, it is important to understand how we got where we are. If we were short sighted in the past, can we learn from that myopia? Fourth, sometimes without realizing it, we hold on to ideas that are out of date. Some of the ideas that are discussed in this book are popular even though many scientists now question them. This is partly because journalists and teachers themselves do not always keep abreast of the latest developments in science, and they continue to use metaphors such as the balance of nature that no longer reflect the latest thinking. But it is also because old ways die hard. That may put us in a position where we are reasoning from contradictions that we do not recognize. For example, if we try to manage to sustain ecosystem processes, an approach that is based upon the flux-of-nature perspective, yet we continue to envision nature as tending toward stability, the contradictions in our approach could undermine our efforts. We might, for example, try to restore a plant community that we think of as a stable climax community without understanding that that climax might never have existed.

Finally, our heroes sometimes have feet of clay. John Muir and Teddy Roosevelt expressed racist attitudes that we find abhorrent today. Acknowledging this makes it possible for us to understand the context in which those attitudes developed and envision possibilities for change. Sometimes people say that the past is water under the bridge, meaning that we can't do much about it. But. as the historian Patricia Limerick points out, the past may be "water under the bridge," but "the river of time has not stopped flowing. The river continues to flow ..., and every moment presents an opportunity to find a fresh, and better, way of living in that flow of time" (Limerick, 2010:xi).

Now that I have tried to convince you of the importance of being aware of our assumptions and values, I should acknowledge my own. I am a white, female, American, retired ecologist. I grew up in a large city on the East Coast but have spent most of my adult life in the semi-rural West. I believe in science, and I tend to favor explanations that make sense within the framework of Western science. But I am aware that scientists have their blind spots, and I think scientists could generally benefit from a large dose of humility. I tend to be suspicious of conventional wisdom whether it is in science or politics. I am drawn to the natural world (my friends call me a "plant nerd"), but I also have a strong commitment to human rights and cultural diversity. I am more interested in analyzing environmental issues in terms of their historical context than in explanations that focus on human nature as a root cause of our problems.

Study Questions

To give you a chance to test your understanding of this material, a comprehensive list of study questions is available at www.cambridge.org/weddell. Answers are provided for some of the questions; others do not have a right or wrong answer and are designed to help you express your own opinions. Selected questions are referenced throughout this text.

Acknowledgments

In writing this book, I have been fortunate to receive help from people who were not only wise, smart, interesting, and competent but also a pleasure to work with.

For years, Linda Cook urged me to take this on. I am glad I finally listened.

I am grateful to Aleksandra Serocka at Cambridge University Press for her expertise, efficiency, guidance, and endless patience with my questions and delays as she shepherded me through the early stages of the project. In addition, I thank Jenny van der Meijden at Cambridge and Preethi Sekar and the copy editors at Lumina Datamatics. The editorial and technical assistance and the professional attention to detail they provided were indispensable.

Stuart Braman, Linda Cook, Michael Jennings, Lee Josephson, Mary Noland, Vivienne Solís Rivera, and Leah Wallach read parts of the manuscript and gave me valuable feedback. I am grateful to each of them for their comments.

Leah also kept me on track when I needed some prodding. Thank you Leah.

I owe a special debt to Edward Josephson, who read the whole thing and greeted each chapter as if it were a treat. With his keen legal mind, vast knowledge of history and the natural world, and probing questions, Ed consistently took me down interesting pathways by bringing up terrific examples and always asking how they came to be. He was the ideal critic.

I also thank Ina Gravitz, who helped me understand the process of indexing, and Diana Witt, who provided her professional expertise during the early phase of generating the index.

E. Kirsten Peters, Joan Rudd, and Leah Wallach shared their experiences with the process of writing. I value their insights about writing everything from newspaper columns to murder mysteries to memoirs to textbooks.

As always, I am grateful to the members of Washington State University's library staff. They invariably went the extra mile to do the myriad things that librarians do, ranging from helping me access material that I needed to dropping off books in a snowstorm. I am also grateful to the staff of the Plummer Public Library for being there when I needed Internet access while on vacation in north Idaho.

It has been a delight reconnecting with Emily Silver after many years. Her watercolor painting "A Curve in the Smith" was everything I had hoped for in a book cover.

I thank Eva Strand for the skill and ideas she brought to the task of creating maps showing the locations of examples discussed in the book.

Thanks to W. Frank Ray at Cactus Computer and Internet and Brian Augenstein at AugyTek for technical support.

Friends and family provided support, tea, zoom chats during the pandemic, good discussions, and company along the way. MaryAnn Boehmke, Nancy and Stuart Braman, Zoe Cooley, Beth DeWeese, Karen Gray, John and Tess Fields, Debra Gaw Josephson, Lee Josephson, Caroline Landrum, Judy Meuth, Katherine O'Rourke, Qiuping Peng, Julia Piaskowski, Lisa Polenberg, and Alfie Wishart shared my joys and frustrations.

Special appreciation goes to my late husband Jim Weddell, whose love, support, and companionship through the decades made me a better person. And who, even as his health failed, was always willing to offer suggestions for navigating the challenges I encountered. His partnership was invaluable.

And of course, Angie Weddell, Wes Weddell, Matt Lam, and Kat Bula were a superlative support team. Thank you for your patience, good cooking, understanding, and humor during hard times and good times.

It took a village to complete this book, and each of these people provided crucial assistance. Any mistakes, of course, are my own.

Introduction

Getting and Evaluating Information for Making Decisions about Conservation

Sources of Information about the Natural World

Getting information that will help us to make decisions about how to conserve the natural world involves piecing together information from many sources. Each kind of information has advantages and disadvantages. The principal types of information that are useful for scientific investigations of the natural world are considered below.

Controlled Experiments

One route to gaining information we need is through controlled experiments in which testable propositions, termed *hypotheses* (singular: *hypothesis*), are formulated, and information is gathered to test them. The investigator makes predictions about what will happen under certain circumstances if a particular hypothesis is true, and then determines whether those predictions are fulfilled. If the predictions are not fulfilled, the hypothesis is falsified.

To do a controlled experiment, scientists compare the responses of two or more identical groups to different treatments. If the responses to the treatments are accurately measured and the results are correctly recorded, then any differences in the responses of the different groups may be evidence that the treatment had an effect.

It is important, however, to be aware of potential *bias*, a consistent tendency to deviate in one direction from a true value. Bias occurs when one outcome or answer or piece of evidence is selected or encouraged over others. Biases might be caused by using faulty instruments or they might result if an observer allows their perceptions to be influenced by preconceived notions. Bias may be introduced deliberately, or it may be unintentionally caused by unacknowledged assumptions or by faulty sampling.

Even evidence that we think of as objective may be biased. For instance, we tend to regard photographs as objective records of reality. We say "seeing is believing" and "one photograph is worth a thousand words." But the photographer makes decisions, consciously or unconsciously, about what to place in the frame, how much light to let into the picture, and how sharp to make the focus. These influence our response to a photo. If a photographer of wolves selects only wolves that are attacking prey, focuses on bloody wounds they inflict, and doesn't allow a lot of light into their photos, that is likely to get an unfavorable response from viewers.

Because researchers at Michigan State University and the Michigan Department of Conservation suspected that nutrition affected the ability of fawns to survive their first winter, they tested the hypothesis that the amount of protein in the diet of white-tailed deer fawns during their first autumn would affect the rate at which they gained weight (Ullrey et al., 1967). The research team conducted a controlled experiment in which 45 captive white-tailed deer fawns were assigned at random to receive one of three diets for 98 days after they were weaned. The three diets differed only in the percentages of corn and soybean meal they contained, which resulted in three different levels of crude dietary protein. At the beginning of the trial and 70 and 98 days after the start of the trial, all fawns were weighed, and rates of weight gain were calculated. The results showed a statistically significant effect of dietary protein on rates of weight gain in both males and females, with fawns that received more dietary protein gaining weight faster than those that got the lowest level of protein.

The use of statistics allows investigators to evaluate rigorously any differences between experimental groups. The term "significant" has a specific meaning in statistics. The *statistically significant* effect of dietary protein on rates of weight gain meant that there was a low probability (which is stated as a calculated P level) that the observed differences were due to chance alone rather than to the different treatments. Sample size is taken into consideration when significance levels are calculated in order to account for the greater likelihood that differences in small groups will be due to chance alone.

Regardless of whether a controlled experiment gives the expected results, an experiment should be repeated, or *replicated*. This is especially useful if the repeat experiment is done by different investigators. If the different groups of investigators get the same results, that further bolsters the original conclusions. If not, then the researchers on each team should look for reasons for the discrepancy, perhaps formulating new hypotheses about what is going on and designing experiments to test them.

Getting unexpected results is not a bad thing in science. In fact, unexpected results are a major driver of scientific progress because they spur scientists to try to find out why things didn't turn out as expected. Because it provides a framework within which results can be compared to predictions, and conclusions can be modified as new evidence becomes available, science has the potential to be self-correcting.

Controlled experiments assume that the different treatments are the only difference between treatment groups. In the study of the effects of dietary protein on fawn weight gain, the three treatment groups were similar at the start of the experiment. They were the same approximate age, had been reared under the same conditions, and were weaned at approximately the same age. During the experiment, the three experimental groups were kept under identical conditions except for the different diets they received.

In studies of captive wildlife, ensuring that all experimental subjects are kept under the same conditions is relatively straightforward. That is a lot harder to do in field experiments. This is illustrated by an experiment about the effects of predation on prey populations in Australia. Drought is a major factor limiting populations of small mammals in Australia, but prey populations typically recover quickly after they collapse during a drought. Wildlife biologists at CSIRO, Australia's national science agency, designed a controlled experiment to test the hypothesis that predators could prevent populations of rabbits from rebounding after a crash due to drought (Newsome et al., 1989). To do this, the researchers killed cats and foxes on study plots at Yathong Nature Reserve (Figure P.3) that had low populations of rabbits, and then compared estimated rabbit populations before, during, and after the predators were removed from those plots to estimated rabbit populations on similar plots where predators had not been removed. Fourteen months after the start of the experiment, estimated rabbit populations had rebounded on the treated plots (the plots where predators were removed) but not on the plots from which predators had not been removed. The researchers concluded that predation on the rabbits had delayed recovery from the drought-induced population crash.

The concept of testing the effects of predators on prey populations by conducting removal experiments is simple, but putting it into practice is not. In this study, fencing to keep mammalian predators from moving onto the treatment plots was impractical, so predators were shot throughout the experiment. It would have been even more challenging to cover the study area with something that would keep out bird predators. Fortunately, earlier studies had suggested that birds of prey would have little effect on rabbits in this setting, so they were not considered in the experiment.

It is relatively easy to measure changes in captive animals, like the fawns in the experiment described above, if good instruments and observers are available, but obtaining reliable measurements in the field is more complex. The most thorough way to determine population size is to count all individuals in an area. However, it is rarely possible to do this in studies of wild animals. Consequently, it is usually necessary to estimate populations of animals by making an approximation that is based on a sample. Many sampling methods – each with advantages and disadvantages – are available. When it is not possible to observe individual animals directly, it may be necessary to rely on counting signs of activity, such as calls, feces, nests, or burrows and to use this information as an indicator (*index*) of abundance. This technique is only useful if the number of individuals that produce a given amount of sign is known. In the predator-removal study, researchers used counts of active burrow entrances to estimate rabbit populations, as well as a second method – counts of rabbits observed with a spotlight at night along sampling transects.

Sometimes when researchers cannot answer a question through field experiments, they can get part of the answer they need by doing experiments in artificial settings and extrapolating the results to field conditions. For instance, populations of the American black duck declined in eastern North America at around the time when *wetlands* (habitats between terrestrial and aquatic environments) in that region became more acidic due to acid precipitation (Section 5.2.1.5). If someone wanted to test the hypothesis that acidification of the wetlands in black duck habitat caused a decrease in the growth rate of ducklings, there would have been substantial obstacles to doing the sort of controlled field experiment that might answer this question because such an experiment would risk serious negative impacts to the experimentally acidified habitats and the ducks within them. Instead, Dr. Barnett Rattner and his colleagues at the US Fish and Wildlife Service constructed six artificial wetlands where they studied this question. Three of these wetlands were randomly selected to receive a treatment of sulfuric acid, and the other three were left as controls

(Rattner et al., 1987). A captive black duck hen with three or four ducklings was placed in each wetland for 10 days. Initially, the ducklings in all groups were of similar age and weight. At the end of the trial, the ducklings reared on the acidified wetlands weighed significantly less than those reared on the untreated wetlands. Despite the limitations on experiments to test the influence of acidification on black ducks, this study provided evidence of an effect. However, because the experimental system used captive animals and constructed wetlands, we cannot be certain that the results applied to the wild.

Well-designed controlled experiments shed light on many important questions. They are particularly useful when questions of policy (What level of pesticide application should be permitted? Should predators be killed? Should naturally started fires be put out? Should drainage of wetlands be allowed? What about mining in the Amazon?) are at stake, because in matters where we need to evaluate alternative courses of action, this methodology defines a standard for evidence. However, the design, methods, and underlying assumptions of scientific studies should always be carefully evaluated.

The principal advantage of controlled experiments is that they reduce the workings of the world to a collection of understandable variables that can be manipulated. Fortunately, there are statistical techniques that allow us to evaluate in a single study the effects of many variables and the interactions between them. The conceptual simplicity of controlled experiments is also a disadvantage, however. Ecological systems are complex. In nature species interact with each other and their environment in myriad ways that are not easily mimicked by controlled experiments.

In addition, there are many situations in which it is impossible or unethical to do controlled experiments. Fortunately, we can sometimes take advantage of natural experiments to get information that is useful for testing hypotheses.

Comparative Studies (Natural Experiments)

If we want to know the effects of potentially harmful treatments – such as pesticides, radiation, oil spills, or acid rain – on populations of wild organisms, there are substantial practical and moral obstacles in the way of doing controlled experiments like the ones described above. This is particularly true when we study past environments (where it is not possible to experiment) or rare or sensitive species and ecosystems (where it is not ethical to deliberately cause exposure to something harmful).

The *ozone layer* in our upper atmosphere reduces the amount of harmful ultraviolet radiation that reaches the Earth (Section 5.2.1.3). This layer became depleted during the 1970s, and the thinning of the ozone layer that resulted is thought to be responsible for a variety of ecological changes ranging from altered food chains to cataracts and blindness in some species. Controlled field experiments to evaluate the effects of reduced ozone are not possible, because there is only one Earth.

Fortunately, there are other ways of getting relevant information. One approach is the comparative study, in which conditions are compared in two or more situations that differ in place, time, or another variable but are alike in many other respects. For instance, we might compare similar events among closely related organisms or in similar habitats or in the same place at different times. When we want to assess the effects of inadvertent environmental perturbations, such as the thinning of the ozone layer, studies of conditions before and after the change are useful. If a pronounced change in the value of a variable – such as a decrease in the amount of ozone or an increase in the concentration of carbon dioxide (CO_2) in the atmosphere – is observed after a certain date, investigators can search for events that preceded and might have caused the change.

This method can shed some light on what might have caused thinning of the ozone layer, but it too has limitations. First of all, in most studies of this type, the early data were not gathered in the same way as more recent data. Second, there are likely to be multiple variables that changed during the period of interest, and this will complicate interpretation of the data. Third, there are usually time lags between a cause and its effect, but most often we don't know how long those lags are. Whatever caused ozone thinning might have begun changing a few years or a few hundred years before the resulting change in atmospheric composition was noticed.

Usually, we do not foresee the consequences of our actions, so we do not plan before-and-after comparisons ahead of time. If someone had suspected a hundred years ago that the ozone layer might wane, they could have tried to gather data on conditions before that happened (although they wouldn't have had the technology to do this very well) and compared it to data gathered subsequently. On the other hand, if someone had foreseen this change far in advance, perhaps people would have taken steps to prevent it or slow its course.

So, scientists are often left scrambling to conduct the first phase of an unplanned comparative study. To do this, they may scour historical records and earlier studies to glean information about prior conditions. This kind of information is very useful, and it underscores the importance of keeping accurate records because one never knows what use data will be put to in the future. But frequently, this type of information was gathered using methods that differ from the ones we would choose. Investigators might have to use data from many different sources, or data that were gathered using different methods and by workers with different degrees of expertise and training and different ideas about what is important. Since this type of information was rarely compiled with the questions that concern contemporary researchers in mind, often the relevant information was simply not recorded. Consequently, researchers have to make inferences from scraps of information.

As with controlled experiments, the results of comparative studies can be used to test hypotheses. We can state hypotheses, make predictions derived from our hypotheses, and then look at evidence from the past to find out if our predictions are correct. Care must be taken, however, in interpreting the results of this type of study. When trying to disentangle cause and effect in the past, we often face situations where several variables changed simultaneously. *Correlation*, the association of variables with each other, does not equal causation. Comparative studies may identify certain factors that occurred together in time, and statistical methods can be used to evaluate the significance of these associations (that is, the likelihood that they are due solely to chance), but this does not prove a causal relationship between them. There may be other factors that changed at the same time and that actually caused or at least contributed to the effects we are interested in. (In Box 2.2, we will encounter an example of an unplanned before-and-after study in which it was difficult to determine causation because multiple variables changed simultaneously.)

Models

Scientists often seek to understand the behavior of systems under conditions that cannot be observed directly. This may be because the system is too small (an atom) or too large (the Earth's atmosphere) to observe directly or because the phenomena of concern took place in the past or are still going on (climate change). In such situations, scientists often construct a *model*, a concrete or abstract representation of a system, that can be used to predict how a system behaves under specified conditions. A scientific model may take many forms: a physical structure, a description, an equation, an analogy, or a theoretical projection. A *simulation* is a type of model that predicts the changes a system undergoes given certain starting conditions and assumptions. Computers are very useful for this type of modeling because they allow researchers to manipulate many variables and to perform calculations rapidly under a wide range of scenarios (Section 10.3).

Models and simulations are used a great deal in conservation. If we want to predict what effects a proposed policy will have on habitats or populations, models are very useful. How long will the world's tropical forests last if we continue clearing them at current rates? What will the average summer temperature be in London in 2,050 CE if we cut our production of greenhouse gases in half? How long will it take for a population to become extinct if its current population trends continue? If we introduce six wolves into an area of suitable habitat, what size will the wolf population attain in 20 years? We cannot answer these questions directly, but we can measure responses under certain conditions and use this information to predict the outcome under other conditions. If our predictions are borne out, we can develop simulations to predict parts of the system in more detail. If not, we can revise our simulations in an effort to come up with better predictions.

Although models and simulations usually represent systems that are not amenable to experimentation, experiments may be useful for examining how certain parts of a system work. For example, you might wish to conduct experiments to test the responses of plants and animals to several treatments for cleaning up spilled oil. The information obtained from the experiments could then be used to modify your model of how long it takes for an ecosystem to recover from oil spills under different condition. A model's predictions should be repeatedly tested against reality, and the information generated in this way should be used to refine the model in order to make it more realistic.

Models are particularly useful where the risks of doing experimental studies are unacceptable, as in the case of research on rare organisms. Field studies inevitably involve a degree of disturbance to wild populations, while laboratory experiments require the removal of some individuals from the wild (and possible stress or mortality from handling). Both these outcomes should be avoided when dealing with sensitive populations. Models are one way to avoid these negative impacts.

Like other methods of getting information, simulations and other kinds of models have limitations as well as advantages. A model always incorporates certain assumptions about how a system behaves. We should therefore keep in mind the assumptions on which models are based. One reads a great deal these days about debates over models that predict global changes in environmental conditions, population growth, and resource availability. Much of the debate focuses on different assumptions about how the system in question behaves.

Scientists are influenced by their values and assumptions about the natural world, which guide their decisions about what to study and how to interpret their findings. These values and assumptions are derived from many sources including intuition, conviction, faith, ideology, experience, and other intangible states. In choosing what to study (in other words, what to focus on), the scientist is like the photographer choosing how to frame their shot. We need to be on the lookout for cases where models incorporate assumptions about how the world works that are at odds with the data that are used to construct the models (Botkin, 2012).

Models necessarily oversimplify the behavior of the systems they portray. However, a model that incorporates a lot of the important factors influencing a system and contains realistic assumptions about how the system changes is likely to do a good job of predicting that system's behavior. An oversimplified model with unrealistic assumptions will not.

Natural and Historical Records

Ecologists often look to the past to get information that will help them to understand the present or plan for the future. If we want to know how much the climate varied in the last 2 million years, how often grassland fires occurred in Australia before policies of fire suppression were instituted, where wetlands used to occur in China, what the former geographic range of the snow leopard was, or what the extinction rate of native mammals of Canada was before Europeans arrived, we must study the past.

Natural records include (but are not limited to) ice; soils or sediments; tree rings; fossils, pollen, and artifacts preserved in sedimentary rocks; packrat *middens* (piles of accumulated objects); and the tissues of long-lived individuals. Museums maintain collections of specimens such as skeletons, study skins, and dried plants. The value of such natural records depends on whether information we are interested in was preserved.

Processes that occur in pulses often produce layered records that are very useful. Periodically deposited sediments and rings or layers that result from variations in the growth rates of wood, bone, fish scales, or coral are examples. These form where alternating cold and warm seasons produce marked differences in the seasonal growth rates of living tissues. In temperate climates, trees produce distinct annual rings, hibernating mammals deposit bone, and fish scales have layers that are correlated with periods of growth.

The position of material in a sequence of layers may provide information on its relative age. By comparing growth rings in the trunks of individual trees that have overlapping life spans, scientists can date tree rings over periods that are longer than the lifespan of an individual tree. When this information is combined with the position of tree scars that resulted from fires which occurred at known dates, chronologies can be constructed that cover thousands of years.

Plants and animals are useful indicators of environmental conditions because every species has a specific range of environmental conditions it can tolerate. Evidence of muskrats or cattails in the past indicates that surface water was present, and the past presence of cacti indicates that there was a hot, dry environment. (This only works if we find records of them where they lived, not if they were transported somewhere else after they died.)

The time span covered by natural records ranges from years to millennia depending on the type of record. Fossils cannot distinguish a year or even a decade within the fossil record, but they can give us information about what was going on millions of years ago. Tree ring chronologies do not go back that far, but they can allow us to pinpoint the year when an event occurred.

Some natural records are more likely to be preserved than others. In other words, the samples passed down to us by natural records are biased. For example, packrat middens are found only in rocky terrain. The absence of packrat middens in sandy soil does not mean that packrats never lived there; it means only that if they lived there, their middens were not preserved. In addition, the record of the past that natural processes provide is often too short or too fragmented to tell us what we want to know. Or the record may be extensive but not provide information for the places and time periods we are interested in.

Historical records are made by people. Journals, maps, notes, photographs, genealogies, censuses, books, newspaper articles, interviews, sketches, paintings, legal transcripts, and recordings can be valuable sources of information about historical ecology. Repeat measurements over time, known as *time series data*, are useful for reconstructing historical trends and evaluating variability in those trends. Weather stations, stream gauges, astronomical observatories, and satellites record time series data.

Documents are an inexpensive, easy-to-use source of information about the past. However, the value of historical documents depends on their condition and whether the information that is preserved is representative. Such documents provide valuable windows to the past, but the viewpoint of the observer must be taken into consideration. The decision about what to record is always subjective, and historical documents reflect the recorder's assumptions about what was important. Because *anecdotes* (personal accounts) recorded in historical documents present the specifics of a particular time and place, they represent a small sample, but the details they capture are useful for understanding the larger context in which events occurred.

The usefulness of historical documents also depends on the accuracy of the recorded information, which in turn depends on the observational skills, memory, meticulousness, and honesty of the person who recorded the information and also on the technical capabilities of the equipment used. The time span covered by historical documents is relatively short (usually decades or centuries), but such documents often allow us to pinpoint when events occurred to the nearest month, week, day, and sometimes even hour.

Scientists have combined natural and historical records to reconstruct changes in vegetation in the Cerro Grande grasslands in northern New Mexico during the twentieth century (Figure P.2). Tree ring chronologies going back to the year 1480 and repeat aerial photographs between 1935 and 1979 showed that forest cover expanded and grassland shrank in that region during the twentieth century. Researchers used this data along with weather records and records of changes in grazing and fire management to evaluate the causes of this shift in vegetation. Subsequently, managers with the US National Park Service made use of this information when they planned nearby restoration programs (Swetnam et al., 1999).

Oral Traditions

In many cultures information about the past is preserved in oral traditions such as narratives, songs, poems, or sayings that are passed down through generations. These sometimes describe events such as volcanic eruptions that occurred thousands of years ago, or they may transmit information about ecological relationships and insights about the effects of management. Oral traditions also embody traditional knowledge, attitudes, and insights about phenomena in the natural world.

The Indigenous Maori people in the Waikato region of New Zealand (Figure P.3) have at least 19 ancestral sayings that pertain to New Zealand flax, or *harakeke*, a culturally important plant of freshwater wetlands and coastal habitats. Some of these sayings relate to ecological relationships of the flax. One expresses the relationship between flax and the *kākā*, an endangered, nectar-eating parrot that pollinates it: "Your flax bush … has nurtured the fledgling, and the full-grown *kākā*." Others describe the environmental conditions that favor the growth of flax ("When the flax plants are plentiful, it is a sign of much rain"; "the flax is nourished by the dead leaves that fall around its base") or provide instructions for management ("Clear away the overgrowth so that the flax will put forth many young shoots"). These sayings provide information useful for restoration of wetland ecosystems impacted by drainage, invasions of non-native species, and fragmentation (Wehi, 2009).

Data Recorded by People without Formal Training in Science

Not everyone who contributes to scientific endeavors has professional training in science. Citizen scientists and parataxonomists are two examples. *Parataxonomists* (Section 12.4.2) – local, often Indigenous, people who are experts in identifying and classifying local flora and fauna on the basis of their observations and experience rather than academic training – are sometimes employed to collect, identify, and preserve specimens for further study by Western scientists. *Citizen scientists* are voluntary amateurs who participate in the collection of information that is integrated into a database for use in large, scientific studies. Participants contribute data about phenomena such as the identity, locations, and abundance of animals or plants, *phenology* (climate-related phenomena such as flowering or migration), measurements of water chemistry, or astronomical observations from around the world.

In other contexts, Indigenous peoples have assembled data that challenged scientists' conclusions about wildlife abundance (Section 12.6). Data produced by Indigenous mapping projects as part of territorial claims have also provided baseline information for assessing ecological change (Nietschmann, 1994).

These different kinds of information about the natural world are not mutually exclusive. They can be used to complement one another, with each method suggesting fruitful areas of inquiry that can be pursued using other tools. Regardless of which tools we use to study the natural world, information should be evaluated carefully.

Evaluating Information about the Natural World

If something is presented as a scientific fact, ask yourself, is it science? There are many ways of doing science and many areas of scientific study. But in spite of this variability, science has some core characteristics. Science should involve observation, making predictions that are based on evidence, testing those predictions, critically analyzing results, and revising conclusions when reality doesn't conform to expectations.

It is also important to question the way information is presented. Whenever we encounter material that is presented as evidence, we should ask the following questions regardless of whether or not we agree with the information:

- What are the main points of this work?
- What kinds of evidence are used to support the authors' arguments (anecdotal, descriptive, comparative, experimental, written, oral)?
  - Was the sample size adequate?
  - If the evidence comes from an experiment, were there good controls?
  - Were all relevant factors considered?
  - If the evidence comes from historical documents, what factors might have colored which information was recorded and how it was presented?
- Does the evidence that is presented support the authors' conclusions?
- What are the authors' assumptions?
- Is the evidence that is presented consistent with your understanding of the subject?
- Who are the authors?
  - Do they have any professional credentials?
  - If not, do they have other qualifications?
  - Are they from a group that has historically been denied access to conventional communication outlets?
    - Although professional credentials generally reflect expertise, it is unwise to assume that someone who is well known should always be believed or that unknown sources are never reliable. Evaluating information requires judgments about when to accept material that is presented and when to question it.
- How is the information made available? Is it published in a professional journal?
  - Most scholarly publications go through a process termed *peer review* in which submitted articles are evaluated by others in the same field before being accepted for publication. *Gray literature* is research material that has not been through peer review. Many government agencies and *non-governmental organizations* (NGOs) publish gray literature.
  - Is this material in a publication that is trying to push a particular viewpoint?

- ○ Is it in a publication or from a website that is trying to sensationalize its subject?
- ○ Is it on social media?
- Do the authors have an interest in advocating a particular policy or theory?
- How well do the data support the authors' conclusions?
  - ○ Does this work contain contradictions?
  - ○ Are the authors' arguments logical and consistent?
  - ○ Are there other possible interpretations of the information that is presented?
- What types of material do the authors use to support their point of view?
  - ○ Are sources of additional information provided?
  - ○ If so, are those sources reliable?
- Do the authors discuss any evidence that does not support their conclusions?
- Do the authors consider any alternative explanations of their data?
- Since this work was written or posted, has new information come to light that is relevant to the subject or that suggests other possible interpretations?
- What additional research could shed light on the topics that are discussed?
- How do the authors convey their point of view?
  - ○ Is information presented in a misleading way?
  - ○ Do the authors use emotional or sensational language or photos to try to influence their readers?
  - ○ Do the authors use disrespectful language to discredit those with whom they disagree?
  - ○ Do the authors use unfair, irrelevant tactics to demonize their opponents? Some common examples of such cheap shots are:
    - · "Everyone who is anyone knows that …"
    - · "Examples of this are too common to be worth mentioning …."
    - · "Anyone who would believe such a thing is an idiot …."
    - · "Only [members of some unpopular group] believe that …."

The bottom line here is that if we want to increase our understanding of the natural world, there is no substitute for careful observation, respectful dialog, and critical thinking that is grounded in honesty and humility.

Part I

Maintaining Populations of Featured Species:
A Utilitarian Approach to Conservation

1

Historical Context

Beginnings of Formal Utilitarian Conservation

In this chapter, we will look at the historical roots of formal conservation. We will see how the disciplines of wildlife management, forestry, range management, and soil science arose in response to threats to living natural resources that followed intensive exploitation, habitat alteration, and the introduction of non-native species. These disciplines are primarily utilitarian in their approach. They focus on the exploitation of economically valuable species to protect a long-term supply. To exploit something (a natural resource or a person's labor) is to use it, but the term often carries an implication of excessive use, unfair use, or use without appropriate compensation. In this book, the term *exploitation* as applied to the use of resources is meant to be synonymous with utilization, without a connotation of exorbitant or inappropriate use, although we will see many examples of unregulated or excessive exploitation that resulted in depleted wild plants and animals.

1.1 Royal Reserves and Sacred Groves

Many rulers of ancient and medieval societies issued decrees regulating the use of wild plants and animals. Such rules are sometimes held up as the "earliest traces of a conservation conscience" (Alison, 1981), but that designation assumes that societies which lacked authoritarian rulers were incapable of conservation. It ignores customary or informal arrangements regulating who, what, when, where, and how wild plants and animals could be used. Rulers were not the sole, or arguably even the most important, source of early conservation. In early human societies, community norms ordered people's relationships with the natural world. These norms included the delineation of revered places and customary taboos dictating what uses were allowed within those places (Box 1.1). A patch of forest, mountain, river, spring, cave, or other features of the environment might receive such a designation. Piecing together a picture of these ancient places of spiritual significance requires a multidisciplinary approach including information from sacred texts, historical records, archaeological artifacts, natural records such as pollen cores, and the oral traditions and traditional ecological knowledge of Indigenous people, which often describe practices understood to date from time immemorial.

Kings, emperors, sultans, and czars decreed limitations on peasant uses. These regulations were aimed at reserving certain resources for rulers, aristocrats, and the clergy.
During the Middle Ages, monarchs in many parts of the world, including Britain, France,
Japan, Africa, and Java, reserved some forms of resource use, such as hunting, for the elite
(Olson, 1984; Grove, 1990). For example, after the Norman invasion of Britain, common
grazing lands were enclosed, and the conquerors marked off large areas of land for the
"pastime or 'game'" of hunting by the king and some members of the gentry (Cox, 1905:5).
Privatization of land for enclosure was encouraged under feudalism. Local people had
rights to some kinds of resource use, however, such as collecting firewood and medicinal
plants, harvesting grass, pasturing animals, and sometimes cutting timber (Peluso, 1992;
Williams, 2003).

By the sixteenth century, England witnessed the rise of a country gentry that regarded
owning land as a capital investment. To obtain a return on that investment, it was deemed
necessary to improve the land. Marshes were drained to create farmland, and common lands
were enclosed. Land improved in these ways could be defined by abstract boundaries. This
trend led to a growing privatization and commercialization of the landscape (Ingrouille,
1995). *Bogs*, freshwater wetlands consisting mainly of partially decayed vegetation known
as *peat*, were regarded as the product of laziness ("want of industry") in "barbarous" countries such as Ireland and Italy. William King, an Anglican clergyman in Dublin, wrote in
1685 that "An act of Parliament should be made … that who did not in such a time, make
some progress in draining their *Bogs*, should part with them to others that would" (King,
1685:955). Similar changes in much of the rest of Western Europe followed.

Box 1.1

Examples of Sacred Groves in Madagascar and India

Two examples of natural features that have been revered and protected for centuries or millennia
are described below.

Sacred groves in spiny forests of southern Madagascar: In interviews with researchers, Indigenous people of the Androy region of southern Madagascar (Figure 1.1) have described practices
relating to the protection of sacred forests dating to the time of their ancestors. For centuries and
perhaps millennia, people of this hot, dry region of spiny thickets (a type of vegetation characterized by plants adapted to low and variable rainfall) have recognized several kinds of sacred
places. Harvesting honey, a food with medicinal and ceremonial uses, was regulated within honey
groves by means of taboos. Stricter taboos regulated access to and use of ancestral spirit sites
associated with pre-burial ceremonies, burials, and funerals (von Heland and Folke, 2014).

Sacred grove of the goddess Janni in the Western Ghats of India: Today many sacred groves
are found in the Western Ghats, a chain of hills along the western edge of the Indian peninsula
(Figure 1.1). One of these is dedicated to the goddess Janni. Worship of Janni in this grove may
have begun as worship of an early fertility goddess, perhaps dating to a nomadic hunter-gatherer
culture before village settlement. Although about 5 m of rain falls on the grove during the monsoon season from June through September, rain is sparse during the rest of the year. Trees and
woody vines dominate the vegetation, but the understory is sparse. In the past, taboos prevented
almost all extraction of vegetation from the grove (Gadgil and Vartak, 1975).

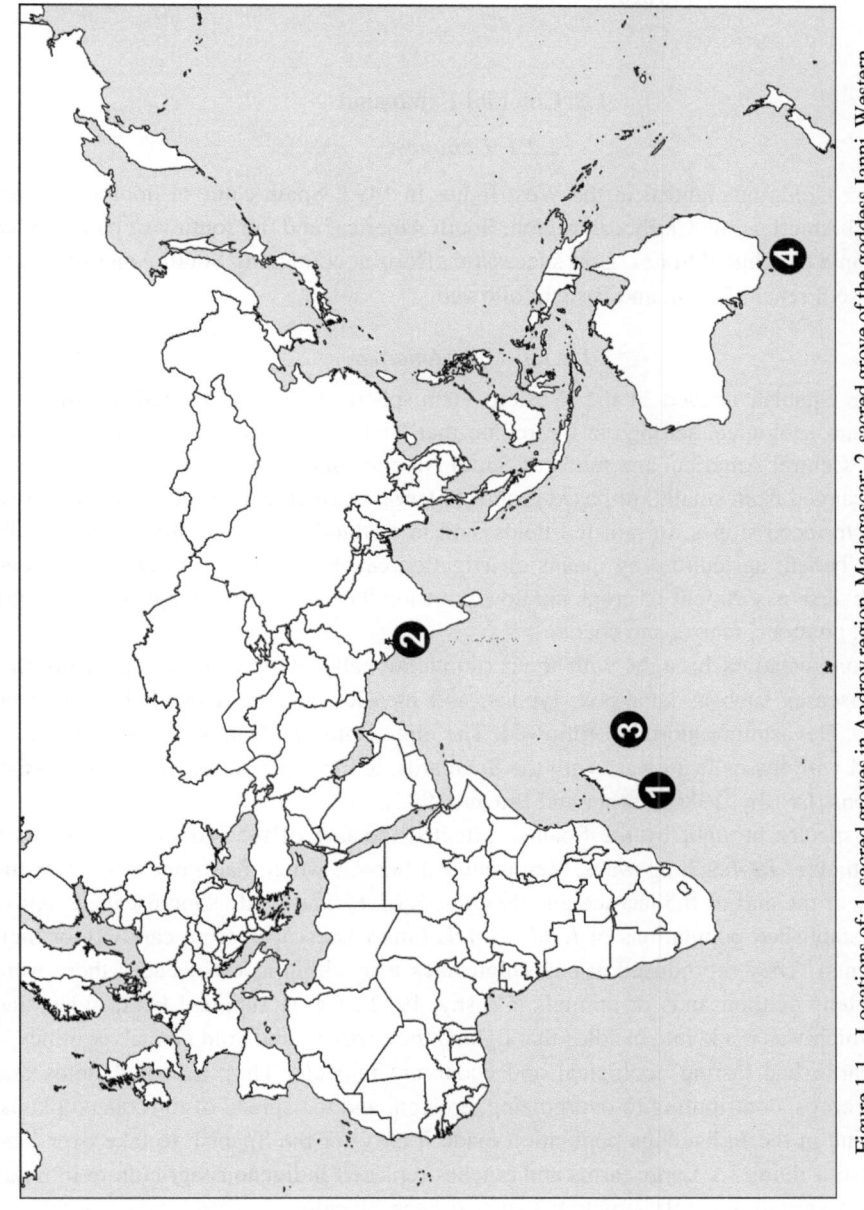

Figure 1.1 Locations of: 1, sacred groves in Androy region, Madagascar; 2, sacred grove of the goddess Janni, Western Ghats mountains, India; 3, extinct giant tortoises, Mauritius; 4, extinct Tasmanian wolf, Tasmania. Map created by Eva Strand using Esri, DeLorme World Countries Generalized Data & Maps for ArcGIS 2013, with permission.

When Europeans arrived in the lands that later became the USA and Canada, these ideas influenced the ways in which they interacted with the Indigenous peoples they encountered. They considered the lands from which the Indians harvested wild plants and animals to be unimproved and unowned.

1.2 Colonial Expansion

1.2.1 Conquest

Soon after Columbus landed in the West Indies in 1492, Spain came to dominate much of Central America, the Caribbean region, South America, and the southwest part of what later became the United States. Less successful efforts at colonizing South America by the Portuguese, French, Dutch, and British followed.

1.2.1.1 Latin America

When the Spanish arrived in the Western Hemisphere, they encountered a variety of cultural and ecological settings in the region that later became known as Latin America (Mexico, Central America, and much of South America and the Caribbean). Social organization ranged from small, kin-based groups to large, hierarchical societies. Indians grew crops on terraced slopes, in rain-fed fields, and in wetlands. They modified water availability to benefit agriculture by means of irrigation canals, dams, dikes, and ditches and grew a diverse assortment of crops unknown outside the Western Hemisphere, including tomatoes, potatoes, maize, and cocoa.

The conquistadors brought with them (unintentionally at first) microorganisms that caused diseases such as smallpox, typhus, and measles, to which the Indians had no immunity. Devastating mortality followed. The high death rate from introduced diseases, combined with losses from wars with the Spaniards, led to a drastic decline in the Indian populations (Crosby, 1986; Turner and Butzer, 1992).

Other species brought by the Spanish affected the fate of the Indians in less direct but nonetheless far-reaching ways. Reintroduced horses, which had gone extinct in the Americas at the end of the last ice age (Section 6.4.1.1), and cattle soon escaped captivity and established populations of *feral animals* (animals escaped from captivity or their descendants). They reproduced at such high rates that within a few decades there were herds of tens of thousands of animals (Crosby, 1972). Cattle supplied beef, hides, and tallow, which was made into candles that lighted the underground gold and silver mines.

The cattle had lasting ecological and economic impacts. They trampled fields and uprooted crops, contributing to overgrazing, erosion, and the spread of introduced plants. The decline in the Indigenous population made it easy for the Spanish to take over land and to justify doing so. Large farms and ranches replaced Indigenous agriculture in fields and within agricultural infrastructure that had been abandoned. Traditional agricultural knowledge was lost except in a few remote locations (Sluyter, 1996).

The Spanish conquistadores exported precious metals as well as products from native organisms. In Europe beans from cacao trees were used to make cocoa, which became a popular beverage. Both *indigo*, a blue dye derived from a shrub that grew along the Pacific coast of Central America, and *cochineal*, a bright red dye extracted from scale insects that grew on cactus plants, were produced on plantations. The ecological impact of this production was modest compared to the plantations of introduced sugar cane, which required clear-cutting and used great amounts of wood for boiling down the sugar juice (Myers and Tucker, 1987).

1.2.1.2 North America

Indians in North America also traded with Europeans. In the northeast, they provided the Dutch, English, and French with beaver pelts – which were in demand for felt hats – as well as furs of other mammals. The resulting fur trade spread westward across North America. Because beaver ponds were easy to find, beaver were especially vulnerable to exploitation. When their populations plummeted as a result, many plants and animals associated with the wetlands created by beaver ponds declined (Box 11.5).

The Indians' hunting territories were part of a multifaceted system of property rights that varied from region to region. In southern New England, for example, Indian villages had collective sovereignty to the territory they used throughout a year. In addition to owning the animals they obtained in their hunting territories, families owned the crops they produced and products they gathered from the land (Cronon, 1983).

Europeans arrived with very different ideas about land ownership. The colonists believed that development of the land's resources legitimized ownership. According to their values, land belonged to those who labored on that land or who employed others to labor on the land. They viewed Indigenous hunting as recreation, not legitimate economic activity, and they considered the lands from which the Indians harvested wild plants and animals to be unimproved and unowned (Hurt, 1987).

Habitats were transformed as a result of this belief that landowners should increase the value of their land by using it to produce crops or livestock. The settlers cleared forests for farming and to provide fuel and timber. Americans consumed far more firewood in the New World than they had in Europe, where wood was scarce (Cronon, 1983). At first, this exploitation depleted forests in the eastern USA; later it spread to the Great Lakes region. As the supply of timber dwindled, forests even further west were cut. This process came to an end only when westward expansion was stopped by the Pacific Ocean. Changes in the intensity of resource use, in habitat structure, and in plant community composition followed each new wave of settlers.

Many Indigenous tribes in the United States signed treaties in which they ceded vast areas of their traditional territories but retained rights to fish, hunt, and gather in those territories (Section 9.6). In most cases, however, those rights were not respected.

1.2.1.3 Asia, Africa, and Oceania

The changes that occurred when colonists from Portugal, England, France, Belgium, the Netherlands, and Spain arrived in much of Asia, Africa, and Oceania were similar in many respects to those that followed colonization in the Western Hemisphere. The specifics differed with the varied cultural and ecological contexts, but some general patterns emerged.

Again, the ecological changes that occurred with colonialism resulted in large part from European assumptions about what constituted legitimate uses of nature. Exploitation of selected resources intensified, while customary uses were restricted. For example, colonial administrators in Kenya did not recognize native Maasai ownership of their lands because by European standards nomadic pastoralism did not improve the land (Collett, 1987). (The words nomad and nomadic are often used to imply aimless wandering, but nomadic peoples actually follow well-defined routes, although the pattern may vary from year to year depending on weather and available vegetation.)

Customary resource management systems were eroded, and profound ecological changes – including the depletion of resources critical for Indigenous peoples and the arrival of non-native plants, animals, and diseases – followed. Large areas of communally owned land were converted to plantations. Yet, as timber and wild plant resources declined and forest cover shrank, colonial concern about the need for conservation mounted and led to measures such as reserves designed to protect forests.

For the most part, the diseases that European colonists brought to Africa and Asia did not cause serious problems for Indigenous peoples. Rather, the reverse was sometimes true. The diverse pathogens and parasites in tropical Asia and Africa limited European expansion into those regions. Because the colonized lands had not been isolated from Europe in the way that they had in the Western Hemisphere, African exposure to European pathogens did not cause the kind of depopulation that occurred in North and South America. *Rinderpest*, a devastating virus of livestock and wildlife that was brought to North Africa by Italians at the end of the nineteenth century, was an exception (Section 10.1.5.2).

In contrast, the native peoples of pre-contact New Zealand and Australia had not been exposed to European diseases before Europeans arrived, and therefore they were vulnerable. Smallpox was particularly devastating to the Aborigines of Australia. For the Maori of New Zealand, tuberculosis, respiratory infections, and sexually transmitted diseases reduced the population (Crosby, 1986).

Rabbits were released on mainland Australia on Christmas Day in 1859 by a member of an organization dedicated to establishing European plants and animals abroad. They multiplied rapidly, as rabbits usually do. By overgrazing and suppressing the regeneration of many native plants, rabbits damaged soils and contributed to the decline and extinction of many native species in Australia (Williams et al., 1995). Those problems were exacerbated by introduced predators, such as foxes and feral cats (Caughley and Gunn, 1996).

1.2.2 Changes in Resource Use: Intensification and Criminalization

As with the fur trade in North America, the intensity of wildlife exploitation increased when Europeans colonized Africa. East Africans had a long history of trading ivory from elephant tusks, but in the latter half of the nineteenth century, the level of trade escalated. Africans began supplying ivory to commercial operators to meet British and American demand for knife handles, piano keys, combs, and similar items. As ivory became scarce, missionaries, colonial administrators, and entrepreneurs traded buffalo and other wildlife to raise revenue for their African enterprises (MacKenzie, 1987).

Throughout the colonies, Indigenous resource uses were criminalized while colonial exploitation of trees and wildlife increased. Authorities in Africa and India forbade traditional hunting but permitted sport hunting. The turn of the nineteenth century in Africa witnessed the elevation of a mystique around hunting for sport, which came to be known as the Hunt. Colonial hunters and wealthy international adventurers killed to obtain skins, trophies, and horns as emblems of prestige (or, later, for museum specimens), yet the Hunt was said to embody ideals of sportsmanship (Manore, 2007). Game laws excluded Africans from hunting, although their skill as trackers was indispensable for a successful Hunt, and their labor as porters was essential to large safari parties. Theodore Roosevelt described the porters on his safari to Africa (which killed 512 animals) as "strong, patient, good-humored savages" (Roosevelt, 1910:94).

In South and Southeast Asia, protection of forest resources, rather than game, was the pre-eminent concern of colonial governments. Much communally managed tropical forest in India, Indonesia, and Indochina, as well as islands in the Indian Ocean, was brought under the control of the Dutch, British, and French East India companies to facilitate commercial timber production. Mixed forests were converted to plantations with single-species *stands* (collections of trees) of commercially valuable species such as ebony and teak. Local uses, except for harvests of some forest resources for customary subsistence, were forbidden. Unauthorized users became "poachers," "squatters," or "timber thieves" (Grove, 1990; Gadgil and Guha, 1995; Peluso and Vandergeest, 2001).

1.2.3 Colonial Reserves

As the effects of intensive resource use became evident, colonial scientists and administrators began to worry. They were uneasy about the fate of economically valuable resources, but their concerns also grew out of a growing recognition that high levels of resource exploitation, especially hunting and deforestation, had serious consequences.

In 1875, the professional hunter W. H. Drummond praised southeastern Africa as "the finest game country in the world," but lamented that "day by day, almost hour by hour, and with ever increasing rapidity, the game is being exterminated or driven back." He especially feared that the "wanton and wasteful wholesale destruction" of elephants for ivory could not "last much longer" (Drummond, 1875:viii, 220, 221).

By the late seventeenth century, scientists employed by the East India companies feared that colonial economic policies had harmful environmental effects. In addition to being concerned about timber famine and wildlife extinctions, they worried that soil erosion due to deforestation, especially at the headwaters of streams, was causing local climate change. These combined fears spearheaded conservation centered on designating forest reserves to reduce erosion and safeguard water supplies (Grove, 1992).

Similar patterns occurred in North America and Europe, where concerns about resource depletion contributed to the rise of conservation. We turn our attention to those changes next.

1.3 Changes from 1800 to 1950

1.3.1 Altered Habitats

During the initial colonial period in eastern North America, forests were considered a deterrent to progress. Clearing the forests was considered a prerequisite for taming the wilderness and improving the land. Later, during western expansion, trees were viewed as a resource for commercial exploitation. Little thought was given to future supplies. "The common assumption was that trees, like Indians, were an obstacle to settlement, and the woodsmen were therefore pioneers of progress" (Udall, 1963:67).

When the timber industry developed, a cut-and-take mentality of making a quick profit prevailed. Because forest land was cheap and harvesting trees did not require much investment, the timber business attracted entrepreneurs. With so much competition there was no incentive for conservation. Companies made quick profits and moved on, leaving large piles of dead wood behind. When that material dried, it was highly flammable, and when it burned it did so with unprecedented intensity (Hays, 1959). The capacity for forest regeneration was nil. Devastating forest fires followed in the wake of careless logging. In 1871, the Peshtigo Fire in upper Wisconsin and parts of Michigan consumed 1.5 million acres and killed thousands of people.

Similarly, minerals and oil were extracted using methods that involved making quick profits. In California in the 1870s, hydraulic mining of gold washed tons of soil and gravel downslope, causing problems for people living in the valleys below and irreversible ecological impacts upslope (Udall, 1963). In mining as in timber harvest, money could be made by those who extracted the resources before anyone else did, regardless of the waste or ecological damage that followed.

As settlers proceeded westward, many habitats, including forests, prairies, and wetlands, declined because they were converted to croplands or used for grazing. As livestock replaced native herbivores, problems from overgrazing developed. Cattle and "buffalo" (bison) had quite different ecological effects. Bison herds grazed an area intensively and then moved on, allowing the vegetation to recover, but livestock grazing is prolonged, especially if the animals are fenced in. Settlers were unfamiliar with the dry climate of the Midwest and failed to appreciate the potential impacts of overgrazing and loss of plant cover in such a setting. Furthermore, because the impacts of grazing are gradual, most people did not recognize what was happening until the effects on native vegetation were dramatic, soil erosion had become severe, and alien weeds were entrenched. (In a similar vein, colonial administrators and post-independence national governments in Africa did not recognize the importance the seasonal movements of large grazers for sustainable grazing regimes (Section 10.1.1.2).)

By the 1920s, thousands of homesteaders had moved to the Midwest to farm. Using mechanical tractors, they uprooted the native sod and planted crops. The severe droughts that occurred in the following decade led to an ecological and humanitarian crisis. Crops and livestock died, wind blew away the topsoil, and dust storms darkened the sky, resulting in a period that came to be known as the Dust Bowl. Unable to grow food or find jobs, hundreds of thousands of farmers and ranchers left their homes and migrated west in search of work in the fields of California, a phenomenon that was recorded in photographs, books, and songs.

Some species such as the coyote benefited from the changes in habitats. However, the effects of habitat modification, increased exploitation, and predator control contributed to the decline and sometimes the disappearance of many native species.

1.3.2 Population Declines

1.3.2.1 Prairie Dogs and Ferrets

In open habitats of the midwestern USA, colonies of prairie dogs (small rodents related to ground squirrels) dug extensive underground burrow systems and lived in colonies covering as much as 100 ha. Because these rodents fed on grasses, ranchers viewed them as potential competitors of livestock. By the early 1900s, elaborate public and private poisoning programs were directed at prairie dogs (Nowak and Paradiso, 1983). As a result, their distribution and abundance declined, as did the black-footed ferret, a type of weasel that depended on them and became extinct in the wild in 1987.

1.3.2.2 Predators

Prairie dogs were not the only animals that competed with the settlers' livestock. Many species of predatory land mammals and birds were systematically killed in efforts to minimize conflicts with livestock and people (Section 4.3.1). Hawks, owls, wolves, foxes, bears, and wild members of the cat and weasel families declined throughout the USA and Canada (as well as Europe and Russia) because of predator control programs.

1.3.2.3 Bison

An estimated 50 million bison roamed the plains of North America when Europeans arrived. Because of their great numbers, bison were easy for market hunters to find and to kill. Much of the carcass was usually wasted; often only the tongues and hides were taken (Hornaday, 1887).

The elimination of bison was, in part, motivated by politics. The persistence of their massive herds was not consistent with the dominant vision of how the frontier should develop.

Because of their numbers and mass, these animals had a pivotal role in the ecology as well as the material and spiritual culture of the Great Plains Indians. The bison provided a prey base for predators; influenced nutrient cycling through their grazing, defecation, and urination; and modified the physical structure of the vegetation by trampling and wallowing. They also provided food, clothing, and implements for Native Americans and featured prominently in rituals and beliefs.

But the bison ate prairie plants that ranchers wanted for their cattle. In 1874, Representative O. D. Conger of Michigan argued against a bill in the US Congress that would have limited the killing of bison, contending that the herds were incompatible with settlement because they competed with sheep and cows. He suggested that the bill granted a "privilege" to the wild, savage Indian that is not given to the poor civilized settler .... [The buffalo] eat the grass. They trample upon the plains upon which our settlers desire to herd their cattle and their sheep .... They range over the very pastures where the settlers keep their herds of cattle and sheep to-day. They destroy that pasture. They are as uncivilized as the Indian.

(Congressional Record, 1874:2107)

Some government officials wanted to reduce bison populations for the specific purpose of subjugating native peoples of the plains. The explicit connection between eliminating bison and Indians was expressed in 1874 by Congressional Representative James Garfield of Ohio (who was later elected president of the USA) commenting on the same bill:

The best thing which could happen for the betterment of our Indian question ... would be that the last remaining buffalo should perish .... So long as the Indian can hope to subsist by hunting buffalo, so long will he resist all efforts to put him forward in the work of civilization .... The Secretary of the Interior said that he would rejoice, so far as the Indian question was concerned, when the last buffalo was gone.

(Congressional Record, 1874:2107)

Congress did pass the bill, but President Grant failed to sign it. By 1890, there were fewer than 1,000 bison in North America. Most of those were in Canada.

The bison had declined in the face of market hunting. However, there were also other circumstances that contributed to the bison's decline. Competition with introduced horses, diseases transmitted to bison by livestock, and drought all had negative impacts on bison populations (White, 1991).

Like its American relatives, the European bison, or wisent, barely escaped extinction. Hunting combined with conversion of forested habitat to cropland contributed to the wisent's decline. By the early twentieth century it survived only in Central Europe, and by 1919 the species was extinct in the wild (although a small number of individuals survived in zoos). It has since been reintroduced in the wild (Nowak and Paradiso, 1983).

1.3.2.4 Marine Mammals

Marine mammals (whales, sea otters, seals, and walrus) also declined in the face of intense exploitation.

Seals. Seals and walrus come ashore to breed, where they are vulnerable to exploitation because they concentrate in large groups and cannot move quickly. Prior to the eighteenth century, many coastal peoples exploited seals in the North Pacific, the North Atlantic, and the coast of the Mediterranean Sea for their meat, blubber, skins, bone, and – in the case of the walrus – tusks. In the folklore of Scotland and the islands of the North Atlantic, *silkies* – legendary creatures that are seals in the sea and transform into people on land – figure prominently.

In the late eighteenth century, a thriving trade in seal pelts developed. By the end of the nineteenth century, many commercially hunted species of seals had declined markedly in both the northern and the southern hemisphere, and entire *rookeries* (breeding colonies) had been eliminated (Box 3.3). At that point two factors came into play that probably saved several seal species from extinction. First, as it became harder to find

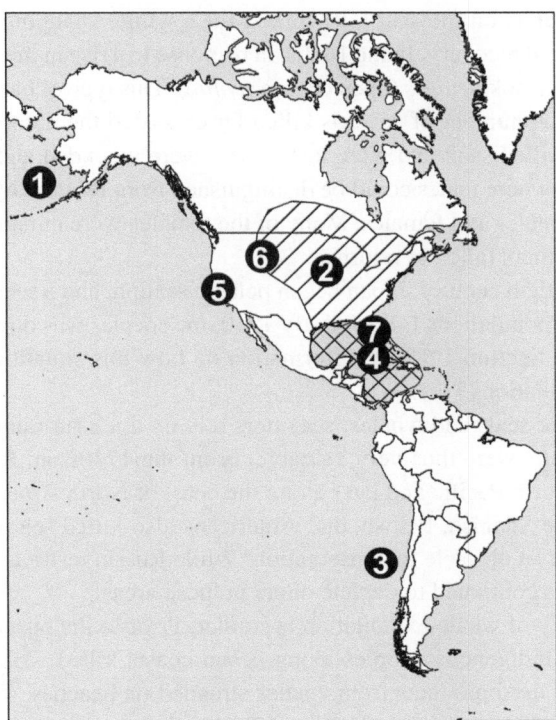

Figure 1.2   Locations of: 1, northern fur seal rookery, Pribilof Islands, Alaska; 2, extinct passenger pigeon, North America (hatched area); 3, extinct Chilean sandalwood, Chile; 4, range of extinct Caribbean monk seal, Gulf of Mexico and Caribbean Sea (cross-hatched area); 5, extinct Xerces blue butterfly, San Francisco; 6, Yellowstone National Park; 7, Pelican Island National Wildlife Refuge, Florida. Map created by Eva Strand using Esri, DeLorme World Countries Generalized Data & Maps for ArcGIS 2013, with permission.

seals, commercial exploitation dwindled, a phenomenon known as *economic extinction* (*see Q1.1*). In addition, the demand for seal oil shrank as alternative sources of fuel were developed.

The fate of the northern fur seal illustrates the dynamics of intensive commercial exploitation. This species breeds on the Pribilof Islands, a chain of islands stretching across the North Pacific (Figure 1.2). After the Russian explorer Gerassim Pribilof arrived at the breeding colonies in 1786, many small companies began to kill fur seals for their pelts. The number of animals they took is not known, but between 1786 and about 1820 the fur seal population declined precipitously (Baker et al., 1970; Gentry, 1998).

Measures to limit exploitation were rarely initiated by private commercial interests, but the Russian American Company, which held the concessions for sealing on islands where northern fur seals bred, was an exception. Sometime around 1834, the company began prohibiting the killing of females on some of the islands where they were harvested. Under this early management plan, the herd increased from about 300,000 animals to over 2 million in less than 50 years.

Although limitations on the seal kill worked for a while, changing market conditions prevented a prolonged recovery. In the 1860s, in response to a rise in the price of pelts, commercial sealers undertook *pelagic* (open ocean) *sealing*. This type of harvest was extremely wasteful, because the number of animals killed far exceeded the number retrieved. Many animals that were killed sank and were lost; others were wounded and died later. Unlike land-based sealing, where males could be distinguished from females by their size, pelagic sealing killed both males and females. Many of the females were nursing pups, which also died when their mothers failed to return.

In the early twentieth century, Japan began pelagic sealing, and a second serious decline in northern fur seal populations followed. By 1910, the species was down to about 10% of its 1867 level. (See Section 1.5.3 for information on how this situation was addressed to restore fur seal populations.)

Sea otters: Unlike seals and whales, sea otters rely on thick fur rather than blubber for insulation. Their pelts were thus very valuable. From the 1740s on, Russians hunted sea otters, first in the North Pacific and later along the coast of North America from Alaska to California. Later the Spanish, British, and Americans also killed sea otters. International competition created an obstacle to conservation. While Russia restricted its hunts in some places, other nations continued to deplete otters in those areas.

Whales: The story of whale exploitation is similar. Prior to the onset of whaling on the open ocean, many Indigenous peoples along ocean coasts killed whales for subsistence from small boats or used products from whales stranded on beaches. (Some cultures continue traditional whale hunts today (Section 12.6.1).) The remains of combs, keys, knife handles, and other objects made from the bones of whales thousands of years ago have been found in Scotland and other islands of the North Atlantic.

When Europeans arrived in New England, they established a prosperous commercial whaling industry. Before the age of modern chemistry, Industrial whaling provided many products. Whale oil was a valuable source of light until petroleum became available. *Ambergris*, a substance formed in the intestines of sperm whales, was used as a fixative for perfumes. *Spermaceti*, a waxy substance found in the large reservoir at the front of the sperm whale's head, was valued as an industrial lubricant. Whale ivory was obtained from the teeth and jaw bones of toothed whales such as the sperm whale and the narwhal.

Blue, fin, humpback, gray, right, minke, and bowhead whales are baleen whales. They lack teeth but have fringed plates (*baleen*) that hang from their upper jaws (Figure 1.3A). When a baleen whale opens its mouth, water flows in, and the soft tissue of the lower jaw expands (Figure 1.3B). The whale then closes its mouth part way and raises its tongue. This allows water to squirt out while the baleen acts as a giant sieve, filtering food organisms from the water.

Baleen provided a flexible, springy, strong material that was used for many items made today from steel or plastic (Figure 1.3). Also known as *whalebone*, baleen formed the stays of women's "whalebone" corsets.

In the eighteenth century, a profitable commercial whaling industry developed off the island of Nantucket, Massachusetts. When a whale was sighted from a whaling ship, a few sailors (most of whom were Wampanoag Indians forced to work because of debt-servitude)

A   **B**

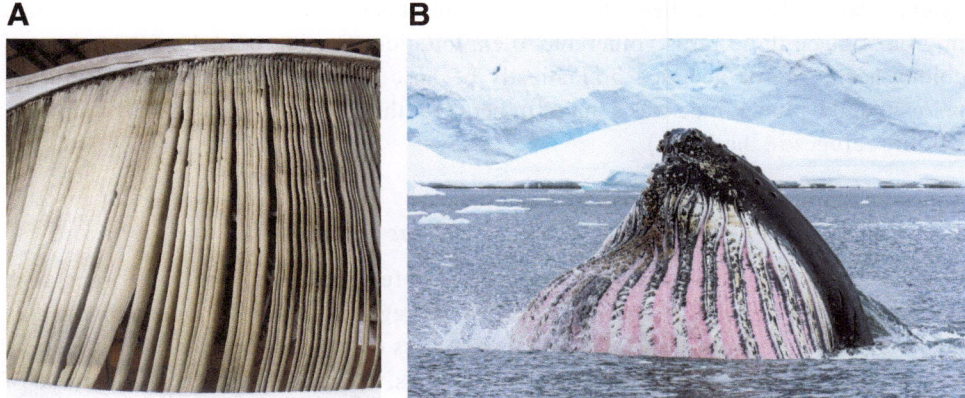

Figure 1.3   (A) Baleen hanging from the upper jaw of a baleen whale. The plates act as a giant filter to strain small marine organisms from seawater. Credit: Kevin Schafer / Moment Mobile / Getty Images. (B) Humpback whale feeding with mouth expanded to take in water. Credit: Adam Cropp / Moment / Getty Images.

pursued it in a small boat (Philbrick, 2001). If the sailors succeeded in harpooning the whale, it thrashed about, dragging the small boat and its crew along, an experience that became known as a "Nantucket sleigh ride." Often the men drowned before the wounded whale died. Sperm, bowhead, and right whales were taken using this method. Those species could be handled from small boats because their carcasses float.

In 1864, a Norwegian captain invented a cannon-powered harpoon that exploded after it entered a whale. This Foyn gun was much more effective at killing whales than a hand-held harpoon. Armed with this technology, whalers became more efficient at finding, killing, and processing whales. As a result, populations of all the large whale species plummeted to low levels. Several species became so rare that whalers rarely encountered them, but the pursuit of other kinds of whales continued.

In 1931, in response to the decline and near extinction of large whales brought about by unregulated exploitation, a Convention (treaty) for the Regulation of Whaling was signed by 26 nations. Fifteen years later, the convention stablished the International Whaling Commission (IWC) to regulate commercial whaling in order to protect the supply of whales for commercial whaling (Section 7.3.1).

Most nations that conducted commercial whaling in the past now abide by the IWC's moratorium on commercial whaling, but some exempt themselves from it, and others continue killing whales for research, which is not regulated by the IWC.

1.3.2.5 Colony-Nesting Birds

In the last half of the nineteenth century, women's hats containing ornate feathers – and even whole, mounted birds – were fashionable in North America and Europe. As a result, a thriving trade in feathers, known as the plume trade, developed. Large numbers of many species that nested colonially – such as herons, grebes, terns, ibises, and egrets – were killed for their plumage. Like the bison, their tendency to concentrate in groups made these birds vulnerable to exploitation (Section 6.5.1.1). Populations of colony-nesting birds declined dramatically because of this exploitation. (We will see below that the plume trade was a major impetus for the establishment of the first national wildlife refuges in the USA.)

1.3.3 Extinctions

Some species were not as lucky as bison, beaver, and fur seals. Certain characteristics, such as limited geographic distribution and defense strategies that worked against nonhuman predators but not against people, made these species especially vulnerable to extinction. Market hunting, efforts to control predators and pests, habitat alteration, and the arrival of non-native species were usually the circumstances that precipitated the declines, however.

1.3.3.1 North America's Passenger Pigeon

In the middle of the nineteenth century, the passenger pigeon may have been the most abundant bird on earth. It was present throughout much of eastern North America (Figure 1.2) in such great numbers that it was difficult for people to imagine its disappearance. The birds traveled in massive groups, accounts of which are so amazing that they might seem exaggerated if they were not so well documented. The ground underneath roosting flocks appeared to be covered with snow because of a thick layer of droppings, and it was common for tree trunks and limbs to break under the birds' weight. In 1806, the ornithologist Alexander Wilson estimated that a flock he saw in Kentucky contained over 2 billion birds. In 1813, the naturalist John James Audubon reported seeing a flock on the Ohio River that obscured the sun at midday (Trefethen, 1975).

By forming such large groups, it is likely that the passenger pigeon was able to overwhelm animals that preyed on it. Although predators were attracted to the pigeons' roosting and breeding aggregations, there were so many birds that predators could not eat them all, and large numbers of pigeons always escaped. This is termed *predator saturation*. Furthermore, the flocks never stayed in one place long enough for predators to increase to levels that would exert sustained pressure (Blockstein and Tordoff, 1985).

Although the enormous flocks were an effective *adaptation* (a trait that is favored by natural selection (Section 6.1.1)) to predation by wild animals, the flocks made the birds especially vulnerable to human predators. The passenger pigeon was a market-hunter's dream. It took little effort to find and to kill large numbers of birds. Professional hunters trapped them in baited nets capable of killing hundreds of birds at a time. Local hunters shot nesting birds and took young from the nest, sometimes killing virtually all the young from a nesting colony.

Birds could be shipped by rail to markets in distant cities. In 1878, hunters in Michigan took over a million birds from the last large nesting colony. In another example of economic extinction, commercial killing stopped when the birds became too scarce to hunt profitably, but unfortunately by that time it was too late for the species to recover. The last individual died in the Cincinnati Zoo in 1914.

The decline of the passenger pigeon was remarkably abrupt. Hundreds of millions of birds persisted into the 1870s, but by the 1890s they were very rare, and two decades later they were extinct. How was it possible for a species that had been so abundant to go extinct so quickly?

To answer this question, it is necessary to understand how the ecology of the passenger pigeon made it vulnerable in spite of the enormous size of its population. This species occurred only in the deciduous forests of northeastern North America. It fed on acorns, beechnuts, and chestnuts, which are produced by *mast trees*, trees that produce fruits and seeds erratically. Every few years, the different species of mast trees throughout a region produced abundant seed crops all at once. This is thought to be adaptive for the tree species that produce mast because the passenger pigeon and other seed-eating species such as squirrels could not consume all the seeds that were produced in a single mast season, so some survived and germinated. Furthermore, just as the squirrels and pigeons could not eat all the seeds of the trees they fed upon, the predators could not eat all the squirrels and pigeons.

Because the availability of these foods was unpredictable, passenger pigeons had to search wide areas to find food. The large size of their flocks may have allowed them to scan vast expanses of the landscape. When some of the birds detected food, they would call to the other flock members. Eventually, when the pigeon populations were reduced by a combination of hunting and habitat loss, their ability to locate food was probably compromised, even though the flocks were still large. It seems likely that their dependence on erratically available food supplies which could only be found by enormous aggregations of birds made passenger pigeons uniquely vulnerable to the combined effects of exploitation and deforestation (Bucher, 1992).

Like the bison, the passenger pigeon might have been intolerable to settlers. It is doubtful whether the landscapes of the eastern and midwestern USA could have supported passenger pigeon flocks along with intensive agriculture and urbanization.

1.3.3.2 Giant Tortoises on Mauritius

Mauritius (Figure 1.1) is a group of oceanic islands in the Indian Ocean east of Madagascar. When the Dutch claimed it in 1598, they found two species of slow-moving, long-lived land turtles: a large, high-backed, flat-shelled tortoise and a smaller, domed tortoise. Both were *endemic* to Mauritius and nearby islands (Griffiths et al., 2010). (The meaning of the term endemic in ecology is different from its meaning in medicine. In ecology, a species that is endemic to a region is not found anywhere else. In medicine, a disease that is endemic to a place is regularly found in that place but is not necessarily restricted to it) (*see Q1.2*).

Ships from several maritime nations stopped on Mauritius regularly to stock up on food, and the Dutch established short-lived settlements for the purpose of harvesting ebony trees. According to an account by a Portuguese missionary who visited the island, the tortoises could "easily carry a man on their back for some time. They are ugly and deformed creatures, whose carapace, hard though it is, can nevertheless be shot through by a bullet, as we have had occasion to test out" (Cheke and Bour, 2014:47).

Before people arrived, adult tortoises had no predators. Consequently, they did not flee from or avoid people who came to the island. Because they were easy to find and kill, the tortoises were heavily exploited for their meat and their fat, which was valued for making candles and was considered "superior to the best butter" (Cheke and Bour, 2014:48). Thirty to 40 animals were needed to obtain a pint of grease. Often only the fat was used, and the carcasses were left to rot or were fed to pigs.

Introduced pigs devastated the tortoise populations, eating their eggs as they were laid. When cats were introduced, predation became even more intense. Successful tortoise reproduction had probably ceased by the end of the 1600s, but some individuals might have lived for many decades. It is unlikely that any survived on Mauritius past the 1620s.

1.3.3.3 Chilean Sandalwood

The Chilean sandalwood was a tree species endemic to the San Fernández Islands (Figure 1.2). This small archipelago of three volcanic islands in the South Pacific became known to the Western world in 1574 when the Spanish explorer Juan Fernández encountered it on a voyage west of Chile. Subsequently, a Scottish naval officer was marooned on one of the islands from 1704 to 1709, an experience which may have provided the inspiration for the novel *Robinson Crusoe*.

We know from herbarium specimens and remnants of wood collected at known locations that the Chilean sandalwood was present on the two largest of the San Fernández Islands. British and American whaling ships often stopped there to make repairs and take on firewood. As a result, native trees on the lower slopes declined, and severe erosion developed. Chilean sandalwood was probably quite common until the eighteenth or nineteenth century when it was depleted by logging.

The reddish heartwood of this plant was used for handicrafts and sculptures and was exploited for its fragrant oil. It was last recorded in 1908 by a Swedish botanist who photographed what may have been the last living member of this species. That plant is believed to have died by 1916, although pieces of Chilean sandalwood were still in circulation as late as 1996 (Stuessy et al., 1998).

1.3.3.4 Tasmanian Wolf

The Tasmanian wolf or Tasmanian tiger was a carnivorous *marsupial* (pouched mammal) of the Australian region. Although it is not closely related to dogs and true wolves, the Tasmanian wolf's teeth, build, and feet are remarkably doglike (Figure 1.4). Its scientific name means "pouched dog with a wolf head." This similarity in unrelated species is known as *convergent evolution* (Box 12.5).

Tasmanian wolves were widespread in Australia and New Guinea until about 3,000 years ago, but subsequently they became extinct everywhere except on the island of Tasmania (Figure 1.1). It is likely that their demise in those regions was due to competition with the dingo (Australian wild dog), which was introduced by Aboriginal hunters about 10,000 years ago (Archer, 1974). Dingoes spread throughout most of the Australian region, but they did not become established in Tasmania.

When Europeans arrived in Tasmania and began raising sheep, predator control programs were initiated. Tasmanian wolves were shot, trapped, and poisoned. Between 1888 and 1909, the government paid bounties for over 2,000 individuals, and others were killed

Tasmanian tiger

Figure 1.4   Tasmanian wolf. Antique engraving. Credit: mikroman6 / Moment / Getty Images.

for private bounties or for income from selling their pelts. Habitat loss, competition with dogs, and disease may have contributed to their decline (Caughley and Gunn, 1996). By 1905 the Tasmanian wolf population had declined markedly, and by the 1930s the species was extinct (Nowak and Paradiso, 1983).

1.3.3.5 Caribbean Monk Seal

The Caribbean monk seal once occurred mainly in the Caribbean Sea and the Gulf of Mexico, but its range also extended south to South America and north to the coast of Georgia (Lowry, 2015) (Figure 1.2). It was one of three species of monk seal. Like other seals, monk seals periodically came ashore on coastlines and islands to rest and breed. Unlike other seals, however, monk seals lived mainly in tropical waters. Authorities disagree on whether the Caribbean monk seal was rare before Europeans arrived or was abundant but subsequently declined because of exploitation.

Humans and sharks were the only major predators of the Caribbean monk seal. Prior to contact with Europeans, Indigenous peoples sometimes used this species, but it was not a major target of exploitation. Soon after Columbus encountered Caribbean monk seals in 1494, Europeans began killing them for food. By the late 1600s, many seals were being killed for their skins and oil (much of which was used to lubricate machines on sugar plantations). Still later they were killed by fishermen concerned that they might be competing with them for fish, and this species was also taken for museum specimens and public displays. It was rare by the late 1880s, although Caribbean monk seals persisted at least until 1952, when a small colony was sighted at a group of coral islands between Jamaica and Honduras. Subsequent searches of the region did not locate any individuals, although there were unconfirmed sightings from time to time (which were most likely a different species).

1.3.3.6 Xerces Blue Butterfly

The Xerces (ZER sees) blue butterfly occurred only in and around San Francisco, California (Figure 1.2), where it was restricted to well-drained soils of coastal sand dunes. Within its restricted geographic range, it had narrow food preferences. Typically, this species occurred near mat-forming patches of a plant known as deerweed, beneath Monterey cypress trees. Females laid eggs on deerweed or yellow bush lupine. After the eggs hatched, the larvae (caterpillars) fed on those plants until they formed pupas. After 10 to 11 months, adult butterflies emerged from the pupae. Although adult butterflies can fly, caterpillars cannot, so the larvae could not move from one host plant to another. For this reason, they were dependent for food on the host species on which the eggs were laid. This put the species at risk.

As the city of San Francisco developed, habitat for the Xerces blue and its host plant shrank. By 1919, these butterflies were seen in an area only 21 m wide by 46 m long. The last known specimens were collected in 1941. Many subsequent searches of its habitat failed to turn up any more individuals of this species. The immediate cause of the Xerces blue's extinction was disturbance due to urban development, but the underlying circumstances that made it vulnerable were its narrow *geographic range* (the region over which it was distributed) and specialized food requirements.

Data from field and laboratory studies of the Xerces blue indicated that, like many of its close relatives, the Xerces blue formed a close association with ants. However, the species became extinct before it was possible to get additional information about this interaction.

These examples illustrate how the depletion of wild plants and animals and their habitats set the stage for conservation aimed at addressing the resulting problems.

1.4 Diagnosing the Problem

Habitat modification and species declines fostered the development of conservation sentiment during the last half of the nineteenth century in the USA and in many parts of the Western world. Among those who worried that economically important resources would soon be exhausted, fears of a timber famine, along with concerns about the fate of water supplies and wildlife, were prevalent (Hays, 1959). In his influential book, *Man and Nature: Or, Physical Geography as Modified by Human Action*, originally published in 1864, George Perkins Marsh warned of "desolation … unless prompt measures are taken to check the action of destructive causes already in operation" (Marsh, 1965:201). Similar dire warnings appeared in numerous popular articles with titles such as *Timber Waste: A National Suicide* (Pisani, 1985). In addition to those who warned that resources were being used at a rate that could not be sustained for long, there were others, such as John Muir (Preface) who feared that the beauty of the natural world would be destroyed. Those voices emphasized the importance of protecting places with aesthetic and spiritual value.

The immediate causes of these problems seemed straightforward: high rates of use, waste, and lack of regard for preserving or restoring valued habitats. Better management of living natural resources was clearly needed to stem the tide of declining populations and habitats. The dominant view was that economically valuable species should be managed to support efficient economic growth, although a minor theme that emphasized strict preservation also emerged. We consider that thread first.

1.5 Response

1.5.1 Reserves

Some reserves prohibit or severely curtail the removal of resources within their borders, whereas others allow managed extraction. The first category includes national parks (which generally exclude most forms of consumptive resource use) or other areas that are set aside to protect their scenic values or the habitats and populations within their borders.

1.5.1.1 Strict Protection within Reserves

In 1864, US Senator John Conness introduced a bill to the US Senate proposing that Yosemite Valley in the Sierra Nevada mountains be granted to the state of California. In his comments to the Senate, Conness argued that the lands under consideration constituted "perhaps, some of the greatest wonders of the world" (Congressional Globe, 1864:2300). When President Lincoln signed the bill, Yosemite became the first place that the US government had set aside to preserve scenic beauty for "public use, resort, and recreation" (Yosemite Valley Grant Act of 1864).

Yosemite was not the nation's first national park because it did not become part of the national park system until 1916. That distinction belongs to Yellowstone National Park, which encompassed parts of the Wyoming and Montana territories (Figure 1.2) that were set aside by Congress "as a public park or pleasuring-ground for the benefit and enjoyment of the people" in 1872 (Yellowstone National Park Protection Act of 1872). The Indigenous peoples whose territories encompassed Yosemite and Yellowstone were prevented from using those lands in the designated parks, an issue that received little attention from conservationists for about a century (Spence, 1999).

Like Yosemite, Yellowstone National Park was set aside to preserve scenic and geologic wonders. Both were beautiful places at high elevation locations that were regarded as economically worthless. As far as Congress was concerned, setting them aside did not entail any economic sacrifice. In Senator Conness's recommendation that Yosemite be preserved, he argued that it was "for all public purposes worthless" (Congressional Globe, 1864:2300). (Recall from the Introduction, however, that within a few decades Congress would change its mind about the value of Yosemite's resources.)

The bill that set aside Yosemite Valley also authorized private individuals to apply for leases to build and operate tourist accommodations within the park. Ironically, John Muir, the wilderness enthusiast who championed the natural values of Yosemite, disapproved of some forms of enjoyment of nature. He wrote of yearning to live in the Sierra Nevada mountains "like the wild animals, gleaning nourishment here and there from seeds, berries, etc., sauntering and climbing in joyful independence of money or baggage," and disparaged the "glaring tailored tourists … that frightened the birds and squirrels" (Muir, 1911:4,79).

The designation of lands for the benefit of the public raised thorny questions. Can the public enjoy natural places without harming them? What commercial development of those places should be permitted? Are some forms of enjoyment of natural places more valid than others?

1.5.1.2 Regulated Resource Use within Reserves

After the Civil War, concern increased about whether America's living natural resources could sustain the levels of harvest to which they had been subjected. Beginning in the 1890s, substantial areas of public land in the USA were set aside for regulated resource use within their borders. The resources of concern included timber (national forests), wildlife (national wildlife refuges), and livestock forage (lands managed by the Bureau of Land Management).

In 1903, Pelican Island, off the coast of Florida, became the first national wildlife refuge in the USA (Figure 1.2). It was set aside by President Theodore Roosevelt to protect birds that were hunted for the plume trade. Additional land was rapidly added to the refuge system. By the end of 1904, Roosevelt had designated 51 refuges.

Because the word "refuge" implies a safe place or sanctuary, people are often surprised to learn that hunting and fishing are allowed in many national wildlife refuges. The earliest national wildlife refuges in the USA were indeed intended to serve as sanctuaries for birds targeted by plume hunters. Many of the refuges that were subsequently set aside, however, were established for game species such as *waterfowl* (ducks, geese, or swans) or ungulates, with the explicit purpose of providing hunters and fishers with a supply of game during regulated seasons.

The Bureau of Land Management (BLM) is another federal agency in the USA that administers large areas of federal land on which extractive activities, including grazing, timber harvest, and mining, are permitted. This agency was established in 1946 to take over the administration of *rangelands* (lands that are unsuited for cultivation but produce forage for livestock or wildlife). The BLM manages more than half the federal lands of the United States.

1.5.2 The Rise of Conservation Science

The designation of public lands for regulated use created a need for scientific information to guide conservation within those lands. The need for information to guide water management was particularly pressing in the American West, where most public lands were located. Scientific study of those lands began in 1879. Like colonial scientists in the tropics, Western scientists and irrigators soon realized that forests were critically important for protecting water quality.

1.5.2.1 Forestry

The first American national forests were designated in 1891. Before leaving office, President Harrison set aside 526,000 ha of forest reserves in the American West. Early forest managers drew upon but modified an older tradition of European forestry. Europe's scientific approach to forestry involved intensive management of scarce, privately owned resources. For example, in nineteenth century Germany, stands of spruce were planted and thinned, resulting in *monocultures* (holdings dominated by a single species) of small, symmetrical trees of the same species and age.

American resource managers recognized that European-style intensive management of public forests would not be accepted by the American public. But although American foresters rejected the European approach as too unnatural, they too saw themselves as analogous to farmers. In their view, agriculture and natural resource management differed only in the degree of domestication of the product. Both manipulated critical factors to enhance production. This was to be done in a regulated fashion that would ensure a continuous flow of products.

In 1898, Gifford Pinchot, a wealthy New Englander with European training in forestry, took office as the US Department of Agriculture's chief forester. Scientists and foresters in the US government shared a commitment to rational, efficient management of water resources and forests to promote economic growth. As a result of Pinchot's advocacy, the concept of regulated use was adopted as the guiding principle of national forest management. Forests were to be managed to maximize the amount of wood that could be produced while ensuring a future supply. The goal was *sustained yield*, the harvest of a renewable resource at a level that could be maintained over time. For yield to be sustainable, the rate at which plants or animals are harvested must not exceed the level at which they grow back, and the harvest must not harm other species or ecosystem processes. If these assumptions are met, then sustainable harvest is possible (though it is not guaranteed) unless conditions change. Historically, the implied assumption of no effects on non-target species has rarely been tested.

In response to wasteful practices of the day, and convinced that commercial exploitation of forests was inevitable, Pinchot argued that "the job of the forester was not to stop the ax but to regulate it" (Pinchot, 1947:29). "The purpose of Forestry, then, is to make the forest produce the largest amount of whatever crop or service will be most useful and keep on producing it for generation after generation of men and trees" (Pinchot, 1947:32).

Under Pinchot's leadership, a cadre of idealistic, college-educated young men set out to save America's forests from timber barons through a program of scientific, rational, and efficient harvest management. Pinchot and these young men saw themselves as reformers, curbing corrupt special interests and safeguarding the nation's forests for future use. By studying the growth rate and abundance of each tree species, foresters would scientifically determine appropriate harvest levels. Regulated timber harvests would replace rapacious looting of the forests. Their passion was shared by Pinchot's boss and close friend President Theodore Roosevelt.

According to scientific forestry of the day, disturbances that damaged valued species should be eliminated. One of the young forest managers in the early twentieth century was a midwestern lover of the outdoors named Aldo Leopold. Like his colleagues, the young Leopold focused on protecting resources from whatever destroyed them. Excluding fire from forests meant saving timber. In an essay written in 1920, Leopold stated that "Piute [Paiute Indian] forestry," the practice of setting frequent, low-intensity fires, would "ultimately destroy the productiveness of the forests on which western industries depend for their supply of timber" (Leopold, 1920:12). Similarly, Leopold's contemporary William Greeley wrote in 1920 that the best way to manage forests was "to keep fire out of the woods" (Greeley, 1999:33).

Leopold and Greeley dismissed the idea that Indians had any insight into forest dynamics. "It is, of course, absurd to assume that the Indians fired the forests with any idea of conservation in mind" wrote Leopold (Leopold, 1920:13). Likewise, Greeley disparaged the idea that the Indian "fired the forests regularly ... because his nature lore taught him that this was the way to prevent the 'big' forest fire" (Greeley, 1999:34). These views on light burning reveal the assumption that Native Americans were incapable of managing their environment wisely (*see Q1.3*). This mindset had far-reaching consequences.

1.5.2.2 Game Management

The discipline of game management in North America was launched in 1933 when Aldo Leopold, at that time a Professor of Forestry at the University of Wisconsin, published a text on the subject (Leopold, 1933). (The term *game management* eventually came to be called wildlife management, and the term wildlife itself expanded over the decades. At first, it was used to refer to hunted species; later it came to mean all terrestrial vertebrates. In current usage, the term wildlife can refer to all forms of wild organisms, including animals, plants, and microorganisms, although it is commonly used to mean animals, especially mammals and birds).

Leopold defined game management as "the art of making land produce sustained annual crops of wild game for recreational use" (Leopold, 1933:3). The comparison to agriculture was explicit: "Like the other agricultural arts, game management produces a crop by controlling the environmental factors which hold down natural increase, or productivity of the seed stock" (Leopold, 1933:3). Like foresters, wildlife managers emphasized regulated use of resources through manipulation of the environment to maximize the productivity of desired species and minimize undesirable species. (However, Leopold later came to believe that predators play an important ecological role (Section 4.3.3).)

1.5.2.3 Range Management and Soil Science

Range management differs from the management of forests or game because it concerns domestic animals that are produced from wild forage, whereas foresters and game managers regulate harvests of wild plants and animals directly. But range management is similar to those disciplines in that it was formed to address utilitarian concerns. Like early foresters and wildlife managers, the first range management professionals emphasized maximizing productivity, specifically the productivity of wild crops. A US Department of Agriculture bulletin published in 1926 stated that "more and better forage, as well as the maximum production of beef, wool, and mutton, is a primary object of grazing management" (Sampson, 1926:1).

Like range management, soil science addresses the manipulation of a natural resource (soil) for the purpose of producing domesticated species (agricultural crops). Farmers in many parts of the world had been making observations about agricultural practices that affected their soil for millennia, but it was not until the nineteenth century that formal study of soils developed (King, 1907). Scientists from Germany, Russia, and Denmark synthesized concepts and methods from geology, chemistry, physics, and microbiology in order to study the formation and development of soils and to classify soils according to their physical properties and potential uses. In 1933, the US Soil Erosion Service (later renamed the Soil Conservation Service and even later the Natural Resources Conservation Service, NRCS) was formed in response to the soil erosion crisis of the American Dust Bowl. The NRCS works primarily with private landowners, providing assistance regarding soil erosion, fertility, and other technical issues related to soil conservation (Section 3.4).

1.5.3 Regulations

Jurisdictional problems confound efforts to manage species that move around. In the USA, if a plume hunter killed birds in violation of a state's laws and transported the feathers to another state, neither that state nor the federal government had jurisdiction. This situation changed with the passage of the Lacey Act in 1900, which outlawed interstate shipment of any wild birds or mammals or their parts or products (including feathers and eggs) that had been taken in violation of state laws.

Management of mobile wild animals that move across international boundaries is even more complicated. In 1911, the USA, Imperial Russia, Japan, and Britain (for Canada) signed a treaty regulating commercial harvests of northern fur seals. Under the provisions of this treaty, the USA and Russia, which contained breeding islands, shared the sealskin take with Canada and Japan, which agreed to stop pelagic sealing. After the treaty was signed, fur seal populations rebounded, reaching a high in 1941 (Bailey, 1935; Gentry, 1998).

Efforts to conserve migratory birds ran into similar problems. (Migrating animals make predictable annual round-trip *migration*s, in contrast to human migrations, which are usually one-time events.) By the late 1890s, populations of North American waterfowl were down to very low levels because of a combination of drought and market hunting. But, since waterfowl migrate long distances between their breeding and their wintering grounds, governments were reluctant to pass protective legislation limiting hunting within their own borders. People reasoned that if they did not shoot the migrants, someone in another country, state, or province would.

An attempt to solve this problem in the USA with federal legislation was challenged in the courts as a states' rights issue. But before the Supreme Court decided the case, the problem was circumvented in 1916 by the Migratory Bird Treaty between the USA and Canada, which established restrictions on the taking of migratory game birds such as waterfowl. The USA subsequently signed similar treaties with Mexico, Japan, and the Soviet Union.

We have seen in this chapter how historical conditions gave rise to utilitarian conservation primarily aimed at managing supplies of economically valuable species. The next three chapters describe the central concepts of this approach (Chapter 2), and the principal strategies for accomplishing its goals (Chapters 3 and 4).

2

Central Concepts

Populations, Succession, and Ecosystems

In Chapter 1, we saw how intensive exploitation of wild plants and animals laid the foundation for formal conservation. In response to this situation, professionals in the young disciplines of wildlife management, forestry, range management, and soil science adopted a utilitarian approach. In this chapter, we will consider the principal concepts underlying utilitarian management, which tends to focus on certain phenomena in nature, including population regulation in resource-limited systems, changes in plant communities over time, and ecosystems. The first two of these concepts are grounded in the assumption that the natural world tends toward equilibrium.

In Part II and Part III of this book we will consider developments that occurred as the limitations of the utilitarian approach became apparent. The conceptual framework of utilitarian conservation was not entirely discarded, but utilitarian concepts were refined in ways that involved more nuanced understandings of populations, resources, vegetation change, and ecosystems.

2.1 Populations

A *population* can be defined as a group of a kind of organisms occupying a defined area during a specific time. For example, we may want to refer to the population of people in Germany in 1910 or in 2020, the population of beetles on a log, spruce trees in a forest, fish in a lake, kangaroos in Western Australia, or elephants in Kenya. The rate at which a population grows depends upon how many individuals are added to and removed from it during a given period of time. Members can be added to a population by birth or by *immigration* (permanent movement into a population), and members can be removed either by death or by *emigration* (permanent movement out of a population). If birth and immigration exceed death and emigration, the population will grow, and if more individuals die or leave than are born or join, it will decline.

2.1.1 Population Growth and Regulation

It is easier to get information about death and birth than about emigration and immigration, so *demographers* (scientists who study changes in population size) usually focus on mortality and reproduction when analyzing population processes. The potential rate at which a species can increase, under ideal environmental conditions, is its *biotic potential*. The biotic potential of a species is a genetically determined characteristic of that species. It depends on such things as the number of eggs laid (*clutch size*) or young born (*litter size*) during each reproductive event, the frequency of reproduction, the age at which individuals first reproduce, and the age at which reproduction ends. The age when reproduction begins has a particularly strong influence on the rate of population growth. Populations of animals (including people) grow much more rapidly when reproduction begins at a young age.

Biotic potential is high for some organisms, such as bacteria and dandelions, and low for others, such as people and elephants. We will see in Section 6.5.1.1 that biotic potential is one factor that influences the likelihood that a species or subspecies will become *endangered* (likely to become extinct) or *threatened* (likely to become endangered).

Organisms die from many causes. Individuals are killed outright by enemies that eat them, uproot them, or parasitize them. Mortality is also caused by accidents, or it may result if competition between members of the same species or with members of other species prevents individuals from obtaining resources they need for survival. Finally, organisms die if they are both unable to tolerate environmental conditions (for example, if it is too cold, too hot, or too wet or dry for them) and unable to move elsewhere. An otherwise suitable habitat may become intolerable because of unusual short-term fluctuations in weather (such as a drought), long-term environmental changes (an ice age or global warming) (Section 10.3), high levels of physical disturbance (wave action or winds), or catastrophes (earthquakes, volcanic eruptions, storms, or floods). When such a change occurs, a species will decline and eventually become extinct if mortality exceeds reproduction and its members are unable to emigrate from the unsuitable location to a place where conditions can be tolerated.

Factors that affect mortality also influence reproduction. The reproductive rate of a population under a specific set of conditions depends upon a species' biotic potential as well as on the survival of its members. In addition, reproductive rate is affected by physiological state. Nutritional status, disease, body size, and parasite loads can influence reproductive rates. Individuals that are in poor quality habitat or that are prevented from obtaining resources by competitors may survive but fail to reproduce or may produce few young or young that are unlikely to survive.

2.1.1.1 Exponential Growth

Under ideal conditions – where there is no crowding, resources such as light, water, and nutrients are abundant, appropriate habitat is available, and no substances that inhibit growth or reproduction are present – organisms are capable of increasing to extremely high numbers. Some bacteria can divide to produce two cells every 20 minutes if conditions are favorable. Starting with a single bacterial cell, over 250,000 cells can be produced in just six hours. In the example in Table 2.1, the population doubles in each generation, and a generation lasts 20 minutes. This pattern, in which something is multiplied by a constant in a given time interval is termed *exponential* growth (*see Q2.1*).

Table 2.1 *Exponential population growth in bacteria.* A single bacterium that divides every 20 minutes is capable of producing over a quarter of a million cells in six hours

Time (minutes)	Population size	Population size expressed as an exponential function of 2
0	1	2^0
20	2	2^1
40	4	2^2
60	8	2^3
80	16	2^4
100	32	2^5
120	64	2^6
140	128	2^7
160	256	2^8
180	512	2^9
200	1,024	2^{10}
220	2,048	2^{11}
240	4,096	2^{12}
260	8,192	2^{13}
280	16,384	2^{14}
300	32,768	2^{15}
320	65,536	2^{16}
340	131,072	2^{17}
360	262,144	2^{18}

When the size of an exponentially growing population is graphed as a function of time (with the number of individuals on the vertical (y) axis and time on the horizontal (x) axis), a characteristic curve is produced (Figure 2.1).

A population that grows exponentially builds up slowly at first, but later it increases extremely rapidly. In the example shown in Table 2.1, the population at the start of each interval is twice what it was at the beginning of the previous interval, so the amount by which the population grows keeps getting bigger and bigger. Doubling a starting population of 2 individuals brings the total up to only 4, but at 120 minutes doubling the population of 64 individuals brings it to a size of 128, and so on until a doubling after 340 minutes leads to a population size of a whopping 262,144 at 360 minutes. The column on the right in Table 2.1 shows the population size at each time expressed as an exponential power of 2. A population size of 4 is expressed as 2^2, and if that is multiplied by 2 again the result is 2^3, or 8.

Bacteria have very short generation times and high reproductive rates. Exponential growth can also take place in organisms that reproduce more slowly. In such cases, it takes longer to reach a large population size, but eventually growth becomes very rapid. Charles Darwin (Section 6.1) pointed out that even elephants, which are notoriously slow breeders,

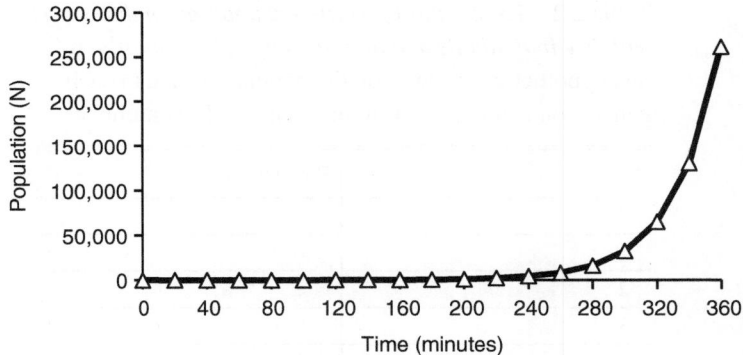

Figure 2.1 Exponential population growth. A characteristic exponential growth curve is produced when the number of individuals (N) in a population is multiplied by a constant (in this case, 2) in a given time interval.

have the potential for explosive population growth. Although different organisms increase at different rates, any type of organism can theoretically increase exponentially.

Obviously, ideal conditions are rarely encountered in nature except, perhaps, when a species is introduced into an environment in which it has no enemies. In that situation, there may be a temporary period when resources are abundant and no predators, parasites, diseases, competitors, or human antagonists are present. Although unlimited population growth is rare, exponential growth can also occur when conditions are not ideal. Even a population with a relatively low reproductive rate can build up to a level where it is quite large. For example, a population that starts with five individuals and increases by 15% annually will reach a size of 331 in 30 years, an increase of over 6, 500% (if there are no deaths) (Table 2.2).

The following legend from ancient Persia illustrates the way that exponentially increasing numbers increase slowly at first but then increase very rapidly.

A clever courtier … presented a beautiful chessboard to his king and requested that the king give him in return one grain of rice for the first square on the board, two grains for the second square, four grains for the third, and so forth. The king readily agreed and ordered rice to be brought from his stores. The fourth square of the chessboard required eight grains, the tenth square 512 grains, the fifteenth required 16,384, and the twenty-first square gave the courtier more than a million grains of rice …. The king's entire rice supply was exhausted long before he reached the sixty-fourth square. Exponential increase is deceptive because it generates immense numbers very quickly.

(Meadows et al., 1974:36–7)

In 1798, the Reverend Thomas Malthus published an *Essay on the Principle of Population*, in which he noted that the human population has a tendency to increase exponentially and suggested that unchecked population inevitably leads to resource scarcity, conflict, and mortality (Malthus, 1798). (See Sections 5.3.1 and 9.3.1 for discussions of contemporary Malthusian views about the role of population growth in generating current environmental problems.)

Table 2.2   *Exponential growth of a population that starts with five individuals and increases by 15% annually.* In this hypothetical population the population size in each generation is rounded off to the nearest whole number

Year	Population
0	5
1	6
2	7
3	8
4	9
5	10
6	12
7	13
8	15
9	18
10	20
11	23
12	27
13	31
14	35
15	41
16	47
17	54
18	62
19	71
20	82
21	94
22	108
23	124
24	143
25	165
26	189
27	218
28	250
29	288
30	331

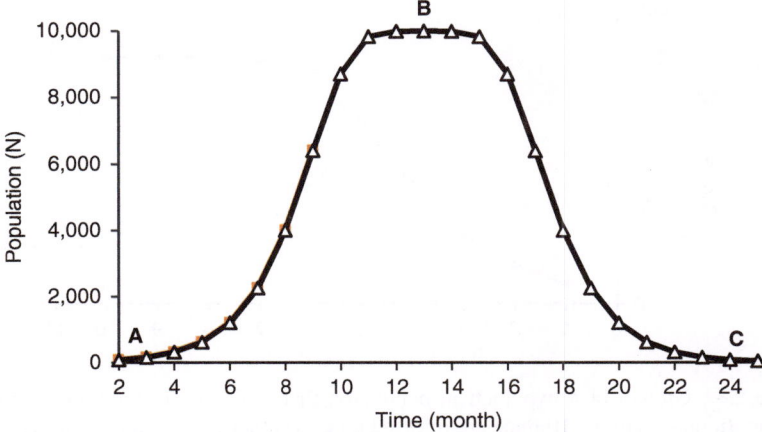

Figure 2.2   Population growth in a closed system. (A) Initial phase of exponential growth; (B) plateau phase; (C) decline phase.

These ideas influenced Darwin, who saw population pressure as the driving force for evolution. In biology, *evolution* refers specifically to inherited changes in populations of organisms that lead to the appearance of new forms. In everyday speech, however, we use the word evolution to mean something different. If I say that my attitude about something evolved, I mean that it changed gradually. That is rather different from what the term means in science.

Malthus was right that populations cannot continue growing indefinitely. However, the exponential growth curve is not a realistic model of how populations behave for very long. To return to the example involving bacteria, if organisms are in a closed flask and no new resources are added, the population will undergo an initial phase of exponential growth (Figure 2.2A), but as resources are used up, the population will level off and reach a plateau phase (Figure 2.2B). Eventually, if additional resources do not become available, resources will be exhausted, and the population will decline (Figure 2.2C).

2.1.1.2 Eruptions

In nature, populations do not exist in closed systems such as the hypothetical flask depicted in Figure 2.2. Organisms that are introduced into habitat where resources are abundant and enemies are few may at first experience a period of rapid population growth analogous to the bacteria introduced into our hypothetical flask. Unless more resources become available, the graph of population dynamics in the introduced species is likely to resemble Figure 2.2C, culminating in a crash.

This occurred after 29 reindeer were introduced on St. Matthew Island in the Bering Sea (Klein, 1968). Nineteen years later, an estimated 6,000 reindeer lived on the island, but after the following winter, all but about 42 animals had died. Similar events occurred after moose dispersed naturally to Isle Royale in North America's Lake Superior. This pattern, in which a population initially increases suddenly and then declines markedly, is termed an

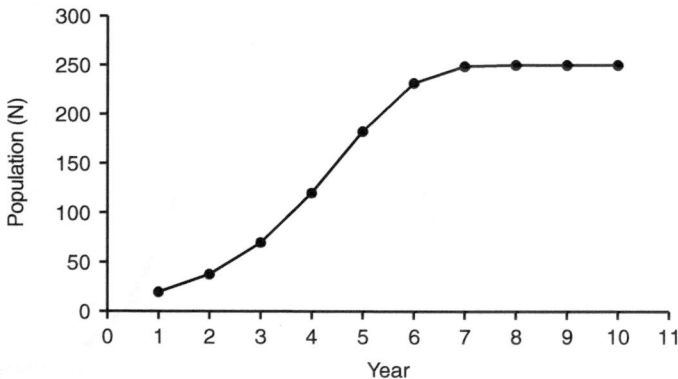

Figure 2.3   Growth of a hypothetical population that is limited in a density-dependent fashion. In this example, 10 mice are released on a hypothetical island that has a carrying capacity of 250 mice.

eruption (or sometimes, an *irruption*). Eruptions have been documented for many members of the deer family including elk (known as wapiti in Eurasia), white-tailed deer, and reindeer. The crashes that occurred during the final phase of these eruptions were attributed to starvation after an explosively increasing population depleted its food supplies. In some cases, biologists assumed that this was the case on the basis of insufficient or biased data.

2.1.1.3 Density-Dependent Growth

In the example in Figure 2.2C, the organisms are in a closed system where resources are limiting and are eventually used up. If a fresh supply of resources is added from time to time, however, the population might remain at the plateau phase instead of dying out (Figure 2.3).

The number of individuals of a given species that can be maintained in a particular environment is sometimes termed the *carrying capacity* (*K*) of that environment (Leopold, 1933). According to this model of population growth, if a population drastically exceeds the carrying capacity of its environment, environmental damage may result, and if the damage is severe enough carrying capacity may be permanently diminished.

When a population stabilizes at carrying capacity, additions to the population are balanced by losses. In other words, the population is at equilibrium. This kind of population growth is regulated by processes that depend on population density; it is *density-dependent*. In a density-dependent population, population growth slows if density increases and vice versa. Mortality will be low and/or survival will be high when population density is low. This means that a population that is subject to density-dependent regulation will increase relatively rapidly at low densities and will increase slowly or not at all at high densities (*see Q2.2*).

Density-dependent processes can affect reproduction and mortality. In a dense population, growth may slow because of competition between members of the same species (*intraspecific competition*). Plants that are closely spaced will die if they fail to obtain enough water, light, or nutrients to survive. In animal populations, high population density may reduce access to food or other resources, or it may lead to injuries from fights. It might result in increased stress that physiologically weakens individuals and leaves them susceptible to diseases. Contagious diseases operate in a density-dependent fashion because they spread from one individual to another more rapidly in a dense population than where individuals are widely spaced (a phenomenon that became obvious throughout the world in 2020 during the Covid-19 pandemic). Individuals may also be more vulnerable to predation when they are densely clustered because it may be easier for predators to find them.

If individuals do not die from the effects of high population density, they may nevertheless fail to reproduce, or they may produce fewer healthy offspring or reproduce less frequently because of stress or resource shortages associated with high population density. The reverse may occur at low densities in populations that are regulated in a density-dependent fashion. Reproduction may increase (Box 2.1), mortality may decrease, or both may occur when density is low.

Management to control the size of wild populations is often desirable either for the purpose of limiting unwanted species or else to keep a valued species from exceeding the carrying capacity of its environment. Density-dependent responses can also make control efforts inefficient and present practical difficulties for managers interested in keeping down the densities of pest species (Section 4.5.1).

Box 2.1

Effects of Population Control on Feral Horses at Assateague Island

Two herds of feral horses inhabit Assateague (AHS a tig) Island, a long, narrow strip of sandy beaches, salt marshes, and pine forests that runs parallel to the Middle Atlantic coast of the USA (Figure 2.4). Horses have inhabited the island for over 350 years. They were made famous with the publication in 1947 of the children's book *Misty of Chincoteague* (SHIN ke teeg).

Both herds live on land managed by federal agencies, but the two herds are managed quite differently. This difference in management creates a natural experiment (Introduction). At the south end of the island, the US Fish and Wildlife Service has intensively managed the herd within Chincoteague National Wildlife Refuge by removing some of the foals born each spring in an effort to prevent overpopulation within the limited habitat available at the refuge. At the north end of the island, horses inhabit Assateague Island National Seashore, which is managed by the US National Park Service. In keeping with that agency's policy of nonintervention, managers do not attempt to control population growth in that herd.

Jay Kirkpatrick, a wildlife biologist, and John Turner, Jr., a physiologist, designed a study to test a hypothesis about density-dependent population regulation in the herds (Kirkpatrick and Turner, 1991). They hypothesized that by removing foals in order to artificially lower the horse population at the wildlife refuge, biologists might actually be triggering a density-dependent increase in reproduction. They predicted that if that were the case, the reproductive rates of mares would be higher in the herd at the wildlife refuge (from which foals had been removed) than in the herd at the national seashore (where there were no removals). In this natural experiment, the treatment was population control (removal of foals), the managed herd at the wildlife refuge

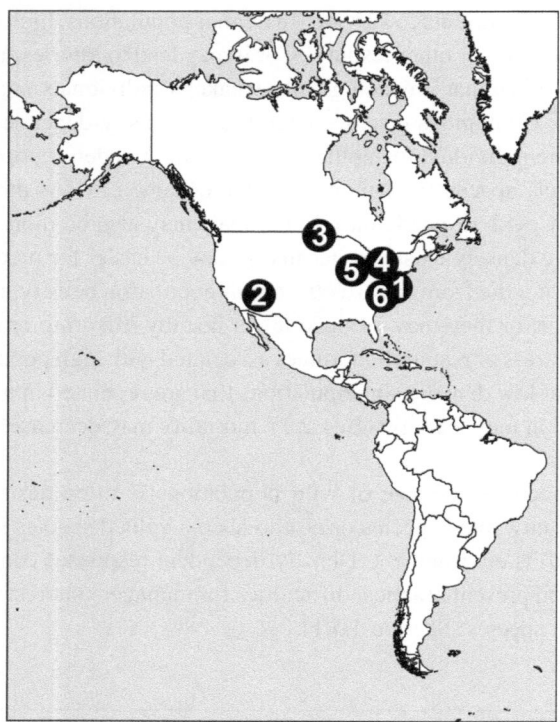

Figure 2.4   Locations of: 1, Assateague Island; 2, Kaibab Plateau, Arizona; 3, Agassiz National Wildlife Refuge, Minnesota; 4, Welland Canal, Canada; 5, Lake Michigan; 6, Duke Forest, North Carolina. Map created by Eva Strand using Esri, DeLorme World Countries Generalized Data & Maps for ArcGIS 2013, with permission.

was the treatment group, and the unmanaged herd at the national seashore was the control group (Introduction).

To find out whether a compensatory increase in reproduction occurred in the refuge herd, and if so, what caused that increase, the research team combined data from laboratory analyses of horse urine and feces collected in the field with data from behavioral studies. To do this they followed the fates of 88 sexually mature mares, each of which had its *home range* (the area within which an individual animal occurs) confined to either the national wildlife refuge or the national seashore. All the sample mares could be individually identified from unique markings. To determine pregnancy rates, urine, feces, or both were collected from each individual in the field and analyzed in a laboratory. Later the pregnant mares were observed in order to determine whether they were nursing foals. This information allowed fetal loss rates to be calculated. The results are summarized in Table 2.3.

Two-thirds of the 48 sample mares from the refuge herd, which was subjected to population control, became pregnant, whereas the average pregnancy rate for the herd of 40 mares from the unmanaged national seashore herd was only 35%, a statistically significant difference. Not surprisingly, given their higher pregnancy rate, the sample horses from the refuge also produced more foals than the sample from the national seashore. This difference was also statistically

Table 2.3   *Reproductive characteristics of feral mares at Assateague Island National Seashore and Chincoteague National Wildlife Refuge*

	Chincoteague National Wildlife Refuge	Assateague National Seashore	Difference
Management	Foals removed	Foals not removed	
Total mares tested	48	40	
Pregnant	66.6%	35.0%	significant
Number of live foals born	32	13	significant
Proportion of mares that produced foals	62.5%	32.5%	significant

significant (Table 2.3). Thus, the hypothesis that the reproductive rate of the population that was subject to population control (the refuge herd) would respond in a density-dependent fashion through a compensatory increase in reproductive rate was supported by Kirkpatrick and Turner's findings. The mechanism responsible for that increase was a higher rate of pregnancy in the refuge population, where population density had been artificially lowered.

Not all populations are regulated in a density-dependent fashion. Biologists who focus on insect populations have long argued that too much attention is paid to density-dependent regulation and, consequently, to resource limitation and competition. In the 1950s, two Australian scientists, H. G. Andrewartha and L. C. Birch (1954), suggested that unfavorable weather, physical disturbances, or environmental catastrophes typically operate in a *density-independent* fashion to keep populations well below the level at which resources become limiting. This point of view, however, is not the dominant one among utilitarian resource managers. Rather, they tend to emphasize density-dependent processes.

Andrewartha and Birch were influenced by the fact that fact that they studied insects, which often experience widespread weather-related mortality that will kill the same proportion of the population regardless of whether the population is dense or sparse. We will see in Section 10.1.1.1 that density-independent mortality can pose some challenges for managers who are used to assuming that mortality operates in a density-dependent fashion.

2.1.2 Interactions between Populations

Populations do not exist in isolation. Every species interacts with other species that prey on it, parasitize it, compete with it, protect it, provide it with necessary resources, or influence it in other ways. When one organism benefits at the expense of others, the interaction is termed *antagonistic*. Predation, herbivory, and parasitism are types of antagonistic interactions. In *mutualistic* interactions, both organisms benefit from an interaction.

Any close relationship between different species is termed *symbiosis*. Thus, antagonistic and mutualistic interactions are examples of symbiosis; however, the term symbiosis is often used, especially in popular writing, to refer only to beneficial interactions.

An assemblage of interacting populations is termed a *community*. Within a community, every species has a specific functional role; this is termed its *niche*. (A description of a species' niche includes the resources it utilizes, the type of habitat it lives in, specific features of its habitat, the seasons when it is active, and so on.)

2.1.2.1 Competition

Early utilitarian conservationists were interested in competition for a couple of reasons. As noted above, they paid attention to populations that were regulated by density-dependent processes such as competition among members of a population. Second, wild species that competed with people for resources got the attention of resource managers. Predators were in this group because they compete with people for wildlife or livestock. So were weedy plants that competed with crops for light, water, or nutrients and birds that fed on crops (Sections 4.1 and 4.5).

Even valued game species were viewed negatively if they potentially competed with livestock for forage on rangelands. This concern gave rise to many studies that documented overlap in the use of a resource (usually food) by wildlife and livestock. For example, researchers G. D. Pickford and Elbert Reid at a US Forest Service experiment station compared the diets of sheep and elk on subalpine (Section A.1.2) summer range in eastern Oregon by estimating the percentage of each plant species that was grazed by sheep in the summer of 1937 and by elk three summers later. They reported that a dozen species of grasses, *sedges* (grass-like plants that usually have triangular stems), and wildflowers made up more than four fifths of the forage used by sheep and elk and concluded that "competition is keen between these two animals for the choice forage plants" (Pickford and Reid, 1943:330). This situation, they suggested, posed a risk for subalpine habitats in the region because of overgrazing. "So long as these conditions persist" they wrote, "the most desirable forage plants cannot possibly retain their important place in the range vegetation, nor can soils remain stable" (Pickford and Reid, 1943:332) (*see Q2.3*).

The underlying assumption in this type of study was that if two species use the same resources then one must have a negative impact on the other. However, the fact that different species use the same resource does not prove that the supply of that resource limits the abundance of either or both species. Studies testing the hypothesis that different species compete with each other, in the sense that one species negatively impacts the survival of the other in areas where both are present, were not generally undertaken until the middle of the twentieth century (Section 10.1.2).

2.1.2.2 Predation

Since predatory mammals and birds kill and eat many of the same animals that people feed on, utilitarian managers have historically tended to view predators as economically detrimental. The assumption underlying this perspective is that predators limit the size of prey populations. The idea that more prey would be available if fewer prey were eaten by predators seemed self-evident. For many years, hunters, ranchers, and resource managers simply accepted that this must be the case on the basis of intuition or anecdotal evidence, a position which both reflected and contributed to the idea that species should be judged as good or bad on the basis of their utilitarian values. In their enthusiasm for this view, utilitarian managers have sometimes focused on predation as a limiting factor while downplaying the effects of other factors (Box 2.2).

If the story of the predators and deer on the Kaibab Plateau was oversimplified, what would be a more rigorous way to test the hypothesis that predation regulates prey populations? To address this challenge, it is necessary to find out if a population is smaller when it is preyed upon than when there is no predation. Comparisons between populations of herbivores in environments with and without predators are a source of data on this question. Controlled field experiments are one option for testing hypotheses about the effects of predators on prey populations. The concept is straightforward (although the logistics of

---

Box 2.2

Did Removing Predators from the Kaibab Plateau Cause an Eruption of the Deer Population?

In 1893, parts of the Kaibab Plateau in Arizona (Figure 2.4) were designated as a forest reserve; other parts were set aside as a game reserve in 1906. Several changes in land use followed. The number of domestic sheep grazing in the area was reduced, deer hunting was prohibited, fires were suppressed, and government agents were employed to kill predators (mainly mountain lions, wolves, coyotes, and bobcats) in the hope of increasing the number of mule deer.

After the reserve was created, the deer population of the plateau grew at first, but subsequently it declined. In 1941, Irvin Rasmussen published the results of research he conducted for his PhD dissertation on the *Biotic Communities of Kaibab Plateau* (Rasmussen, 1941). This publication contained detailed descriptions of the animals and plants of the plateau, including estimates of trends in the deer population from three sources: the forest supervisor, park visitors, and Rasmussen himself. The forest supervisor's estimates, which were probably the most reliable since he was most familiar with the situation, were much lower than those of Rasmussen and the other observers. Nevertheless, Rasmussen used the most extreme data points in his analysis, particularly the estimate that the herd had reached 100,000 animals in 1924. By connecting those data points, he came up with a graph that depicted a population explosion followed by a dramatic crash (Rasmussen, 1941:236) (Figure 2.5). Rasmussen also pointed out that "Total removal [in the Kaibab Plateau] 1906 to 1939 inclusive has been, 816 mountain lions, 30 wolves, 7,388 coyotes and 863 bobcats" (Rasmussen, 1941:246).

Two years later, Aldo Leopold writing in the Wisconsin Conservation Bulletin reproduced Rasmussen's diagram and wrote that the increase in the deer population was followed by "two catastrophic famines which reduced the herd 60 per cent in two winters. By 1939 the herd had dropped to a tenth of its peak size, and the range had lost much of its pre-irruption carrying capacity" (Leopold, 1943:3).

This version of events was subsequently repeated in textbooks. Finally in 1970 Graeme Caughley re-examined the Kaibab deer story. He concluded that "little can be gleaned from the original records beyond the suggestion that the population began a decline sometime in the period 1924–1930, and that the decline was probably preceded by a period of increase." He pointed out that it is not possible to determine in retrospect whether the increase in deer was caused by the decline in predators, because predation was not the only factor that changed, so that "the factors that may have resulted in an upsurge of deer are hopelessly confounded" (Caughley, 1970:56).

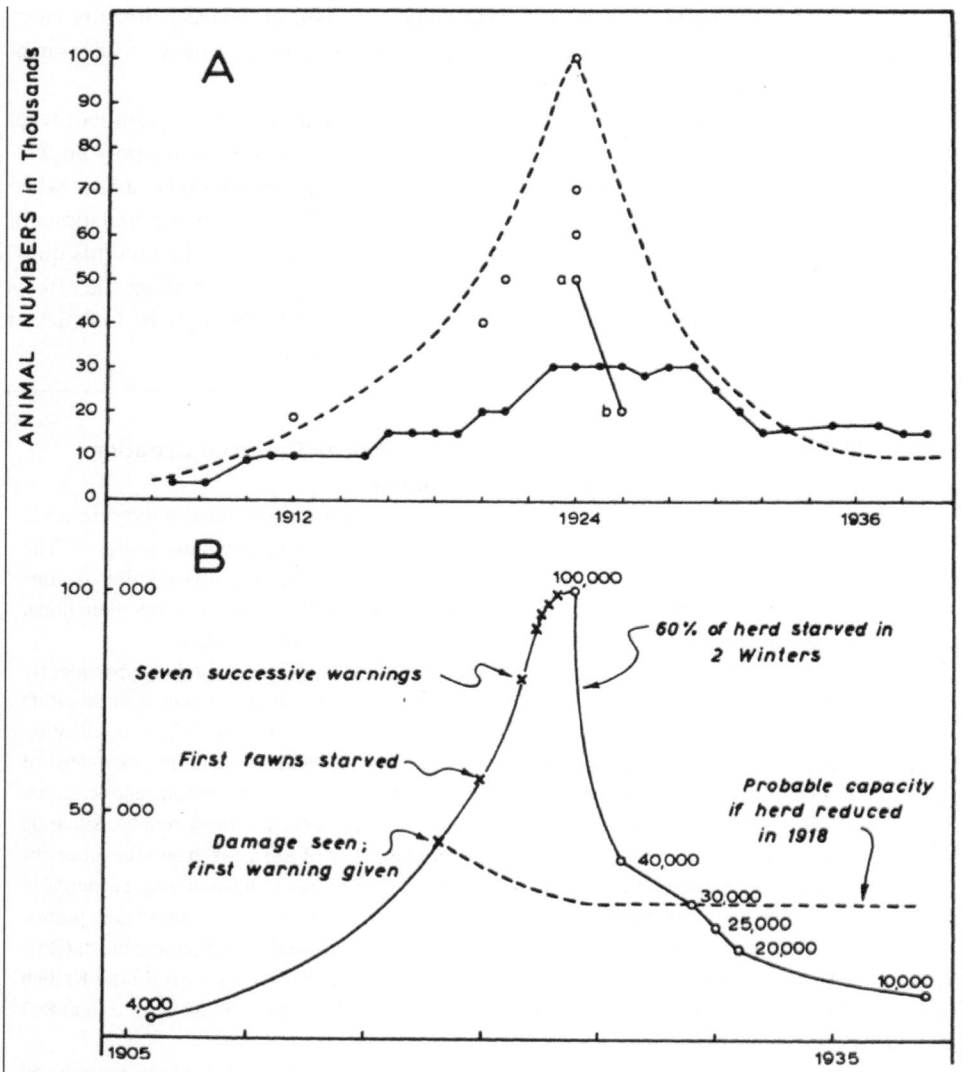

Figure 2.5   Estimated changes in the population of deer on the Kaibab Plateau from 1906–1939 (A) according to Rasmussen (1941) and (B) reproduced by Leopold (1943). Used with permission of John Wiley & Sons, Ltd., from Caughley (1970).

No one has suggested that Rasmussen's interpretation of the Kaibab deer data was consciously biased (in fact he was considered a gifted observer), but this example from the early days of the profession of wildlife management illustrates the shortcomings of before-and-after comparisons conducted without adequate controls, with subjectively chosen data sets, a preconceived interpretation, and confounding variables. Fire suppression and decreased grazing might have resulted in an increase in vegetation available to the deer and contributed to a rise in their population. Although the Kaibab deer data suggest that predators might regulate ungulate populations, this example cannot be considered conclusive evidence, because more than one variable changed simultaneously.

such experiments are challenging): compare populations of a prey species in paired study areas from one of which predators have been removed and another where predators have not been removed. A six-year study of the effects of predation on nesting ducks at Agassiz National Wildlife Refuge in Minnesota (Figure 2.4), used this approach. (Another example was discussed in the Introduction.) Predatory mammals and birds were removed annually from the west side of a marshy study area at the refuge for three years, while the east half served as a control (Balser et al., 1968). At the end of that time, the treatment and control areas were reversed in order to control for any environmental differences between the two areas. Researchers recorded the number of nests and ducklings on the treated and untreated halves. They also assessed predation by placing chicken eggs in artificial nests and recording whether the eggs were attacked. On the areas from which predators had been removed, 7,571 ducklings were recorded, compared to 4,858 ducklings on the sites with predators, a statistically significant difference. In addition, fewer artificial nests were attacked in the predator-removal areas than in the untreated areas. These findings suggest that predation limited the reproduction of duck populations on the refuge. However, although the study demonstrated higher duckling production on the predator-free areas, it did not show that the greater productivity caused a lasting increase in the duck populations. If increased duckling production were followed by higher duckling mortality, duck populations might still not be greater on the predator-free areas.

Studies on muskrats by the wildlife biologist Paul Errington illustrated how this might occur. In his field studies of muskrats in Iowa, Errington found that populations were regulated by social behavior rather than by predation. The number of individuals that older muskrats would tolerate in a given habitat determined the size of a muskrat population. Young animals were driven from their lodges by older muskrats and engaged in "foot-loose and hazardous wandering" as they dispersed through unfamiliar country. They died from intraspecific strife; motor traffic; attacks by farm dogs, mink, and other predators; and a variety of other causes. Errington considered these transient muskrats "a biological surplus, largely doomed through one medium or another" (Errington, 1946:146). He suggested that in muskrats and some other prey species predation does not limit populations because approximately the same number of individuals will die regardless of whether predation occurs. This is termed *compensatory mortality*. If mortality from predation is compensatory, then the same amount of mortality would be expected to occur with or without predation (and therefore removing predators will not have an effect on the size of the prey population because individuals that predators might have killed would probably die of other causes).

To summarize, the effects of predation (including human predation in the form of hunting or trapping) on prey populations are complex and depend on many factors including the behavior of the prey species. Predators sometimes exert a limiting effect on prey populations, but this is by no means always the case because sometimes compensatory population processes counteract losses due to predators.

Either viewpoint – that predators limit prey populations or that mortality from predation merely substitutes for other forms of mortality – is compatible with the objectives of utilitarian resource managers. Both perspectives suggest justifications for harvesting economically valued species: Hunters kill deer that might otherwise be taken by predators; trappers remove muskrats that would probably die from other causes. Similarly, foresters thin forests to remove young trees that otherwise might die from competition.

2.1.2.3 Herbivory

Herbivory is widespread. Horses, cows, deer, elk, and pandas eat plants. So do snails, butterflies, koalas, elephants, rabbits, aphids, tadpoles, hippos, prairie dogs, geese, mice, grouse, kangaroos, rats, and many other animals. In the early years of utilitarian conservation, managers focused on how grazing animals affect rangeland vegetation. They accumulated a great deal of information about the effects of grazing by sheep, cattle, deer, and elk. As part of these efforts, utilitarian managers classified rangeland plants according to their responses to grazing. Plant species that become less abundant in the face of heavy grazing are termed *decreasers*; those that become more abundant in an intensively grazed area are *increasers*. The decline of decreasers is usually due in part to the fact that they are preferred by grazing animals. But decreasers are also more sensitive to grazing than increasers. They recover from being eaten or trampled more slowly than plants that increase when grazed.

Range managers are interested in manipulating the timing and amount of grazing in ways that sustain high-quality vegetation. Of course, what constitutes high-quality vegetation is a value judgment. The choice of strategies will depend, for example, on whether the objective is to provide forage for cattle, sheep, deer and elk, or a mixture of different species or perhaps to protect a *watershed* (an area in which water drains into a given river or other water body) and provide erosion control on steep slopes. Often the manager must decide how to handle tradeoffs between different values.

2.1.2.4 Mutualism

Mutualistic interactions are widespread, and some mutualisms, such as the relationships between crops and their pollinators, have tremendous economic importance. Hence, utilitarian managers are interested in understanding how to promote such interactions. Mycorrhizae (singular, *mycorrhiza*) (from *myco*, meaning fungus, and *rhiza*, meaning root) are common and economically important mutualistic interactions between a plant and a fungus. (*Fungi* are a domain of organisms – including yeasts, molds, mushrooms, and many disease-causing microorganisms – that have cell nuclei, like plants, but cannot manufacture their own food). Many fungi have long, threadlike underground filaments that form close mutualistic relationships with the roots of many plants, including valuable trees and crops. The fungus receives sugars from the plant roots, while the plant receives nutrients that are absorbed from the soil by the fungus and passed along to the plant (Section 11.4.3).

The Clark's nutcracker, a member of the crow family, gets its name from its stout bill and powerful jaw muscles that allow it to crack open the hard coats of pine seeds. Nutcrackers harvest these in late summer or fall and bury them for later use. This activity is especially important for whitebark pine, a tree that grows near tree line in western North America. A single nutcracker can harvest 32,000 whitebark pine seeds in a year. This is far more than the bird would need in order to meet its nutritional requirements. It might seem that this behavior would be harmful to the long-term survival of the pines because the birds take so many seeds. However, the nutcrackers also promote reproduction by burying large numbers of seeds that they don't retrieve, placing them in a favorable environment for germination. Thus, the whitebark pine provides food for the Clark's nutcracker, and the nutcracker contributes to reproduction and dispersal of the whitebark pine. This is crucial because unlike many other species of pine trees, the whitebark pine's seeds do not have papery wings that would allow them to be spread by wind (Tomback, 1982).

When different species interact, one or both species may influence the other in ways that affect its evolution. This kind of reciprocal interaction, in which different species influence each other's evolution, is termed *coevolution* (Section 6.1.2). Utilitarian conservationists described many interactions between species, but although Darwin was interested in coevolution (Box 6.1), how those relationships affect coevolution did not attract a lot of scientific attention until the 1960s.

Like many terms that pertain to evolution, the meaning of coevolution is often misunderstood. In popular media it is used loosely to refer to any kind of interdependence among organisms in nature and often implies cooperation, although actually coevolution often involves antagonistic interactions, which, by definition, are not cooperative.

2.1.2.5 Introduced Species with No Antagonists

In 1958, the British ecologist Charles Elton published a book on *The Ecology of Invasions by Animals and Plants*. Not all introduced species succeed in establishing viable populations, but those that do often increase to the point where they pose a threat to native organisms and communities. "Nowadays we live in a very explosive world," wrote Elton.

It is not just nuclear bombs and wars that threaten us ...: there are other sorts of explosions ... ecological explosions. ... I use the word "explosion" deliberately, because it means the bursting out from control of forces that were previously held in restraint by other forces ....

Ecological explosions ... can be very impressive in their effects, and many people have been ruined by them ....

[W]e are living in a period of the world's history when the mingling of thousands of kinds of organisms from different parts of the world is setting up terrific dislocations in nature. We are seeing huge changes in the natural population balance of the world.

(Elton, 1958:15,18)

Organisms that are accidentally or deliberately introduced and successfully colonize places they did not previously inhabit are considered non-native, alien, or exotic. Once an introduced species becomes established, its increase is often dramatic. If the newcomers do not encounter antagonists – such as predators, grazers, parasites, or competitors – in their new home they often become invasive. (Fictional versions of this phenomenon have been humorously treated in many science fiction movies, such as *Invasion of the Body Snatchers* and *Little Shop of Horrors*.) A successful invader frequently undergoes an explosive, exponential increase in population size and geographic range, with negative impacts on native species and habitats. Introduced species may compete with native organisms, eat them, or parasitize them. Non-native species can also indirectly alter ecological processes, such as nutrient cycling, thereby causing changes in community structure and function.

For an alien organism to become established it must be able to get to its new homeland, and it must survive and reproduce when it arrives. Dispersal is limited by geographic and climatic barriers, but many organisms are aided in their dispersal by the activities of people. We introduce plants, animals, and microorganisms both deliberately and inadvertently. Contaminated batches of grain in rail shipments and ballast water in ships have unintentionally transported dozens of species to new environments. Species that can take advantage of these means of transportation are more likely to invade new regions than species that cannot. This is why beetles, which hitchhike in stored grain, and rats and mice, which stow away on ships, are well represented among introduced fauna.

When they succeed in becoming established, introduced species can have dramatic effects. Elton described invasive viruses, bacteria, plants, insects, crabs, birds, and mammals that moved between the continents, often with serious consequences for native organisms. Box 2.3 describes two examples.

Box 2.3
Examples of Elton's Invaders: Lamprey and Blight

Sea lamprey: Lampreys are known as living fossils because they have changed very little over the past 360 million years. These fish belong to the oldest group of vertebrates. Unlike modern fishes, lampreys have no jaws, no bone, and no scales. They are sometimes called eels because their long, cylindrical bodies resemble eels (which are bony fish and thus not closely related to lampreys). Instead of jaws, adult lampreys have an oral disk with several rows of sharp teeth.

Some species of lamprey, including the sea lamprey, are parasitic. Sea lampreys spend most of their life in the North Atlantic but migrate to fresh water to breed. The immature sea lamprey is a free-living filter feeder, but adults attach themselves by means of their oral disc to the body of a host fish, secrete an anticoagulant that keeps the host fish's blood from clotting, and feed on fluids and tissue that they suck from their host, which usually dies as a result.

For millennia, Niagara Falls on the border between Ontario, Canada, and New York State formed a natural barrier that prevented sea lampreys from entering the Great Lakes except for Lake Ontario. However, in 1829 the completion of the Welland Canal connected Lake Ontario and Lake Erie, giving lampreys access to Lake Erie and, eventually, the other Great Lakes (Figure 2.4). The entry of sea lampreys into these lakes devastated populations of native fishes, especially lake trout, a species of both ecological and commercial importance.

Chestnut blight: The American chestnut tree (which is not related to either the horse chestnut tree or the water chestnut) once dominated the temperate deciduous forests throughout much of eastern North America. Its wood was prized, and its nutritious fruits, which were also called chestnuts, were an important food for wildlife (including the passenger pigeon, Section 1.3.3.1), Native Americans, and Euro-Americans. Around the beginning of the twentieth century, some American chestnut trees became infected by a fungus that was apparently brought to New York City on nursery stock from East Asia. The non-native fungus, which came to be known as the chestnut blight, spread quickly throughout the eastern USA by means of its wind-borne spores. Within a few decades, viable American chestnut trees were wiped out, although live shoots sprouting from stumps persisted on some trees. (Today chestnuts sold in American supermarkets, especially around the winter holidays when roasted chestnuts are popular, are imported from Europe.)

In both these examples, native organisms succumbed to non-native antagonists against which they had no defenses. By contrast, the impacted species that had coevolved with lampreys and chestnut blight in their native ranges had developed some resistance. In the sea lamprey's native range in the Atlantic Ocean, fish usually survive lamprey attacks. Likewise, chestnuts in East Asia had some ability to survive infection with chestnut blight.

Many deliberate introductions, Elton suggested, are motivated by utilitarian concerns, especially the desire to simplify ecological relationships in ways that maximize production and minimize losses:

> with land in cultivation, whether pastoral, ploughed, or gardened, the earnest desire of man has been to shorten food-chains, reduce their number, and substitute new ones for old. We want plants without other herbivorous animals than ourselves eating them. Or herbivorous animals without other carnivorous animals sharing them.
>
> *(Elton, 1958:127)*

2.2 Succession

2.2.1 Changes in Plant Associations over Time

Vegetation often changes over time, sometimes in ways we can predict. Past hunting and foraging peoples tracked changes in dynamic environments, and later farmers undoubtedly understood that plants other than crops would take over their abandoned fields. Formal study of such changes goes back only a few centuries, however.

In 1895, the Danish botanist Eugenius Warming described a sequence of vegetation assemblages on sand dunes in Denmark. His work influenced Henry Cowles, an American graduate student at the University of Chicago. In 1899, Cowles published the results of his graduate research on sand dunes along the shores of Lake Michigan (Figure 2.4). He described a complete sequence of vegetation types that occurred in bands extending away from the edge of the lake.

The environment along the shore of Lake Michigan is characterized by extreme conditions: bright sunlight, fluctuating temperatures, wind-blown sand, frequent storms, and dry, infertile soil. In a belt of vegetation immediately adjacent to the shoreline, Cowles found a "barren" zone, the lower beach, where seedlings could not get established. Farther inland, the environment of the middle beach was still "exceedingly severe," yet it was slightly more hospitable to plant growth because driftwood and other debris deposited by storms provided some shelter from the wind. Under those conditions, a few *annual plants* (plants that germinate, grow, and produce seeds within a single growing season, in contrast to *perennial plants*, which normally live for two or more years) could get established. Fleshy leaves stored water and allowed these plants to withstand the drying effects of strong winds and direct sunlight. "Just as succulent plants inhabit deserts (Section A.1.6) where no other high grade plants can grow," wrote Cowles, "so, too, they are able to withstand the severe conditions of the beach" (Cowles, 1899:116).

Even farther from the shore, Cowles found a zone where grasses and perennial *forbs* (broad-leaved, non-woody plants such as many wildflowers) along with scattered, stunted shrubs modified the physical environment by reducing wind speed, trapping moisture, providing shade and nutrients, and retarding the downslope movement of sand. Other, less hardy, plants were then able to colonize the dune, eventually resulting in the establishment of forests still farther from the lakeshore.

Cowles' description of this sequence of vegetation stages became a classic in studies of the development of groups of plants that tend to occur together, or *plant associations* (which later came to be called *plant communities*). It illustrates key features of *plant succession*, the process by which vegetation changes over time, as understood by Cowles and his contemporaries including the ecologist Frederic Clements (1916). According to this view, changes in vegetation follow a predictable sequence of stages, which are summarized below:

1. *A few pioneer species colonize bare surfaces, initiating a process of primary succession.* A bare substrate such as rock or sand is initially colonized by *pioneers*, organisms that are able to tolerate the harsh conditions on an unvegetated surface. Bare areas can result from geologic activity, the retreat of glaciers, or erosion. Colonization takes place on relatively unweathered rocks (including grains of sand) that were not previously vegetated, so the development of soil during primary succession is gradual. For this reason, *primary succession* (succession on previously unvegetated substrate) proceeds slowly.

The colonization of bare rock and the colonization by annual forbs of the middle beach along the shores of Lake Michigan both illustrate primary succession. Initially bare rock surfaces can be colonized only by certain organisms that have adaptations which allow them to eke out a living in such a demanding setting. Usually, few organisms other than mosses and *lichens* (composite organisms made up of photosynthetic organisms such as *algae* (sing: alga) (seaweeds and other photosynthetic organisms that have cell nuclei but lack true roots, stems, and leaves) or bacteria living in a mutualistic association among the filaments of fungi) can exploit such habitats. Succession in rock crevices proceeds a bit more rapidly because soil accumulates more rapidly in cracks than on flat rock surfaces. Because most rocks are impermeable to water, periods in which water stands on the surface alternate with dry intervals.

Lichens are able to endure prolonged periods of desiccation and are able to carry on photosynthesis at low temperatures. This allows them to be active in winter, when moisture is likely to be available in depressions on rocky surfaces.

2. *Pioneer species modify a site, making it hospitable to other organisms.* As colonization proceeds, pioneers gradually change their environment, making it less favorable for some living things and more favorable for others. Along the shore of Lake Michigan, pioneers on the middle beach modified wind speed, temperature, light availability, and moisture in a host of ways, and when they died, their decomposed tissues added organic matter to the soil.

Pioneer species also colonize sites that have been disturbed by the removal of vegetation, initiating a process termed *secondary succession*. Many types of disturbances, both natural and man-made, lead to secondary succession. Anything that removes vegetation, such as volcanoes, floods, landslides, fires, insect outbreaks, storms, cultivation, logging, and grazing can set the stage for secondary succession. This type of succession proceeds far more rapidly than primary succession. We have already seen one reason why this is the case: Soils on previously vegetated sites are more developed; they are deeper and contain more nutrients. In addition, the soils of previously vegetated areas contain fungi, other microorganisms, seeds, and spores that greatly facilitate revegetation. The soils on such sites are also likely to contain burrows made by earthworms and other animals and the root channels of plants that were there before the disturbance. These passageways aerate the soil and make it more permeable. Finally, disturbances of vegetated sites rarely remove all pre-existing vegetation. The plants that remain moderate the local environment, creating microhabitats with more shade and higher moisture. If those plants are alive and mature, there may be living roots or stems that can give rise to new growth. Consequently, the plant associations that develop after a disturbance are often characterized by a mixture of pioneer species and residual species from prior associations. Understanding the process of secondary succession has important implications for habitat restoration (Section 11.1.3).

3. *A predictable sequence of plant associations develops as each community of organisms changes the environment.* These changes create conditions that allow other species to become established. The species in this new group again alter the site, allowing still other species to colonize it. This process may be repeated several times, so that different plant associations follow each other in a fairly predictable series of stages, like the associations on the dunes along Lake Michigan. Ultimately, a stage may be reached when the species on a site are capable of reproducing under the conditions they create. This final stage is termed the *climax*, and the communities preceding the climax are termed *seral stages* (Box 2.4). Once a climax has developed, succession will, by definition, stop. If that happens, the plant association of will be stable; no further change will occur unless a disturbance returns the vegetation to an earlier stage of succession.

Box 2.4

Utilitarian Management of Forest Succession in Abandoned Fields in North Carolina

When settlers came to central North Carolina, they cleared and cultivated large areas of forest (Figure 2.4). Although agricultural productivity was high at first, fertility quickly declined and erosion increased, after which fields were abandoned and allowed to revert to forest. The result was a mosaic of forest types in varied stages of secondary succession.

In 1931, Duke University in North Carolina designated 1,900 ha for the Duke Forest Demonstration and Research Laboratory (Korstian and Maughan, 1935) as part of its efforts to build a program in forestry. The Duke Forest provided a natural laboratory for studying the development of forests on cleared land.

One year after upland fields were abandoned, two annual forbs were prevalent, but just a year later, a mixture of tall, brushy annuals and perennials dominated. By the third year, native perennial grasses were dominant, but pine seedlings were already present. The grasses maintained their dominance until the pines grew tall enough to form closed canopies, which usually occurred within 10 to 15 years. Although the pines were able to reproduce in the open grass- and forb-dominated communities that developed after field abandonment, they could not reproduce in their own shade. By about 40 years after abandonment, a distinct understory of hardwood trees such as red gum, black gum, sourwood, and dogwood was established, and oaks and hickories of varied ages were present. After 70 to 80 years, pines were less abundant, and oaks and hickories had increased. At similar sites in the region, only a few scattered pines would remain after 150 to 200 years, and an oak-hickory climax would remain (Oosting, 1942).

The Duke Forest was intended to "contribute toward the solution of some of the problems which forest-land owners must face if the forests of this region are to be perpetuated and wisely managed as a renewable natural resource" (Korstian and Maughan, 1935:7). Scientific understanding of succession had important practical applications for that objective. Information about changes in the sequence of plants during succession was applied to lands that were commercially harvested. For example, if the goal was to maximize the commercial production of pines, harvest should take place while they were rapidly growing, which occurred in a seral stage of development, before the upland forest was around 75 years old. After that, the pines were considered "overmature" because their growth had slowed to the point where they had little value as timber.

The concept of a sequence of plant associations leading to a final climax allows managers to think about the potential future vegetation of a site as well as the vegetation that is there at a particular moment in time. For example, commercially valuable trees such as pines can be harvested when they are at a rapidly growing seral stage. The concept of succession has other practical applications as well. Some wildlife species utilize late-successional vegetation, such as old-growth forest, whereas others utilize early successional vegetation, such as clearings and groves of saplings. By understanding these relationships, managers can manipulate habitats to favor species of particular interest. It was not until the last half of the twentieth century, however, that utilitarian managers began to realize that although old forests might not have optimum economic value, they supported many ecologically important species and habitats (Section 11.4.2).

2.2.2 Controversies

Clements (1916) made two additional assumptions about plant succession, both of which were challenged by some plant ecologists. First, he asserted that every region has one and only one climax, which he termed the *climatic climax*. According to Clements, within a specific regional climate, succession would always tend toward the establishment of the climatic climax. Under typical conditions (deep, fertile soils on gently sloping topography) in any given climate region only one plant association could develop. Hence, the final stage in the vegetation of a region could be predicted. Clements "confidently affirmed that stabilization is the universal tendency of all vegetation under the ruling climate." He recognized that most of the Earth's vegetation was not in a climax state, but he argued that most deviations from the expected climax were due to the activities of people: "Man alone can destroy the stability of the climax during the long period of control by its climate" (Clements, 1936:256). Clementsian ecologists held that the preponderance of early successional plant associations was due to human-caused disturbances such as cultivation and deliberate burning. Natural disturbances played only a minor role in this view.

Second, Clements viewed plant associations as comparable to individual organisms. According to this interpretation, the climax association that develops at the end of a successional sequence is a stable entity. The species that make up a plant association are interdependent, and associations are fundamental units of nature: "Each climax formation is able to reproduce itself, repeating with essential fidelity the stages of its development ... comparable in its chief features with the life-history of an individual plant" (Clements, 1916:3). To Clements, the comparison between plant associations and organisms was not simply a metaphor. He believed associations were objective units in nature.

Both these assertions were controversial. First, the British plant ecologist Sir Arthur Tansley and others contended that because the environment of any climate region was far from homogeneous, the stable vegetation of a region was never uniform (Cowles, 1911; Tansley, 1920, 1935). Because of factors other than climate, "different kinds of stable vegetation are developed and remain in possession of the ground, to all appearance as permanently as the climatic climax" (Tansley, 1935:292).

Some plant ecologists also rejected Clements' argument that plant associations were true organisms, although they did consider the comparison between plant associations and organisms to be a useful, if subjective, analogy (Tansley, 1920, 1935; Cooper, 1926). Instead, these ecologists considered plant associations to be fairly loose collections of species. They argued that plant associations resulted from assemblages of organisms responding to the environment individualistically, according to species-specific ranges of tolerance (Gleason, 1917, 1926). In their view, a plant association should be regarded as a collection of individual plant species with similar, but not identical, environmental requirements, rather than as a highly integrated unit.

If associations are true organisms, then all the species in an association should respond in the same way to the environment. This hypothesis can be tested by obtaining quantitative data on the distribution of different species across a variable environment. Two European scientists, the Russian Leonid Ramensky and the Polish scientist Josef Paczoski, did just that (Maycock, 1967; McIntosh, 1983). In the 1920s and 1930s, they presented data on how different plant species responded to slight variations in the environment. Their results showed that different species did not respond as if they were a unit. Each species became more abundant or less abundant as the environment varied, but they had different responses to that environmental variation. Thus, the hypothesis that the different plant species are interdependent and function together as a single organism was not supported. At the time it was published, however, this kind of evidence did not receive a lot of attention. Clements' views tended to prevail until the middle of the twentieth century (McIntosh, 1983) (Section 10.1.5.1).

2.3 Ecosystems

2.3.1 The Concept

In the nineteenth century, several European and American scientists had studied interactions between organisms and their environment, but it was not until 1935 that Tansley used the term *ecosystem* to denote a system formed by the interaction of a community of organisms with each other and with their physical environment (Tansley, 1935). He suggested that the fundamental concept for ecological study should be

the whole *system* (in the sense of physics), including not only the organism-complex, but also the whole complex of physical factors .... Though the organisms may claim our primary interest, when we are trying to think fundamentally we cannot separate them from their special environment, with which they form one physical system.

It is the systems so formed which, from the point of view of the ecologist, are the basic units of nature on the face of the earth .... These *ecosystems*, as we may call them, are of the most various kinds and sizes.

(Tansley, 1935:299, emphasis in original)

As the last sentence indicates, defining the boundaries of an ecosystem is somewhat subjective. This is because biophysical interactions occur at a variety of spatial scales and because ecosystems are open and interconnected across a landscape. An ecosystem might be thought of as a pond, a marsh, a forest, a rotting log, or the entire *biosphere* (the portion of the Earth that contains life).

The implications of these different theoretical perspectives on ecology are discussed at the end of this chapter (Section 2.4). We will also return to a consideration of those implications in the light of developments in ecological theory at the end of the twentieth century (Section 10.1).

2.3.2 Ecosystem Components

2.3.2.1 Soil

Ecosystems are composed of living (*biotic*) and nonliving (*abiotic*) components. Soil is a link between biotic and abiotic parts of the environment; it is formed when rock weathers and the resulting material mixes with decomposed tissues and products of organisms. Dead plant matter is incorporated into soil and also forms a layer of organic material on the surface. Animals influence soil through their burrowing, wallowing, and trampling. Microorganisms modify soil fertility, nutrient cycling, and properties such as texture and chemistry.

Soil is the basis of terrestrial ecosystem productivity. The terrestrial soil community includes bacteria, fungi, plants, earthworms, and other invertebrates, as well as burrowing vertebrates such as ground squirrels and badgers.

To support vegetation, soil must provide substrates that are suitable for root growth and must supply adequate carbon, nitrogen, phosphorus, micronutrients, water, and oxygen. Well-functioning soil must store and release water necessary at times and in amounts appropriate for plant growth. Soil also has the capacity to store or release greenhouse gases (Section 10.3.2).

Because soil influences the distribution and abundance of wild organisms as well as the productivity of crops, it is of interest to conservationists. Soil quality declines when soil is polluted or compacted, when the community of microorganisms is degraded, or when fertility is depleted because nutrient losses exceed nutrient inputs. Compaction, salinity, excessive evaporation of salts, and irreversible alteration of the soil community can cause soils to lose their ability to absorb water, resulting in soil loss from runoff and erosion. This is of concern because soil forms slowly, but its loss due to wind, water, landslides, or earth-moving machinery can be very rapid.

2.3.2.2 Autotrophs and Heterotrophs

The biotic components of an ecosystem are divided into two categories on the basis of whether or not they can make organic compounds such as carbohydrates, fats, and proteins. In chemistry, the term *organic* refers to compounds containing bonds between carbon atoms. Note that this is different from its meaning in everyday usage, where the word organic can refer to food that is produced without the addition of *synthetic chemicals* (chemicals that don't occur in nature). This can lead to confusion. The word organic is often used loosely as a synonym for natural. But a chemist considers the synthetic pesticide DDT (Section 5.2.1.1) an organic compound because the DDT molecule has carbon–carbon bonds. A farmer would not consider DDT an organic compound, however, because it is made in a laboratory.

Autotrophs (*auto*, self; *troph*, feeding) are organisms that produce their own food. They do not have to feed on organic matter derived from the tissues of other organisms because they are able to manufacture organic compounds using energy either from sunlight (in *photosynthesis*) or from chemicals (in chemosynthesis). Green plants, algae, and some one-celled organisms contain the green pigment *chlorophyll*, which absorbs energy from sunlight. In the process of photosynthesis sunlight drives chemical reactions that form high-energy organic compounds such as glucose from carbon dioxide (CO_2) and water. Some microorganisms, such as certain bacteria found in the soil or in nodules on the roots of some plants convert the element nitrogen into a form that plants can use, a process known as *nitrogen fixation*.

Some bacteria and *archaea* (a domain of single-celled organisms that are similar to bacteria but differ from bacteria in their chemical composition) use the energy from chemical bonds or charged particles, rather than sunlight, to make organic compounds in the process of chemosynthesis. Because they don't require sunlight, these chemosynthetic autotrophs can live in dark ecosystems, such as some caves and deep-sea environments (Box A.1).

The total amount of biomass produced by photosynthesis in an ecosystem is its *productivity*. (This is distinct from what we mean when we refer to the productivity of an animal population, which is its population growth.) Plants use some of the products of photosynthesis in chemical reactions in which the energy in organic compounds is converted into a form that cells can use. The resulting plant is *net primary productivity* (NPP). For terrestrial plants, NPP includes biomass below as well as above ground, but in practice roots are often ignored, and NPP values are based on matter above the ground – stems, leaves, flowers, fruits, and seeds.

Heterotrophs (*hetero*, different; *troph*, feeding) are organisms that cannot manufacture their own food; they obtain energy by consuming autotrophs. Animals, fungi, and some microorganisms are heterotrophs.

Carnivores feed primarily on animal tissue, whereas herbivores feed on plants. Some plants, such as pitcher plants and Venus fly traps, are partially carnivorous. They can trap and digest insects and other small animals. These adaptations allow them to function as heterotrophs in addition to carrying out photosynthesis. Some heterotrophic plants and animals get some or all of their nutrients by parasitizing other organisms. Mistletoes (a group of plants that are used in traditional medicine in some cultures and that are the source of the custom in which people standing under a sprig of mistletoe at Christmastime are expected to kiss) get some of their nutrition from photosynthesis and some from parasitizing photosynthetic plants, including some commercially valuable trees such as Douglas-fir.

2.3.2.3 Trophic Levels

The biotic components of an ecosystem can be classified into a series of functional levels termed *trophic levels*. Autotrophs comprise the first trophic level of an ecosystem, the *producers*. Herbivores and carnivores are *consumers*; they obtain their energy by ingesting producers. Herbivores are *primary consumers* because they feed directly on the first trophic level. Carnivores are *secondary consumers* if they feed on herbivores, or *tertiary consumers* if they feed on other carnivores. Heterotrophs that chemically break down dead organisms from any of the other trophic levels are *decomposers*. Fungi, bacteria, and earthworms are examples of decomposers. They convert the complex molecules of plant and animal tissues into substances that can be used by plants.

On land and in the Earth's surface waters, some of the solar energy input to the biosphere is trapped by photosynthesis and converted into *biomass* (plant matter), some of which passes up the food chain when it is eaten by animals. After dead plants and animals decompose, fungi and plants take up the molecules from the decomposed tissues and the cycle repeats. In the depths of the ocean chemosynthesis sets in motion a similar process.

Relationships between trophic levels can be represented by a *biomass pyramid*, which shows the total dry weight (usually expressed as g/m^2) of the organisms at any trophic level (Figure 2.6). In terrestrial ecosystems, the biomass of each trophic level is greater than the biomass of the next trophic level on the pyramid. The fact that biomass decreases as we ascend the pyramid means that persistent toxins become more concentrated at each trophic level (Box 5.2). For this reason, toxins often accumulate in the tissues of predators.

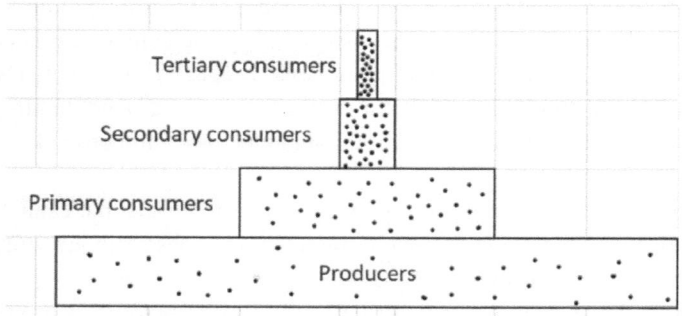

Figure 2.6   Hypothetical pyramid of biomass for a field. Dots indicate molecules of a compound (such as a DDT) that does not break down. Although biomass decreases with each trophic level, the amount of DDT stays virtually constant. Therefore, the DDT becomes more concentrated with each subsequent level of the pyramid. (Horizontal dimensions indicate relative biomass but are not to scale.)

2.4 Implications: Equilibrium and Stability

We have seen in this chapter how ideas from the developing field of ecology provided a theoretical basis for utilitarian conservation. The next two chapters cover the practical application of those concepts to utilitarian goals.

Utilitarian conservation focuses on controlling environmental variables in ways that attempt to keep populations in balance with their environment. Particular attention is paid to population regulation in resource-limited systems and changes in plant associations over time. Both these concepts are grounded in the assumption that the natural world tends toward equilibrium. Ecological studies undertaken in the context of a utilitarian framework have contributed to our understanding of population growth and regulation and interactions between species in some situations. This perspective is useful in elucidating how populations behave when resources are limiting and how vegetation develops when disturbances are few. Many of the strategies for conserving economically valuable species that were developed with that mindset were later applied to the conservation of rare and sensitive species (Chapter 7).

In the world according to utilitarian conservationists, nature could be conserved by favoring beneficial species, controlling harmful ones, and minimizing disturbances. This world was thought to be predictable and stable. Density-dependent changes in survival and mortality would keep populations in balance with the carrying capacity of their environment. Plant associations would develop in orderly sequences that culminated in stable, enduring climaxes. Nature would tend to be in balance. Processes leading to equilibrium would prevail.

However, many scientists and managers later saw weaknesses in this reassuring viewpoint. They argued that many species that did not have obvious economic value, including predators, had critical ecological values and that controlling unwanted species had unforeseen consequences. In this view, abiotic and biotic components of the environment are linked in complex systems; plants react in species-specific ways to their environments rather than as a coordinated group; some plant associations depend on periodic fires and other disturbances; and not all populations are regulated by density-dependent mechanisms. In short, the natural world is more complicated, less predictable, and just plain messier than conservationists in the late nineteenth and early twentieth centuries thought it was.

We will explore the implications of those ideas in Parts II and III. But first, let us look at the strategies used by utilitarian managers.

3

Strategies

Managing Harvests and Habitats for Valued Species

In this chapter, we consider how utilitarian managers regulate harvests and manipulate the amount and quality of resources available to wild species of interest. In the early days of utilitarian management, a rather narrow subset of species was considered worthy of conservation. Section 3.1 illustrates one way of defining valuable species.

3.1 Classifying Species and Habitats on the Basis of Utilitarian Values

In 1885, the US Department of Economic Ornithology and Mammalogy (which at that time was part of the US Department of Agriculture) used data from field observations, experiments on captured birds, and laboratory examinations of stomach contents to classify over 100 species of birds as useful or injurious (Palmer, 1899). Harmful species were those that consumed crops; beneficial ones ate harmful species. In one sense, this work promoted conservation by recognizing that many nongame birds play an important role in the control of insects and weeds. The underlying message, however, was that some species were better than others.

3.2 Managing Harvest

3.2.1 Types of Harvest

Wildlife harvests can be classified according to whether their main objective is to provide recreation, profit, or subsistence and whether or not they are legal. *Recreational harvests* are legally regulated, noncommercial harvests of wildlife or plants. In *commercial harvests*, the objective is to sell a harvested wild product at a profit. *Subsistence harvests* are harvests in which a harvester's family or community depends to some extent upon the products that are obtained. Harvests are regulated by community norms and by laws and regulations.

Many harvests fall in more than one category. Wild plants, animals, and fungi are harvested to obtain food and other products, to observe tradition, to provide recreation, to acquire luxury items for trade or other forms of commerce, and to remove dangerous or destructive organisms. Often wild product harvests fulfill several of these objectives.

Hunting for recreation may also involve tradition and obtaining food or other products. Some species are sought for several reasons. Salmon, for instance, are harvested for recreation, commerce, subsistence, and to maintain cultural practices.

This chapter covers the management of recreational and commercial harvests. Sections 12.6.1 and 12.6.2 describe some examples of subsistence harvests.

3.2.1.1 Recreational Harvests

Hunting and fishing for sport are legally regulated, noncommercial harvests of wild organisms in which recreation is a principal objective. Flowers, herbs, mushrooms, and berries are sometimes harvested for recreation, but usually the term recreational harvest applies to hunting game animals, trapping furbearers, and fishing for sport. *Trophy hunting* is a type of sport hunting in which the primary goal is the recreational killing of an animal that is rare or dangerous or is an unusual physical specimen. In many parts of the world, trophy hunting by tourists provides income for conservation.

In some countries, participants in recreational hunting, fishing, and trapping must buy a license and adhere to regulations about the number of animals that can be killed, their sex or age (or, for fish, size), the methods or equipment that can be used, and the type of lands or waters where harvest is allowed. In addition, hunters and trappers may be required to attend safety training.

Managers take into account the quality of the recreational experience. Sometimes this is more important to participants than the take. The definition of what constitutes a high-quality recreational experience depends on culture and history.

Recreational hunting, trapping, and fishing are emotionally charged issues, with strong feelings for and against. Passionate feelings about related issues such as gun control and animal rights come into play. Table 3.1 summarizes some arguments for and against recreational hunting and trapping.

Many people who oppose recreational hunting and trapping find some forms of these activities more objectionable than others. Disapproval of trophy hunting is high because the goal is the acquisition of a status symbol. Bow hunting and other forms of hunting that use traditional technology are controversial. They require more skill than hunting with modern technology, but these methods may result in more suffering if they do not kill as quickly as firearms.

In general, people with a negative opinion of hunting express the highest degree of disapproval for forms of hunting in which the meat is wasted. Conversely, approval of hunting tends to be relatively high if the meat is used. There is evidence that this is especially true in societies where eating game is common. In Sweden, where meat from recreational hunting can be bought and sold, many nonhunting households consume meat. When a team of wildlife biologists surveyed a random sample of Swedish residents to test the hypothesis that consumption of game meat is associated with favorable feelings about hunting, they found that among nonhunters positive attitudes toward hunting were highest in households that consumed game meat (Ljung et al., 2012).

Attitudes about sport hunting vary with class and region. In Canada, Europe, and the USA, sport hunting is more prevalent in rural areas than in cities. In much of Europe, sport

Table 3.1 *Some arguments for and against recreational hunting and trapping*

	For	Against
Animal welfare	Death from hunting or trapping substitutes for a painful, prolonged death from disease, starvation, or predation. Animals that die from hunting and trapping have better lives than animals that are raised in crowded feedlots.	Hunting and trapping cause trauma and suffering. The products obtained by recreational hunting and trapping are not necessities.
Conservation	Hunting and trapping have a long history of being associated with conservation. Income from the sale of hunting and fishing licenses and from taxes on firearms and ammunition provides funding for conservation. Hunting associations lobby for the protection of wildlife habitats.	Hunting and trapping kill non-target animals including rare and endangered species. Hunters and trappers disturb habitats by driving and trampling in remote areas and leaving trash.
Culture	The ancestors of modern humans were hunters. Hunting is part of our heritage.	Early human hunters exterminated many species.
Ethics	Killing to obtain meat rather than purchasing it promotes awareness of the connection between people and nature.	Hunting and trapping do not promote respect for life or nature because they inflict preventable suffering.
Personal development	Hunting and trapping require physical stamina and mental discipline. Hunting and trapping promote sportsmanship and provide opportunities to develop skill in observing nature. Obtaining and processing meat, skins, and furs from hunting or trapping encourage self-reliance.	Because sophisticated technology for finding and killing animals is available, recreational hunting and trapping do not have to be physically or mentally challenging. They do not foster sportsmanship because they do not involve a fair chase.
Safety	The experience of killing animals makes hunters aware of the danger firearms can inflict and promotes responsible gun ownership. Required hunter safety training promotes responsible gun owners.	Hunters and trappers kill or wound unintended animals including pets. People are killed and wounded in hunting accidents. The widespread gun ownership associated with hunting contributes to gun-related violence, including suicide and domestic abuse.
Values	Hunting and trapping promote personal and social growth by fostering appreciation of nature, hunting traditions, and bonds among family members and friends and between hunters and their dogs.	Recreational hunting and trapping contribute to the idea that nonhuman life forms have no rights, which is at the root of many spiritual and environmental problems in modern society

hunting is associated with aristocratic traditions. In contrast to this situation, hunting in North America has not generally been associated with elites, except for British sportsmen in nineteenth-century Canada, who participated in a type of hunting that embodied values associated with British colonial hunting (Section 1.2.2) (Gillespie, 2007).

Sport hunting in Europe dates back to the Middle Ages, when it was a privilege of the nobility and sometimes the clergy. Hunting in Central Europe is still an exclusive endeavor, with stringent requirements for participation. To join the Czech Hunters' Union, for example, one has to go through 60 hours of education, gain a year of experience, and pass qualifying exams (Newman, 1979). An emphasis on formal customs persists:

This country has very old and rich hunting customs and traditions which have to be regarded as important cultural heritage and observed accordingly. We encourage the hunters to use the historical hunting terminology and to honour the killed game in a proper way. We also encourage them to observe the dress code and show good manners.

(Stepan Muller, Vice Chairman, Czech Hunters' Union, Radio Prague International, 2003, https://english.radio.cz/czech-hunters-union-founded-eighty-years-ago-8073238)

In this cultural context, trophies confer considerable status. Usually, the head or the skin of a game animal is displayed as a sign of the hunter's prowess.

Falconry, a type of sport hunting in which falconers use captive birds of prey (usually falcons, hawks, or eagles) to hunt wild animals, has a long history in the Middle East, Europe, and Central Asia. Because raising and training the birds requires substantial time and resources, falconry was historically practiced by aristocrats, although that is no longer the case.

Due to international publicity about traditional falconry in western Mongolia, tourism in the region has increased (which is turning out to be a mixed blessing) (Box 3.1).

Box 3.1
Influences of Tradition and Modernity on
Eagle Falconry in Western Mongolia

The Kazakhs of the Altai mountain range in western Mongolia (Figure 3.1) use female golden eagles (which are larger than males) to hunt foxes from horseback. In 2014, an online photo of a young eagle huntress named Aisholpan Nurgaiv went viral and inspired an award-winning documentary about her story. The film shows Nurgaiv capturing an eagle fledgling from a rocky precipice, training it, and traveling at the age of 13 to the region's annual eagle festival, where she showcases her skill and bond with her eagle during competitions.

The film includes interviews with elders who disapprove of girls participating in falconry, giving the impression that traditional eagle hunting is exclusively a male activity in Nurgaiv's community. However, Takuma Soma, a historian of Asian falconry, disputes this interpretation, arguing that in Kazakh society of western Mongolia "anybody can capture and own their eagle, and then start hunting without any restrictions" (Soma, 2013:91).

Golden eagles are a symbol of Kazakh identity, and eagle festivals help to transmit the customs and skill associated with falconry. In 1990, after the nation of Mongolia began to transition from a socialist state to a market economy, *cultural tourism* (Section 12.3.2) showcasing traditional Kazakh culture, especially eagle festivals, followed. According to Soma and Battulga Sukhee at the National University of Mongolia, this development has had mixed impacts. The festivals bring in outside income to the region and motivate people to take up eagle hunting. However, some newcomers keep eagles without the knowledge, skill, and commitment necessary to tame and hunt with them, which undermines the transmission of traditions (Soma and Sukhee, 2014).

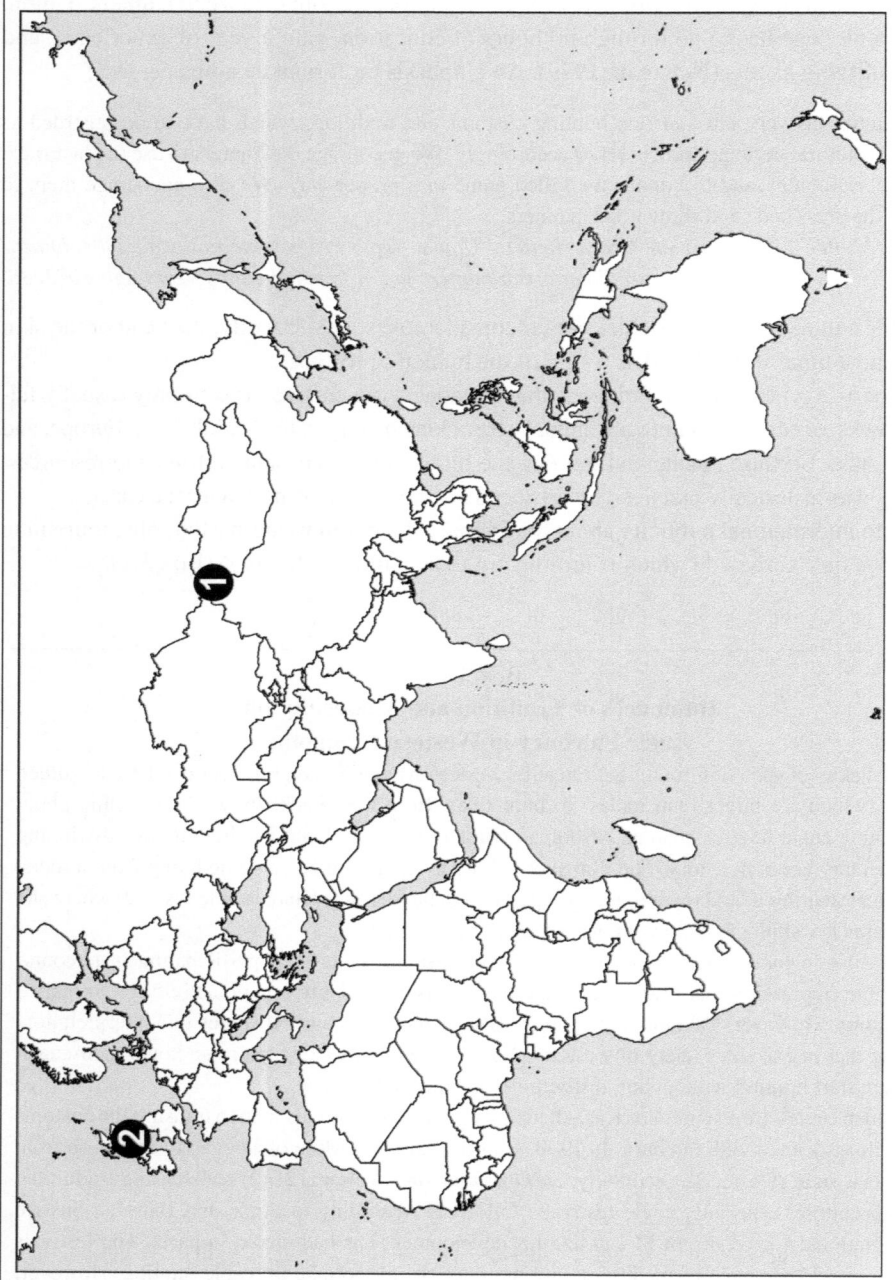

Figure 3.1   Locations of: 1, Altai Mountains, western Mongolia; 2, moors of Scotland and England. Map created by Eva Strand using Esri, DeLorme World Countries Generalized Data & Maps for ArcGIS 2013, with permission.

Table 3.2 *Selected examples of small-scale commercial harvests of wild plants or fungi*

Plant or fungus	Countries	Part used	Used for
Blueberry	USA, Finland, Sweden	Fruits	Food
Brazil nut tree	Bolivia	Seeds	Food
Cactus	Mexico	Stems, flowers, fruit, spines, whole plants	Food, building materials, ornamental plants, psychoactive drugs
Chicle tree	Guatemala	Latex	Chewing gum
Cinderella tree	China	Fruits	Traditional medicines
Ginseng	South Korea, USA	Roots	Dietary supplements
Matsutake mushroom	China, USA	Fruiting body	Food
Pine, cedar, spruce, and fir trees	USA, Canada	Branches	Floral greens, decorations
Rattan palm	Philippines	Vines	Wicker furniture
Porcini mushroom	Finland	Fruiting body	Food

3.2.1.2 Commercial Harvests

In commercial harvests of wild products, the objective is to sell or trade a harvested plant, animal, or fungus or its parts at a profit. Commercial harvests may involve the meat, eggs, bones, antlers, horns, teeth, skins, furs, organs, or bodily fluids of animals. In some places, wild plants and fungi are commercially harvested on a small scale (Table 3.2). These regulated local harvests do not involve elaborate machinery or a large labor force. In some contexts, they provide much-needed food and income for poor women. However, in some cases – such as harvests of matsutake mushrooms – enormous profits are possible from small-scale harvests of wild products.

The effects of commercial wild product harvests vary. Some may be ecologically sustainable and economically viable, whereas others are not (Section 12.3.1.2).

Industrial fisheries harvest fish, shrimp and other invertebrates, and seaweed; timber companies harvest trees commercially. Some of these organisms are also commercially harvested by small businesses such as marine divers, small-scale fishers (Section 12.7), or private owners of woodlots. These harvests are legal when they comply with laws, treaties, and state, and local regulations.

3.2.1.3 Subsistence Harvests

Hunting gathering, and fishing for wild things, often referred to as subsistence harvests if these activities are not for commercial purposes, are important for maintaining traditional cultures as well as for the nutritional and material benefits they provide. In many parts of the world, hunting and gathering wild products for fuel and forage are economically significant forms of subsistence harvest.

Colonial governments enacted laws criminalizing Indigenous harvest of meat and other resources. However, beginning in the 1980s, Indigenous peoples in several parts of the world, particularly the Canadian and American Arctic, entered into agreements with national or international agencies to co-manage subsistence harvests of whales (Section 12.6.1), caribou (reindeer), and the brown bear (Section 12.6.2).

3.2.1.4 Illegal Harvests

Rights to harvest wild products are regulated by customs, laws, and treaties, which are not always compatible with each other. *Poaching* is the illegal taking of wild products. A poacher might be a professional who illegally kills game or fish for profit, an Indigenous hunter who is prohibited from meeting their needs in traditional ways, or a recreational hunter who hunts an abundant game species but does so without a license or takes more than the legal limit or harvests out of season.

Attitudes toward poaching vary depending on cultural context. Where hunting regulations are made by one group of people to govern others who have little or no say in the matter, compliance is likely to be low. In such cases, poachers are often viewed sympathetically by the general population. In Europe, Australia, Africa, and parts of Asia, game management has a long association with protecting wildlife for the wealthy, and poaching was seen as a right of the commoner. Consequently, people were generally more sympathetic to poachers than to enforcers, an attitude that persists in favorable portrayals in legends, children's books, and songs. Local people may favor poaching if they see state control within public lands as serving the interests of an urban elite (Jacoby, 2014).

Poaching is likely to be prevalent where penalties are small compared to potential profits. International trade in wildlife is a multi-billion-dollar enterprise. Enforcement is complex and dangerous. Not all wildlife trade is illegal, but it can be difficult for enforcers to distinguish items that are legally traded from those that are not. For this reason and because of the high profits involved, wildlife trade is a serious threat to numerous endangered species (Section 7.3.1 and Box 7.2) (Rosen and Smith, 2010).

3.2.1.5 Harvests of Bush Meat

In many parts of the tropics, wildlife meat is harvested for subsistence or for sale (Box 12.1). These enterprises are legal in some contexts and illegal in others. Local consumption of wild meat (sometimes termed *bush meat*) may provide income and an affordable source of protein to poor households. But overhunting has caused many local extinctions (Milner-Gulland and Bennett, 2003).

In markets where wild meat is sold, serious risks to human health can arise from handling, consuming, or being bitten by wild animals if disease-causing organisms present in wild animals become capable of infecting people. This happened in 2002, when a virus in bats and a species of wild civet (a medium-sized carnivorous mammal native to Asia and Africa) at a market in China infected a single person. An international epidemic of severe acute respiratory syndrome, or SARS, that resulted in the death of nearly 800 people followed. There is evidence suggesting that a similar but far more deadly scenario may have unfolded late in 2019 and led to the pandemic of coronavirus disease (Covid-19) that killed millions of people and had devastating social and economic consequences.

3.2.2 Managing Harvests in Theory

3.2.2.1 Maximum Sustained Yield

Resource managers often have to decide how many individuals can be removed from a population without depleting it. To do this, they obtain data on population dynamics, use that data to set harvest levels, and try to enforce the resulting regulations. In this way, they attempt to prevent both overpopulation and overharvesting of the harvested species. The basic question is: how many individuals can be removed from a population without compromising its ability to recover by replacing the harvested individuals? Stated another way, if we want to harvest a population without causing it to decline permanently, how much can we take? If we take only an amount that is equivalent to what the population produces (and if the harvest itself does not cause reproduction to decline), then we should be able to harvest repeatedly, without triggering a population decline. We would expect this harvest to be *sustainable*, that is, capable of being continued indefinitely.

The population's *maximum sustained yield* is the largest number of organisms that can be removed repeatedly under existing conditions without causing a population to decline. The maximum sustained yield of a harvested population depends on the growth stage of the population, that is, the relationship of a population's density to its carrying capacity.

Assume that we are dealing with a population that is density-dependent (Section 2.1.1.3). In other words, our hypothetical population is limited by resources. Figure 3.2 presents a model of the predicted effects of harvesting (exploiting) when a density-dependent population is at different points on its growth curve. The number of individuals in the population is shown on the y axis and is represented by N. Time is on the x axis and is represented by t. The subscripts give the value of N or t at different moments in time. For example, time t_2 is one year later than time t_1. Each year, the population of N_1 individuals at time t_1 grows to N_2 individuals at time t_2. The carrying capacity for the environment in this example is reached when the population is approximately one half of the carrying capacity.

Suppose we guess that it would be good to remove animals when the population is high, represented by Figure 3.2C. So, at time t_2 we decide to remove enough animals to return the population to the level it had the year before at time t_1. To do this, we harvest N_2-N_1 individuals. After that, the population continues to grow as it replaces the animals that were removed. It grows with the growth rate of the population at time t_1 in the bottom graph (3.2C), which is the difference between N_2 and N_1 near the top of the curve of the bottom graph. This is indicated by the short vertical lines. If we continue to harvest at that rate, the same number of animals will be added to the population every year, as long as the harvest goes on.

Those vertical lines are pretty short, which indicates that the growth rate of the population is not very high when it is near carrying capacity. That is what we would expect because the growth of a population that shows density-dependent regulation slows down as population density increases. So next, we use our model to predict what will happen if we

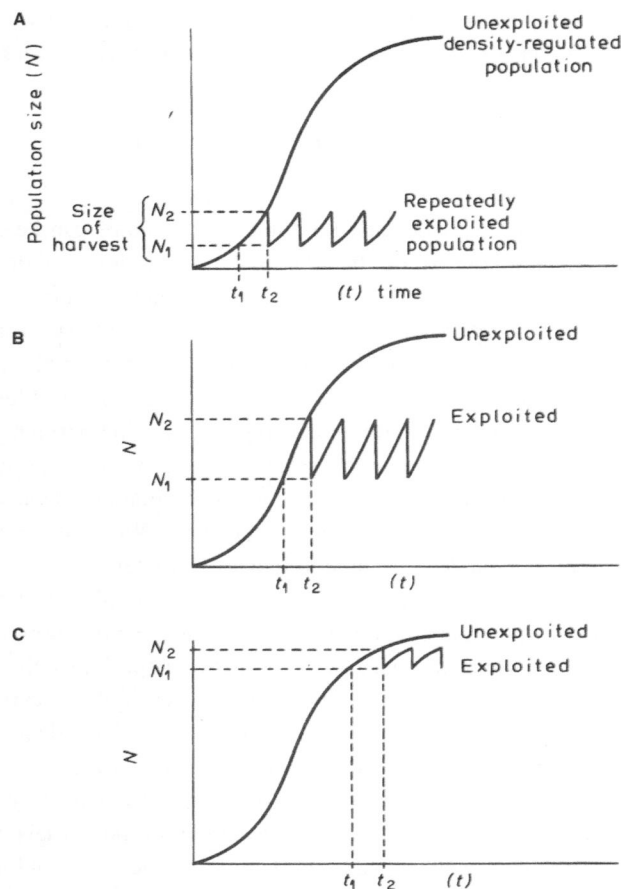

Figure 3.2 Effects of harvest on a population that experiences density-dependent regulation. (A) Effect of harvesting when the population is well below the carrying capacity of the environment. (B) Effect of harvesting when the population is approximately one half of carrying capacity. (C) Effect of harvesting when the population is close to the carrying capacity of the environment. According to this model, the rate of population growth following the removal of animals is greatest when the medium-density population is harvested, that is, when N is about half of the carrying capacity. Used with permission of John Wiley & Sons, Ltd., from Begon et al. (1996).

begin our harvest when the population is very small (Figure 3.2A). Again, the population growth rate will be N_2-N_1 individuals per year (but this time on graph 3.2A rather than graph 3.2C). This time, the predicted harvest is a little larger than it would have been if we had harvested when the population was near carrying capacity; but again, the predicted harvest is pretty low.

We would like to be able to set our harvest limit even higher, and we notice that our hypothetical population seems to be growing quite rapidly in the middle of the curve (Figure 3.2B). So, we look at what the model predicts will happen if we harvest when the population is at about one half of carrying capacity. According to our model of density-dependent population growth, the number of animals that will be added to the population each year will be greatest if we harvest when the population is at about half the environment's carrying capacity (*see Q3.2*). (We can think of this as a Goldilocks solution similar to the children's story in which a little girl finds oatmeal that is neither too hot nor too cold but just right.)

Any of these harvests should be sustainable indefinitely if our assumptions are correct (that is, if the population is regulated by density-dependent processes), if we do not take more than the yearly increment of animals (N_2–N_1) with each year's harvest, and if the carrying capacity of the environment does not change. But if we want to maximize the sustainable take, we should harvest when the population has reached about one half its carrying capacity because that is the density at which the population growth rate (shown as the difference between N_2 and N_1 animals at that population size) will theoretically be at its maximum. If the population we wish to harvest is well below half the carrying capacity, then we should not harvest it at all because of the risk of driving such a small population down to a level from which it cannot recover, and therefore might become locally extinct, is quite high. If the population is much higher than half the carrying capacity, we might choose to temporarily harvest at high enough levels to drive the population down until it is at approximately half the carrying capacity, where the population growth rate will be greatest, and then set our harvest at the appropriate level for that density (Gross, 1969).

We have assumed that removing individuals from a population does not interfere with subsequent population growth. That is, we expect that each time we reduce a population to a lower level, it rebounds with the growth rate predicted for a population at that level. The harvest itself must not disturb the population or otherwise change ecological conditions in ways that affect its growth rate. If this assumption is not met, then losses caused indirectly by hunting must be added to the direct losses when considering the effects of harvest. Game species are usually fairly tolerant of people (this is one reason why they are game species), and so in many cases, disturbance from hunters is not a serious problem for the harvested species (although other species that are incidentally disturbed may be more sensitive).

But even with game species, harvesting may influence population productivity in subtle ways. If disturbance from the hunts causes prey species to increase the amount of energy they spend moving away from hunters and to spend less time feeding, the resulting decrease in energy stores might negatively affect their reproduction.

By-products of managing for harvested species have the potential to cause substantial ecological impacts that are often ignored. For example, roads built to facilitate timber harvests increase the amount of sunlight that reaches the forest floor and can make it easier for invasive species to become established. These indirect impacts should be considered when the effects of harvests are evaluated.

3.2.2.2 Compensatory Mortality

In order to assess the effects of harvest on the size of a harvested population, it is helpful if we know whether the mortality caused by harvest substitutes for mortality from other causes. In other words, is mortality compensatory (Section 2.1.2.2)? If mortality from hunting or other forms of harvest merely substitutes for other causes of death, and

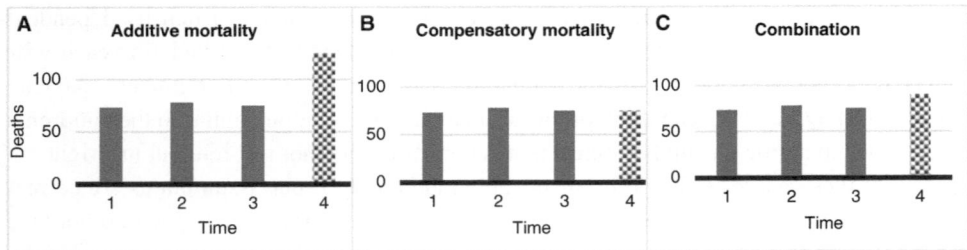

Figure 3.3   The effect of hunting on mortality in a hunted population. Solid bars show number of deaths when there is no hunting. Checkered bars show number of deaths after hunting. (A) In addition to the usual number of deaths, hunting kills additional animals. Mortality from hunting is additive. (B) Animals that are killed by hunting would have died from other causes if there were no hunting. Hunting deaths substitute for other causes of mortality. Mortality from hunting is compensatory. Hunting does not affect the number of animals that die. (C) Hunting mortality is partially additive and partially compensatory. Hunting mortality substitutes for some but not all other causes of mortality. Because hunting mortality partially compensates for natural mortality, total mortality is more than what it would be if mortality were compensatory but less than what it would be if mortality were additive.

harvesting one animal prevents the death of another, then the total number of individuals that die will not increase as a result of a harvest. A graph of the relationship of total mortality to hunting mortality would not show any effect of hunting on the total number of deaths (Figure 3.3B). If hunting mortality is not compensatory, the deaths caused by hunters add to other causes of death, and deaths rise with hunting. This is termed *additive mortality* (Figure 3.3A). If the situation is somewhere in between these two extremes, then there is partial compensation for mortality from harvesting, and total mortality with harvesting increases somewhat but not as much as the total number of harvested individuals (Figure 3.3C) (*see Q3.3*).

In principle, harvest mortality can be compensatory for any harvested organism that is regulated by density-dependent effects. For instance, when many fruits are crowded beneath a parent tree, seedling mortality is likely to be high and to depend on density: the higher the density of seedlings the fewer seedlings survive. In this situation, it may be possible to harvest many seeds or seedlings without having much influence on the tree's reproductive rate (Section 12.3.1.2).

If mortality is truly compensatory, then it should be possible to increase a harvest without increasing a population's losses. This is often a desirable situation from a utilitarian perspective. The other side of this coin, however, is that if a population that experiences compensatory hunting mortality begins to decline, protection from hunting pressure will not be helpful, because natural mortality will substitute for the losses caused by hunting.

Detailed studies of the population dynamics of some large herbivorous mammals such as elk and other ungulates suggest that they exhibit partial compensation for hunting mortality (Caughley, 1985). Hunting mortality often substitutes for some deaths that would have been caused by predation.

Furthermore, if predators have been removed, hunting mortality might substitute for deaths from starvation. Deer can deplete their food supplies, especially in winter. When this happens, starvation usually follows. In addition, economic damage to crops or ornamental plantings becomes likely. Regulated hunting can be a strategy for preventing habitat degradation in such situations.

Early wildlife biologists developed the concept of compensatory harvest mortality from their studies of game animals in temperate environments, but their insight is also useful for evaluating the effects of harvests in some other contexts (Box 3.2).

Box 3.2

Community Harvest of Sea Turtle Eggs in Ostional, Costa Rica

Almost every month, thousands to millions of female olive ridley sea turtles come ashore to lay their eggs on the beach near the village of Ostional, Costa Rica (Figure 3.4). This mass arrival from the sea, or *arribada* (ar ree BA da), is the occasion for a massive harvest of turtle eggs for market. Sea turtle eggs are a popular food in Central America. They are said to enhance sexual potency in men and are also valued for their flavor and nutritional content.

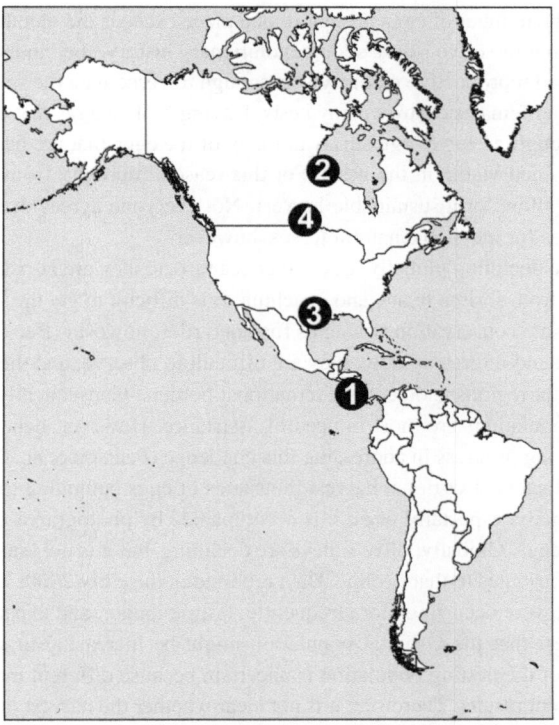

Figure 3.4   Locations of: 1, Ostional, Costa Rica; 2, Churchill, Manitoba; 3, Walker County, Texas; 4, Delta Waterfowl Research Station, Manitoba. Map created by Eva Strand using Esri, DeLorme World Countries Generalized Data & Maps for ArcGIS 2013, with permission.

Turtles have a long evolutionary history; they were alive before dinosaurs appeared on the Earth. The unique anatomy of turtles has contributed to their evolutionary success. Like other vertebrates, turtles have an internal skeleton, but their ribs, backbone, and skin are also modified into a rigid, two-part, external shell composed of a curved shield above and a flat plate below. When in danger, a turtle can draw its legs, head, and tail into its shell. This arrangement protects adult turtles from most predators. (Sharks are the main exception.) But turtle eggs and hatchlings are more vulnerable.

Olive ridleys and other sea turtles spend their life in the ocean, but females come ashore to excavate nests and deposit their soft-shelled eggs on sandy beaches. After laying, they return to the ocean, where they mate again and then return to the beach to deposit more eggs. In most sea turtle species and some populations of olive ridleys, females nest independently of one another, but some olive ridley populations, such as the one that nests at Ostional, and one other sea turtle species, nest in mass synchronized nesting events known as *arribadas.*

Natural mortality of the eggs and newly hatched turtles in *arribadas* is high, especially in the first group of nests. Raccoons and other mammals dig up the eggs, and avian predators such as vultures eat the young. Returning females destroy many eggs when they excavate previously deposited nests before depositing more eggs. In addition, embryos may die if temperatures get too high.

If mortality among eggs and young is so high, how have olive ridleys been able to persist for hundreds of millions of years? The answer to this puzzle lies in predator saturation (Section 1.3.3.1). Because large numbers of females in the *arribadas* nest at one time and each female lays around 100 eggs, the number of eggs present in one place exceeds the amount that predators can consume. Throughout the olive ridley's long evolutionary history, the number of hatchlings that survived and reached reproductive maturity was enough to perpetuate the species.

Because returning females destroy many nests, leaving broken eggs that develop bacteria and fungi which are thought to lower egg survival, many of the eggs that are harvested would probably not have remained viable in the nests. For this reason, mortality from the harvest may be compensatory and allow for a sustainable harvest. Not everyone agrees that mortality from the harvest compensates for inevitable natural losses, however.

Olive ridleys are declining globally. Like other sea turtles, they are at risk from pollution, entanglement in nets from shrimp boats, and poaching. It is difficult to get the kinds of data needed to develop appropriate conservation strategies for sea turtles, however. Because they spend most of their lives at sea and migrate widely, they are difficult to observe, and they face a wide range of threats in different regions and across international borders. Consequently, it is hard to figure out which turtle populations are most in need of assistance. However, genetic and biochemical techniques are making progress in addressing this challenge (Pearson et al., 2017).

The fact that villagers at Ostional harvest thousands of eggs belonging to a declining species generates bad publicity, especially when it is accompanied by photographs of villagers carrying off heavy sacks of eggs. Globally, olive ridleys are declining, but it is not known whether the harvest at Ostional contributes to that decline. Data reported in the early 2000s indicated that the *arribadas* at Ostional were occurring more frequently, lasting longer, and expanding in area. These observations suggest that the Ostional population might be increasing in spite of the harvest. However, the size of the nesting population is uncertain because different methods of estimating it have given different results. Therefore, it is not clear whether the harvest of the *arribada* population at Ostional is contributing to the olive ridley's decline (Cornelius et al., 2007).

In short, although the egg harvest at the Ostional *arribada* is legal, it generates considerable controversy, and much of the controversy centers around the question of whether mortality from the egg harvest is compensatory.

3.2.3 Managing Harvests in Practice

3.2.3.1 Using Population Data in Harvest Management

To manage for a sustainable harvest, it is necessary to assess population trends. In order to do that, it is helpful understand the age and sex structure of a population (Box 3.3).

If it is possible to count all the individuals in a population at several successive time intervals, then managers can determine whether a population is increasing or decreasing and at what rate. They can use this information as a basis for adjusting harvest levels so that fewer individuals are taken from declining populations or more individuals are removed

---

Box 3.3

Management of the Northern Fur Seal Harvest in the Pribilof Islands

Breeding fur seals have several characteristics that historically made it easy for managers to get data on the sex and age structure of their populations, and thus to predict the effects of harvest. Fur seals are *sexually dimorphic* (*di*, two; *morph*, forms), that is, the two sexes can be distinguished. Males weigh several hundred pounds more than females. Fur seals are also *polygynous* (*poly*, many; *gyn*, female), meaning that one male mates with many females (usually between 5 and 10 in a breeding season). Breeding fur seals congregate onshore in *rookeries*. After their winter migration to their feeding areas, the seals return to their breeding grounds in the Pribilof Islands and other islands in the North Pacific.

Older males are the first to arrive. On land, they battle amongst themselves to stake out territorial boundaries. After a few weeks, the females arrive. Almost immediately, they haul out on land, settle on a territory, and give birth. A few days later, each female mates with the male on her territory. The fertilized egg does not implant in her uterus or begin to grow for several months, however. This physiological adaptation, termed *delayed implantation*, allows breeding to take place when the population congregates on land and birth to take place at the appropriate season a year later, when the seals come ashore again. If the eggs began their development right away, rather than being delayed, they would complete their development while the females were at sea. That would be problematic for air-breathing mammals such as seals.

Young, nonbreeding males arrive at the rookery after the adult females and gather at the outskirts in bachelor colonies.

The size difference between the sexes and the spatial segregation of the bachelor males allow managers to study breeding fur seals when they are on land. The numbers of animals in each age and sex class can be determined, and nonbreeding males can be removed without decreasing the reproductive potential of the herd. These circumstances should make it possible to have a closely monitored fur seal hunt because the effects of the harvest on the population can be assessed by comparing successive annual censuses (Baker et al., 1970).

Fur seal populations responded well to the protections afforded by the Fur Seal Treaty of 1911 (Section 1.5.3). After the populations recovered, seal pelts were commercially harvested in the Pribilof Islands for several decades without apparent ill effects on the herd. For many decades, the managed harvest provided a continuing supply of skins without apparently depleting the resource.

The Fur Seal Treaty expired in 1985 and was not renewed, however, primarily because of public opposition to the killing of animals to obtain a luxury. The USA no longer carries out a commercial hunt of northern fur seals (although Alaska Natives hunt seals for subsistence). We will return to the subject of the fate of northern fur seal populations in recent decades in Section 10.1.1.1.

from dense ones. In most cases, it is not possible or practical to count all members of a population, however. Instead, population growth rates must be estimated from a sample of the population of interest (Box 3.4).

Managers also make use of information about the physical condition of harvested individuals. If animals typically lack fat stores or show signs of disease or parasites, or if trees have slow growth rates, managers look for the causes of those conditions. Malnutrition, disease, and slow growth can be signs of high population density, a situation that can sometimes be alleviated by increasing a harvest.

Box 3.4

**Using Mark-and-Recapture Data to Predict the Effects
of Polar Bear Harvest in the Canadian Arctic**

Bears take several years to reach sexual maturity, they produce few young, and the intervals between births are long. Because of this low reproductive rate, bear populations do not recover rapidly after they decline. This means they are vulnerable to overharvesting. It is therefore important to understand the impacts of harvest on these large carnivores.

Polar bears are not easy to study. They are dangerous and occur in inaccessible terrain under rigorous environmental conditions. For these reasons, it is generally difficult to obtain large enough sample sizes to accurately estimate polar bear populations. However, the bears at Hudson Bay in north-central Canada concentrate in a restricted area during the period when sea ice melts, and this facilitates data collection.

From 1977 through 1992, two Canadian wildlife biologists obtained data on the polar bear population near Churchill, Manitoba (Figure 3.4) (Derocher and Stirling, 1995). They trapped bears, immobilized them with drugs, recorded their sex and reproductive status, extracted a premolar (which was subsequently used to estimate the individual's age), and marked the bears with a tattoo on each side of the upper lip and a plastic tag in each ear.

When the population was re-trapped subsequently, some of the bears that were caught had been previously captured and marked. Several mathematical models have been developed for obtaining population estimates from this kind of *mark-and-recapture data*. The basic assumption of these models is that all individuals that are captured initially have the same probability of being recaptured later. If this assumption is correct, and if few marked animals are recaptured, then survival is apparently low. If many of the marked individuals are subsequently recaptured, this suggests that survival is high. However, if the assumption of equal probability of capture is not correct, things get complicated. If some individuals permanently leave the population, researchers know only that these individuals are missing; it is not possible to tell if they died or moved away.

Previous research on the study population had shown that the bears did not often settle in adjoining populations, so the researchers thought that emigration from the population probably did not introduce any serious errors into their analysis.

At the time of this study, Inuit hunters harvested polar bears, and tourist viewing of the bears was popular. The average size of the autumn polar bear population from 1978 to 1992 was 1,000 individuals, and an average of 191 cubs was added to the population each year. Because the population size was high and stable and recruitment of young bears into the population was also high in spite of hunting and tourism, the researchers concluded that the harvest at that time was sustainable.

In exploited populations, the harvest itself can be used to obtain useful data. At hunter check stations, wildlife biologists often obtain information about the age, sex, and physical condition of harvested animals. Similarly, fisheries biologists aboard commercial fishing and whaling vessels obtain data on harvested stocks. For example, the Soviet take of southern right whales between 1951 and 1971 and the Japanese catch of minke whales since 1987 provided substantial data on the distribution, diet, movements, reproduction, and population structure of the harvested species. These research expeditions were highly controversial, however, because in both cases the whales were protected from exploitation by international agreement (Section 7.3.1).

3.2.3.2 Setting Harvest Limits

In practice, setting harvest limits usually involves some trial and error. Usually carrying capacity and reproductive rate are not known precisely, and the mechanisms of population regulation are not fully understood either. Instead of determining carrying capacity and using that as a basis for setting harvest limits, managers often make an educated guess about the level of harvest that a population can sustain, as in Box 3.4. They should then monitor the response of the population and adjust harvest levels accordingly if necessary.

Biological considerations are not always the main driver of harvest levels. A host of non-biological factors, including attitudes and expectations, also enter into decisions about harvest levels. Because hunters like to have high populations of game animals, harvests are usually set at levels that are much lower than what the harvested populations could probably sustain.

Foresters who manage tree harvests attempt to determine how much wood is produced during a given interval and then adjust harvest, so that no more than that increment is removed during an equivalent interval. The crucial management decision is not how many trees can be cut, but how often a stand can be cut, which is termed *rotation length*. To achieve sustained yield, a forester should not allow a forest to be cut until it has reached the same stage of development it had before it was cut. If parts of a large, forested area are cut sequentially, with each stand being allowed to grow back to the stage of development it was at before it was cut, and stands are not cut again until what was removed has been replenished, this procedure should lead to sustainable harvest of timber (but it there may be unanticipated effects on other parts of the ecosystem that are not sustainable).

In managing rangeland for sustained yield of livestock, the utilitarian goal is to avoid overharvesting forage plants. Livestock are the products of this enterprise, but wild plant communities are the resource that managers seek to use sustainably. In principle, this means regulating the levels of grazing so that livestock do not remove plants at rates that exceed their rate of recovery. Grazing pressure can be manipulated by regulating the number of grazing animals as well as the timing of grazing. A number of grazing systems have been developed that aim to minimize the negative impacts of livestock grazing on rangelands. Many of these replace continuous, year-round grazing with a short period of intense grazing followed by a rest period to allow plants to recover.

3.2.3.3 Enforcing Regulations

Even the best harvest regulations work only if they are followed. Compliance depends upon economic, legal, social, and cultural considerations including whether or not the system for setting regulations is respected, whether harvesters believe that their competitors (for instance, sport hunters or commercial operations) will comply, whether they expect to get caught if they violate regulations, and, finally, the ratio of expected gains to expected penalties for noncompliance. Enforcement is difficult if profits are high or the people affected by regulations do not support those regulations. When a resource is harvested by several nations, enforcement is particularly difficult.

Any time the potential short-term economic gains of exploitation are large, there is a danger that scientists' recommendations will be ignored or they may be influenced by special interests. Managers can be subtly or openly pressured into acquiescing to harvest levels that are not sustainable, especially where large profits are at stake, regardless of whether the resource is fish, timber, ivory, whales, furs, medicinal plants, forage, etc.

In addition to managing harvests, resource managers can manipulate habitats with the goal of maximizing the productivity of harvested species. We turn our attention to this topic next.

3.3 Managing Habitats

3.3.1 Modifying the Structure of Habitat Components

The value of a habitat for wild organisms is influenced by its physical structure. Providing or safeguarding structures required by species of interest is a straightforward method of improving a habitat's value for those species. The physical structure of a habitat creates *microhabitats* (small habitats characterized by conditions that differ from those of the surrounding area) and influences the transmission of light, sound, and heat as well as the movements of air and water (Bell et al., 2012).

Cover refers to structural features of an animal's habitat that fulfill functions such as providing visual obstruction or shelter. Different types of cover, such as thermal cover, hiding cover, nesting cover, or resting cover, perform different functions. Plants often provide cover, but soil, rocks, snow, water, landforms, and artificial structures can offer cover too. Cover occurs at many spatial scales. For small animals, a crack in soil or bark may provide cover; grass, boulders, logs, caves, coral reefs, and trees can afford cover for medium-sized to large animals.

Mammals and birds are susceptible to heat stress and therefore require thermal cover, especially when they are young. Environmental features that minimize fluctuations in body temperature provide thermal cover by modifying the local environment in ways that moderate air temperature and the ability to gain or lose body heat.

Late-successional forest provides structures that are critical for some animals. In temperate ecosystems many organisms require dead woody material. Cavity-nesting species such as woodpeckers, small owls, bats, and squirrels are limited by the availability of cavities in *snags* (standing dead and dying trees), and dead wood on the forest floor provide microhabitats used by invertebrates, amphibians, birds, and small mammals. Logs that fall into streams provide coarse woody debris that forms small dams. Pools that form behind these log dams provide spawning habitat for salmon and other fish.

Wildlife conservation aimed at preserving these structures can be at odds with management for other resources. Because managing for decaying wood involves leaving some old trees, which can harbor insects or pathogens, this policy conflicts with management oriented toward maximum timber production.

3.3.2 Modifying the Arrangement and Shape of Habitat Components

The arrangement of habitat patches across a landscape affects habitat use. Organisms move across the boundaries between habitats. These movements may involve permanent relocations, occasional excursions, or regular movements. In many species of plants and animals, young individuals permanently move away from their parents' home range. This *dispersal* may or may not involve movements to or through different ecosystems. In addition, some animals feed in one type of habitat and rest in another on a daily basis. On a seasonal basis, some undertake migrations or less predictable nomadic movements.

Long-distance migrants transport nutrients across hundreds or even thousands of kilometers. Adult salmon ingest nutrients when they feed in the ocean, and they carry those nutrients upstream to their spawning habitat. Terrestrial predators eat these migrants, digest them, and deposit their ocean-derived nutrients as feces in adjacent habitats, fertilizing uplands with nutrients from marine environments hundreds of kilometers away (Hilderbrand et al., 1999).

The borders between contrasting habitats are of particular interest to wildlife managers. From its beginnings, wildlife management was concerned with maximizing edges between contrasting habitats. In his text on game management, Aldo Leopold noted that many species favored by hunters were abundant in places where two contrasting habitats met.

[G]ame is a phenomenon of *edges*. It occurs where the types of food and cover which it needs come together, *i.e.*, where their edges meet. Every grouse hunter knows this when he selects the edge of a woods, with its grape-tangles, haw-bushes, and little grassy bays as the place to look for birds.

(Leopold, 1933:131; emphasis in original)

The influence of the amount of patch border on the diversity and abundance of organisms is termed the *edge effect*. The idea that edges have a positive effect on wildlife abundance is a testable hypothesis. It predicts that: (1) the number of species and (2) the densities of populations will be higher where two habitats meet than in the interior of either habitat. Some early studies lend support to this hypothesis (Box 3.5) (*see Q3.5*). However, we shall see later (Section 5.2.2.5) that in their enthusiasm for edges as a way to increase some kinds of wildlife, managers failed for a long time to realize that creating habitats with high proportions of edge is detrimental to species that require forest interiors.

The relative amount of edge present depends on the size and shape of a habitat patch (Figure 3.5). If two patches are the same shape but differ in size, the smaller one will have a higher proportion of perimeter (edge) in proportion to its area than the larger one

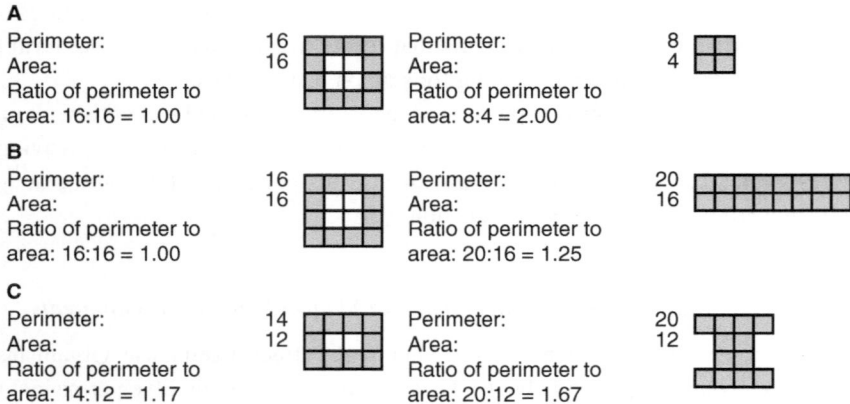

Figure 3.5   The relationship between the size and shape of a patch and the proportion of its habitat that is near an edge (perimeter). To determine the area of a patch, count the number of small squares it contains. To determine the perimeter of a patch, count the number of linear units (edges of a square) it contains. In these hypothetical examples, the influence of the edge is assumed to extend one linear unit from the boundary of the patch into the interior, so shaded squares represent edge habitat. In each row, the patch on the left has a lower proportion of edge than the patch on the right. (A): The two squares have the same shape, but the smaller patch (the one on the right) has a higher proportion of edge than the square on the left. (B): The two patches have the same area, but all of the long, thin patch (the one on the right) is close to an edge. (C): The two patches have the same area, but all of the patch with an irregular area (the one on the right) is close to an edge. Thus, the relative amount of edge habitat is greater in patches that are small (A), elongated (B), or irregularly shaped (C).

Box 3.5

Evaluating Evidence: Testing the Benefits of Edges for Breeding Birds

In 1938, biologist Daniel Lay censused birds in and near forest margins in Walker County, Texas (Figure 3.4) during the breeding season. He walked for 30 minutes (1) along the margins of clearings in pine forests and (2) at least 91 m from the margins, and recorded sightings of birds that belonged to recognized species. In addition, "occasionally, when the unmistakable song of a familiar bird was heard in nearby cover, time was not consumed in stalking or sight of the individual" (Lay, 1938:254). He reported that the margins averaged 16.6 birds of 6.5 species, whereas the woodland interiors averaged 8.5 birds of 4.6 species. Lay concluded that both the forests and the clearings had little value for birds unless they were close to edges:

Obviously a primary essential to the management of woodland areas for wildlife, especially for birdlife, is the provision of clearings with extensive margins .... The interior of a large clearing is as depleted of wildlife as is the interior of the woodland, hence the need for small but numerous clearings ....

   In the management of pine woodland the provision of well scattered (less-than thirty-acre) clearings is distinctly favorable to birdlife.

(Lay, 1938:256)

   When he designed his study, Lay did not consider the effects of breeding bird behavior (*see Q3.5*). Breeding males of many bird species select perches near clearings even if their territories extend far into a forest. Lay introduced a potential bias into his results by overlooking this point. Counting singing males as edge inhabitants might have artificially inflated his estimate of the edge population and underestimated the forest population.

(Figure 3.5A). If the area that can be managed as a given type of habitat is limited, a manager can maximize the edge of a patch by making it have a long, thin shape (Figure 3.5B) or an irregular border (Figure 3.5C). Thus, a manager interested in increasing habitat edges can do so with small patches having irregular margins or thin shapes (*see Q3.6*). This can be done by allowing for or mimicking patchy disturbances that create a mosaic of contrasting habitats. (For information on how these relationships were later applied to the design of nature reserves see Section 8.3.2).

Is there a point at which the patches get so small that the habitat becomes almost uniform? Box 3.6 describes an experiment designed to answer this question by manipulating the size of cattail patches in a marsh.

Box 3.6
Testing the Value for Dabbling Ducks of Cattail Patches with Different Amounts of Edge

Researchers at two Canadian universities and the Canadian branch of the conservation organization Ducks Unlimited designed an experiment to investigate the use of patch edges by ducks in a cattail marsh at the Delta Waterfowl Research Station in Manitoba, Canada (Murkin et al., 1982) (Figure 3.4). They cut a series of perpendicular channels in cattails to create 60-by-60-m plots with different levels of cover removal (Figure 3.6). Using a tractor-drawn hay rake to cut varying numbers of channels in a crisscross pattern, they created plots with patches of three different sizes. Two adjacent areas of uncut cattails were used as controls. Pairs of *dabbling ducks* (ducks of shallow water) and lone males, which were assumed to be paired with a female that the investigators did not observe, were counted on the study plots and control areas periodically. The greatest densities of dabbling ducks occurred on plots with approximately equal amounts of cattails and open water (Figure 3.6B).

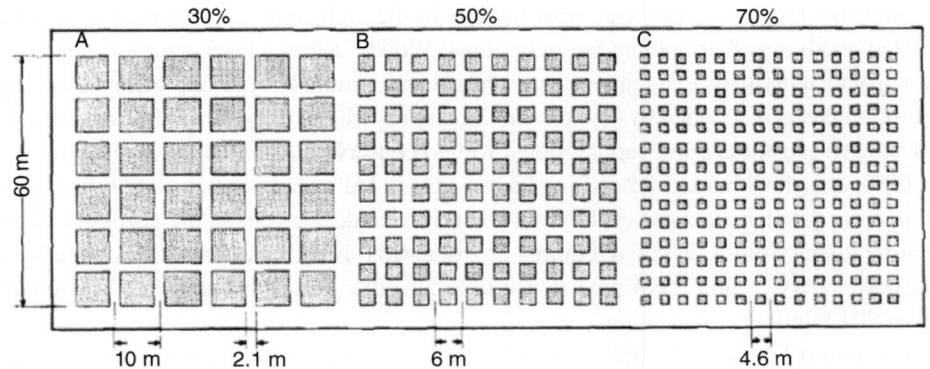

Figure 3.6   Arrangement of test plots used to evaluate the effects of the amount of edge in marsh vegetation on dabbling duck densities. Stippled areas represent patches of cattails in the marsh; clear areas represent open water. (A) Approximately 30% cover removal and 70% vegetation; (B) approximately 50% cover removal and 50% vegetation. (C) approximately 70% cover removal and 30% vegetation. Reprinted from Murkin et al. (1982) with permission of Canadian Science Publishing.

Initially, increasing the number of channels increased the amount of edge between surface water and vegetation (compare Figure 3.6A and 3.6B), and improved the habitat for dabbling ducks. When even more channels were cut (compare Figure 3.6B and 3.6C), duck use declined. The plots from which approximately 70% of the cover had been removed had the highest amount of edge but did not have the most use by ducks. The plot was so finely divided that the vegetation was nearly uniform, at least to our eyes, and presumably the eyes of a duck (Figure 3.6C). The habitat with an intermediate amount of open water (50%) apparently provided the best combination of open water and cattails. (We might call this another Goldilocks solution.)

The results of this study would no doubt have been different if the researchers had looked at the responses of smaller or larger animals. Perhaps a dragonfly would perceive every cattail stalk as a habitat patch. And a deer might see an entire plot as a single patch.

3.3.3 Modifying Succession to Benefit Preferred Species

Habitats can be altered by modifying succession. Although the idea that succession proceeds to a predictable and stable climax is controversial, the idea that the plant communities that occur soon after a disturbance are quite different from those that occur later is not disputed. Managers and ecologists think in terms of vegetation sequences that occur during the course of succession even if they don't think that succession leads to inevitable or stable climaxes. As succession proceeds after either a natural or a human-caused disturbance, early stages are replaced by later ones. Flooding, grazing, logging, burning, spraying herbicides, and mowing will kill or remove vegetation and usually promote seral communities. Similarly, herbivory by large wild or domestic herbivores also affects succession by removing vegetation. Indigenous peoples in many regions of the world have used fire since time immemorial for varied purposes including the creation of early successional habitats conducive to hunting and foraging, but decades of fire suppression dramatically reduced early successional habitats that provided many important resources for Indigenous peoples.

Intensive logging of the American frontier in the eighteenth and nineteenth centuries created conditions that favored large, hot fires. Woody debris left onsite after timber harvest provided dry fuels that ignited easily and caused extremely destructive fires. Utilitarian forest managers responded by suppressing forest fires wherever possible. To them, fire was a destructive agent and fire suppression a way to conserve a commodity (Section 1.5.2.1). In 1910, after fires burned 1.2 million ha of forest in Idaho and Montana and killed nearly 90 people, the chief forester of the US Forest Service declared that "the necessity of preventing losses from forest fires requires no discussion. It is the fundamental obligation of the Forest Service and takes precedence over all other duties and activities" (quoted in Pyne, 2017:260).

The Forest Service's attitude toward fire was grounded in the view of nature championed by Frederic Clements (Section 2.2). Fire historian Stephen Pyne suggests that

the Clementian [sic] concept of "nature's economy" … was a concept especially gratifying to foresters … [that] vindicated fire protection … as a means of assisting the succession of deforested land to forest climax and … as a means of promoting the innate and "natural" drive for successional climax.

That lightning set many fires was really irrelevant: it was well known that nature suffered from waste and entropy, which human engineers could, and ought to, eliminate.

(Pyne, 2017:492)

Not all natural resource managers shared foresters' views on fire. Depending on their objectives, managers in different fields favored different habitats and, consequently, had different views about fire suppression. Wildlife and range managers were more inclined than foresters to use fire as a tool for modifying habitats. Early in the twentieth century, a royal committee recognized the potential value of fire for managing grouse in the moors of England and Scotland (Box 3.7).

The commission's insights were a significant contribution to game management at the turn of the nineteenth century because they were one of the first comprehensive attempts to develop a management plan that considered habitat quality, fire, patch configuration, and socioeconomic context.

Habitat conservation depends on maintaining good soil. Although farmers around the world have long understood that soil can be improved with the addition of manure and plant material, formal soil conservation did not develop until the 1930s.

Box 3.7

Historical Use of Fire to Manage Red Grouse Habitat in the Moors of England and Scotland

The red grouse is a game bird of the *moors* (wet habitats supporting short vegetation) of England and Scotland (Figure 3.1). It inhabits vegetation dominated by heather, a low-growing, evergreen shrub related to cranberries. Since the mid-nineteenth century, many areas of heather on private estates in the British Isles have been managed to produce grouse for hunting.

From 1873 to 1910, a British royal commission studied red grouse in an effort to discover how to minimize mortality from parasitic diseases. The commission members compiled data on dead red grouse that were examined by field observers. One of these was the Antarctic explorer Edward Wilson, who, with the assistance of his wife, dissected and prepared thousands of museum specimens from dead grouse that were sent to him. (This task presented some interesting challenges as he gathered information for the committee: "Every dead grouse found on a British moor was dispatched to Wilson for examination and a good many were suffering from delays in transit .... This was not terribly popular in station hotels" (Anon., 2005–2016)).

Prior to 1850, the owners of large estates had their shepherds burn about one tenth of their holdings annually. As a result, properties were burned about once every 10 years. In the mid-nineteenth century, however, landowners began appointing gamekeepers in response to an increase in the value of grouse as a game bird. The keepers wanted to increase cover to conceal hunters from their quarry, so they burned the moors less often. This, the commission concluded, had caused a decline in the quality of the habitat for grouse because their food had decreased. Therefore, they advocated more frequent burning: "to heighten the average yield of the moor ... the progressive landlord will ... attempt to get the moor into good 'heart'" (Committee of Inquiry on Grouse Disease, 1911:410).

On the basis of their findings, the commission members concluded that burning rejuvenated the food supply for grouse by removing old vegetation and stimulating the growth of young heather. They recognized that the pattern of burning was as important as the amount of moor that burned. Hence, they suggested that by burning long, narrow swaths and alternating burned and unburned strips, gamekeepers could create a mosaic of young vegetation dispersed in a matrix of older patches. This arrangement, the committee thought, would lessen the transmission of parasites between birds by segregating them in separate strips of young heather. According to their report, burning moors in a 15-year rotation would be feasible for most landlords in most years, and would allow three fifths of a moor to be in good condition at any time. Initially, large areas were to be burned in order to make up for years of insufficient burning.

   The report also recognized that attitudes toward burning depended on class. Landowners and the tenants that hunted on their land had different ideas about how to manage the moors. These differences stemmed from different time frames. The landowners could consider their long-term interests, but the tenants' long-term connections to the moor were not secure.

If the landlord knows his own interests his first object must be to burn big stretches ... in order to get the moor into a proper rotation of burning. The tenant, on the other hand, should he be equally well informed, knows that though such heavy burning may be beneficial to the moor in future years, the resulting crop of edible heather will not be increased during his occupancy.

(Committee of Inquiry on Grouse Disease 1911:431)

3.4 Conserving Soil

A naturally potent soil is one of the greatest assets we can have in
setting out to improve conditions for nearly any species of wildlife. It
follows that anything done to conserve fertility and build up humus
and mineral content usually will be a step toward a more abundant
animal life – other things being equal.

(Allen, 1962:26)

In 1935, US senators met in Washington, D.C., to discuss creating a federal agency dedicated to soil conservation. They had a dramatic demonstration of the need for such a body when

The sky darkened as dust from the [Dust Bowl in the] plains arrived .... The Senators suspended the hearing for a moment and moved to the windows of the Senate Office Building. Better than words or statistics or photographs, the waning daylight demonstrated ... that soil conservation was a public responsibility worthy of support and continuing commitment to solve one of rural America's persistent problems.

(Helms, 1990:58)

   The windborne soil from hundreds of miles away that darkened the sky in the nation's capital that day made an impression on the senators. Shortly after the meeting, Congress established the Soil Conservation Service (SCS), followed in a couple of years by legislation that enabled the creation of *conservation districts*. These local bodies carry out federal policy to implement measures for protecting and enhancing soil and water quality. These districts laid the groundwork for interactions between technicians, scientists, farmers, and ranchers in an innovative approach to collaborative conservation. Farmers and ranchers were encouraged to undertake measures designed to reduce erosion and conserve soil, such as planting blocks of trees to reduce wind erosion and plowing along the contours of slopes to reduce water erosion. As of 2021, nearly 3,000 conservation districts in the USA continued to address local issues related to conserving soil and related resources.

3.5 Managing Habitats for Multiple Uses

So far, this discussion treats utilitarian forest, range, and wildlife conservationists as if they were in separate boxes, each concerned only with a single type of resource. In reality, they often try to manage habitats in ways that promote the productivity of multiple resources. This typically involves tradeoffs between different objectives. In the years following World War II, social, economic, and political developments in wealthy countries, including rising population and expanding time for recreation, resulted in pressures to adopt a broad approach to conservation. Americans became increasingly concerned that national forest management was too narrowly focused on timber production. In 1960, the US Congress passed the Multiple Use Sustained Yield Act. Under this policy, federal agencies involved in resource management were to take wildlife habitat, watershed maintenance, and opportunities for recreation into account. The focus was still utilitarian, but by mandating the consideration of other values, the act sought to initiate a new approach to conserving living natural resources.

Managing for multiple uses is easier said than done. Enhancing the productivity of one resource can conflict with managing for other resources. Therefore, choice of an appropriate management strategy depends upon our objectives and priorities. We will return to these points repeatedly.

3.6 Evaluating Harvest and Habitat Management

We have seen that management for sustained yield developed in response to unsustainable uses of plants and animals. Under some circumstances, the strategies discussed in this chapter succeeded in producing a continuous supply of wild products. This approach is based on a conception of the natural world as predictable. It holds that populations tend to stabilize at the carrying capacity of their environment, and plant communities progress toward stable climaxes. Nature is viewed as tending toward balance. Stability and equilibrium are believed to be the rule. Density-dependent responses regulate population growth, and we can anticipate how plant communities will change after they are disturbed.

The risks of managing for maximum sustained yield are fairly low in predictable environments with populations that have high reproductive rates and are regulated by density-dependent processes. This is so because such populations rebound when their density declines. In some environments game species have these characteristics. If they did not, they might well have been overharvested to extinction long ago.

Not all environments are predictable, however. If carrying capacity varies in unexpected ways, the assumptions underlying harvest management may not be met. Where managers do not understand important natural variability, things don't always turn out as expected. By concentrating on maximizing the production of a small number of species and a few types of habitats, utilitarian conservation contributed to simplified communities and had negative consequences for species that were not deemed valuable.

We will return to this point in Part II, but first, let's look at utilitarian management of unpopular species.

4

Strategies

Managing to Minimize Conflicts between Pests and People

In Chapter 3, we saw how populations and habitats can be managed to favor species that are considered valuable. This chapter considers how resource managers using a utilitarian approach attempt to control populations of species that are viewed as pests by some parties because of their effects on wildlife, domestic plants and animals, or people. It begins with the why and how of control, then provides some historical background and explores examples in different contexts.

Pest management affects the conservation of wild organisms in two ways: First, controlling species that are thought of as pests is often part of conservation strategies aimed at minimizing negative effects on valued species. Second, control inevitably has impacts on nontarget species and ecological processes, which need to be understood when evaluating the relationship of control efforts to conservation.

Managing negatively viewed species typically involves applying concepts that are central to utilitarian management, including exponential population growth, density dependence, compensatory responses to changes in population size, and interspecific interactions. Understanding political, social, and economic contexts is also important. This is partly because the definition of what constitutes a pest needs to be understood in a broad context and partly because human behavior affects the behavior of pests and may exacerbate problems.

4.1 What Is a Pest?

You might think of mosquitoes, rats, or dandelions when the subject of pests comes up, but many other living things may be pests in certain circumstances. What you call a pest depends on where you live, how you earn a living, and many other factors. In this book, we consider any species that is perceived as detrimental to people or their interests as a pest, while recognizing that people disagree on which species fall into that category. Pests can be plants, animals, fungi, or microorganisms. Species are considered pests if they cause economic damage (directly, by killing a valued species or damaging property, or indirectly, by competing with valued species) or because they pose a threat to the health and safety of people or domesticated organisms. Often, but not always, pests are not native to the region where they are considered pests.

Pests are sometimes referred to as "weeds." Weedy species generally have high reproductive rates, are tolerant of people and therefore able to live near people, and are good colonizers. Weeds do not have to be plants. To a livestock rancher, the coyote is a weedy species. Weeds have been described as organisms in the wrong place, that is, organisms that have become overabundant (perhaps because they have been transported to a region where their normal enemies are absent) or that conflict with the objectives of people in a given situation. This definition underscores the fact that pest status depends upon context and perspective.

Weediness is in the eye of the beholder. Some animals, such as seals, deer, waterfowl, and bears, are economically valuable and pests at the same time. They are harvested for sport or commerce but cause damage under certain circumstances. We may love our pets, but feral cats and dogs kill wildlife and may attack people or transmit diseases. Tigers (Section 4.8.2) and African elephants are valued by conservationists and considered pests by local people. Grizzlies and gray wolves were once pursued by bounty hunters in North America and Eurasia. Today, they are endangered and protected in parts of those regions. Large amounts of money are spent on recovery plans aimed at rebuilding their populations, but people who are asked to coexist with these large predators have economic and safety concerns. Deer, elk, and waterfowl are valued as game, but when they damage grain fields or orchards, they are pests according to those whose livelihood is negatively affected. Many urban residents enjoy seeing Canada geese and English sparrows (Box 4.2), but others consider them nuisances because of their droppings and noise. Some people enjoy seeing coyotes, lions, tigers, or bears or just want to know that such animals exist in the wild, but those species also have negative effects on property and human welfare in some settings.

Conflicting management objectives complicate these situations. Managers seeking to maintain high populations for sport hunting or for recreational viewing may protect or even artificially feed populations that then build up to levels where they cause problems. In intensively managed German forests of the nineteenth and early twentieth century, populations of artificially fed red deer increased to the point where they stripped bark from young trees and damaged nearby crops (Leopold, 1936a, b; Webb, 1960). In these situations, increasing the level of harvest can be a means of controlling damage, but this is likely to be controversial when the targets of control are birds or mammals that have a lot of emotional appeal (Leopold, 1955).

4.2 Types of Control

The principles underlying management to reduce pest populations are the same as the principles of harvesting for sustained yield, except that instead of removing only as many individuals as can be replenished the objective is to cause more deaths than will be replaced. If control causes deaths and emigration to exceed births and immigration, then the target population will decline. If density-dependent compensation comes into play, however, then control efforts may not make any difference in population size. Consequently, pest control is difficult if increased reproduction or survival compensates for mortality (Section 4.5.1).

Controlling a pest means either reducing its population to the point where it is no longer considered a serious problem or reducing the damage it causes. This can be done by *lethal* (fatal) or nonlethal means. Lethal methods of controlling animals include shooting, trapping, gassing, and poisoning. Lethal control of plants usually involves chemical or mechanical means that poison, uproot, or chop them or otherwise interfere with their survival.

Lethal control of legally protected species often requires permits. Usually, these are supposed to be issued only in situations where economic damage can be demonstrated.

Recreational hunting is sometimes used to reduce population size. For example, it may be desirable to reduce the size of a deer herd that is damaging vegetation. Since deer are polygynous, however (like northern fur seals (Box 3.3)), most of the females in a population mate with just a small percentage of the males. This means that many males can be killed without affecting a herd's reproductive potential if enough breeding males are spared. Therefore, if hunting is to be used as a tool to reduce herd size in a species with polygynous mating, some females must be removed. This is the justification for doe hunts. Although there are sound biological arguments underlying the removal of females to reduce population size, the idea that only bucks should be hunted is deeply ingrained with the sport-hunting public in many countries, and thus public opinion often opposes doe hunts.

Lethal methods of pest control vary greatly in the amount of suffering they cause, depending primarily on how quickly death ensues. Public attitudes toward lethal control depend on several factors. Acceptance is likely to be highest if death occurs quickly, if females with young are not killed, if the targeted species is not viewed sympathetically, if there is minimal mortality to nontarget species, and if the public benefits from control.

Nonlethal pest control involves decreasing reproduction or changing the behavior of the target species. Pest problems can sometimes be addressed by removing problem individuals or altering their behavior, physiology, or environment. Nonlethal methods of control include live-trapping and relocating animals, modifying animal behavior through conditioning, using guard dogs to keep away predators, changing ranching or farming practices, sterilizing problem animals or preventing mating, vaccinating disease-carrying animals, reducing the food supply of problem populations, and using and *lure crops* to direct pests away from places where they cause problems. These methods are designed to minimize wildlife mortality and other environmental impacts of control operations. In a different approach to control, plant breeders develop sterile varieties of ornamental plants that otherwise would tend to be invasive or that have allergy-causing pollen, and veterinarians spay feral cats.

One type of nonlethal control uses chemicals that interfere with normal functions in the pest species. Hormones that regulate growth and reproduction or communication between members of a species are useful for controlling insects by disrupting mating or courtship. A similar approach involves species-specific sex attractants that lure harmful insects into traps.

Nonlethal control sometimes involves modifying the behavior of people to reduce the frequency of encounters with animals or to change conditions that allow pest populations to build up. This kind of control involves educating people about animal behavior in addition to addressing ecological issues. Problems with urban pigeons (Section 4.7.1) and with bears in national parks (Section 4.8.1) illustrate this in different settings.

Methods of control that do not kill the target animal outright may ultimately cause its death anyway. Animals caught in live traps that are supposed to be nonlethal may die if those traps are poorly designed, improperly set, or checked only infrequently. Animals may incur fatal injuries from trying to escape, or they may die from exposure, infection, or predation while they are in the trap. These unintended effects should be considered when selecting a control method.

Biological control, the deliberate importation of predatory or parasitic insects or diseases to reduce pest populations, is a lethal or nonlethal strategy for controlling non-native pests. Disease-causing microorganisms, such as viruses, bacteria, and some fungi, and predatory, parasitic, or herbivorous insects are introduced in areas where targeted plants or insects are present. Biological control may result in the death of the affected organisms, or it may suppress pest populations by slowing or preventing reproduction.

Because this form of control uses coevolved relationships between antagonistic species and avoids the ecological and health risks associated with chemical control (Section 5.2.1.1), it has an intuitive appeal. Biological control has had some spectacular successes controlling insect pests and weedy plants. However, unintended consequences of biological control have led to an intense debate (Box 4.1).

Box 4.1

Evaluating Evidence: Biological Control of Invasive, Introduced Weeds: Powerful Technology for Restoring Balance or a Double-Edged Sword?

Thistles are a group of prickly leaved plants in the sunflower family. In North America, several kinds of introduced thistles, including Canada thistle, Scotch thistle, and musk thistles, are invasive weeds. However, there are also many native thistle species, some of which, like Pitcher's thistle, are rare and/or important constituents of native plant assemblages (Box 7.3).

Biological control has been used for control of some of the non-native thistles. The flower head weevil is a species of beetle native to Eurasia and Africa. Adult females deposit their eggs in the flower heads of some thistles, which prevents them from reproducing by seed. In 1968, flower head weevils were brought to Canada from France and Italy to be evaluated for use in biological control against musk thistles. Before it was released, this introduced species went through extensive screening to determine what thistle species it preferred to attack. In these prerelease assessments, weevils laid eggs in native thistles as well as introduced ones. However, because the weevils preferred to use the target species (musk thistle), and their larvae also did best in musk thistles, the investigators thought that introduced weevils would stick mostly to musk thistles and not damage native species.

After the flower head thistles were released in Canada and the USA, they spread and attacked six species of native thistles. In one species, the weevils damaged less than 1% of the flower heads, but in the other five species, the percentage of flower heads per plant with evidence of weevil damage ranged from 17% to 78%, causing substantial declines in the production of viable seeds (Louda et al., 1997).

In 2004, the journal *Conservation Biology* published two papers that laid out contrasting interpretations of these events. Mark Hoddle at the University of California, Riverside, acknowledged the problem of negative effects on nontarget species, but he argued that biological control is nevertheless "a powerful tool for invasive species management" that could "allow a return to ecological conditions similar to those observed before the arrival of the pest" (Hoddle, 2004:38).

Svata Louda at the University of Nebraska–Lincoln and Peter Stiling at the University of South Florida acknowledged that in some cases, biological control is very successful, but they argued that biological control is a "double-edged sword," with both advantages and disadvantages. In their view, the risks of "introducing self-replicating, self-dispersing, irretrievable biological 'natural enemies' with unexpected ecological side effects" outweigh the benefits (Louda and Stiling, 2004:52).

The disagreement centered on the question of how easy it is to reconstruct a natural community. According to Hoddle (2004:39), introducing natural enemies that an invasive species "left behind" when it arrived in its new homeland is a useful "exercise in community re-assemblage." Louda and Stiling (2004:50), on the other hand, contend that we still do not know enough to predict the outcomes of introducing "alien organisms into new species assemblages in new physical environments, and without the rest of their food web."

One thing that all parties to this controversy agree on, however, is the need for rigorous, well-regulated, case-by-case assessments of the benefits and risks of biological control (*see Q4.1*).

4.3 Historical Background

In Western Europe, Russia, North America, and colonial Africa and Australia, farmers and ranchers viewed and often still view predators negatively. Historically, this attitude was not controversial; it was considered common sense and was shared even by biologists and conservationists until well into the twentieth century. Theodore Roosevelt – who is generally regarded as pro-conservation because of his efforts to regulate timber exploitation and his designation of wildlife refuges – considered the wolf "the beast of waste and desolation" (quoted in Mech, 2012:143). In 1913, William Hornaday, director of the New York Zoological Park and author of the book *Our Vanishing Wildlife*, was concerned about declines in animal populations, including bison (Section 1.3.2.3), but he had no sympathy for large carnivores. Using the emotional language typical of his time, he wrote, "there are several species of birds that may at once be put under sentence of death for their destructiveness of useful birds …. Four of these are *Cooper's Hawk*, the *Sharp-Shinned Hawk*, *Pigeon Hawk* [merlin] and *Duck Hawk* [peregrine falcon]" (Hornaday, 1913:80; emphasis in original). In 1919, in his book *A Colony in the Making. Or, Sport and Profit in British East Africa* Lord Cranworth expressed a similar attitude:

There are at the present time certain animals, such as the eland and buffalo, which are under taint of suspicion of bringing in their train tsetse-fly or other obnoxious parasites, and therefore are inimical to stock raising. Should this suspicion develop into a certainty, these species must disappear from all settled lands ….

(Cranworth, 1912:310)

4.3.1 Bounties

In North America, governmental efforts to control species that were viewed negatively date back at least to colonial New England, when there was widespread agreement that animals that were deemed harmful should be controlled. In 1630, the Massachusetts Bay Colony passed a law providing for payment to anyone who killed a wolf. During the next three centuries, state and local governments in the USA and Europe enacted bounties on mammals and birds believed to be harmful to livestock or crops. Between 1883 and 1930, over 280,000 bounties were paid for dingoes in New South Wales, Australia (Glen and Short, 2000). These measures provided for payment to anyone who presented the carcass or body parts of a designated species such as a wolf, cougar, fox, weasel, hawk, owl, crow, blackbird, English sparrow, or other animal suspected of causing economic damage.

The bounty system had numerous shortcomings. It was wasteful and ineffective, and its ecological consequences were not thought out. During the nineteenth century, states in the USA paid an estimated $250,000 in bounties on birds alone (Palmer, 1899), often with little effect on the target populations. The system also invited fraud. County employees in charge of paying bounties were rarely motivated to be thorough when checking out a sack of decomposing animal parts. If presented with a bag of chicken heads topped off with a few hawk heads, the clerk was likely to pay without examining the evidence or challenging the claim. Children often took advantage of the fact that it could be difficult to identify baby birds. Boys would take the nestlings of protected species and claim they were English sparrows, for which bounties could legitimately be collected (Box 4.2) (Barnett, 2010).

4.3.2 Conflicting Attitudes and Values

By the nineteenth century, pest control had generated considerable controversy. Constituencies with different interests and values arose. Agricultural and livestock lobbies wanted the bounties, but sport hunters were ambivalent. Although they generally supported controlling animals that killed livestock, sometimes they opposed the bounty system because it targeted game animals such as foxes, raccoons, and black bears. Since hunters paid license fees to the states, this put sportsmen in the position of helping to finance the destruction of species that they wanted to hunt (Trefethen, 1975).

One heated controversy regarding a bounty species involved a small bird introduced to the USA from England, an action that caused a dispute that was so acrimonious it became known as the sparrow war (Box 4.2).

These examples illustrate a point that we will come back to repeatedly. Different positions on pest control (and other conservation issues) stem from different values, and those differences are related to status, geography, occupation, and other social and economic factors. How should we weigh the value of a species of sparrow as "amusement and inspiration" versus the economic damage it causes? What about the positive value of a fox as a game species when weighed against the chickens it might kill? Who gets to decide?

Box 4.2
The Sparrow War

The English sparrow or house sparrow is native to Eurasia, North Africa, and the Middle East where it is abundant around human dwellings and in croplands. In the early 1850s, the director of the Brooklyn Institute in New York City imported eight pairs of English sparrows from England (Figure 4.1). He released these in the hope that they would control the canker worm or inch worm that attacked shade trees in urban parks from Boston to Washington, D.C., but they did not persist. Subsequently, Ernest Schiefflin, a wealthy New Yorker and future founder of the American Acclimatization Society, a private organization devoted to introducing European species in North America, released dozens more English sparrows in New York. By the 1870s, the birds were breeding in New York City as well as New England and eastern Canada, and by the 1880s, the species had been brought to South Carolina, Texas, Utah, and California. Once it became established. it spread rapidly, reaching an estimated average radius of 161 km within 15 years (Robbins, 1973:3).

The newcomers ignited a passionate debate. Elliott Coues and Thomas Brewer, two members of the American Ornithologists' Union, a private organization for the conservation and study of birds, engaged in a heated exchange that lasted until Brewer's death. Most members of the organization agreed with Coues that the bird was a serious pest, as did scientists with

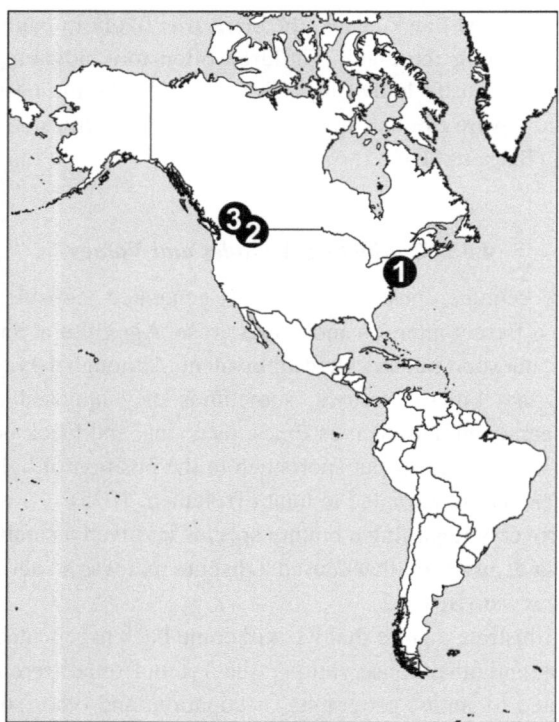

Figure 4.1   Locations of: 1, the first release of the English sparrow in the USA; 2, Glacier National Park; 3, Revelstoke, British Columbia. Map created by Eva Strand using Esri, DeLorme World Countries Generalized Data & Maps for ArcGIS 2013, with permission.

the federal Department of Economic Ornithology and most farmers. But members of local Audubon societies and others agreed with Brewer that introducing the English sparrow was, on balance, a good thing. Thousands of outdoor enthusiasts, sportsmen, and members of conservation groups joined in the fray by writing impassioned letters on the good or bad qualities of the little bird.

The feud between Coues and Brewer ended when Brewer died, but the controversy continued for several more decades. In 1907, the secretary of the Michigan Audubon Society championed the positive value of the English sparrow:

[w]ho can look at this bird with the temperature about the zero mark, hopping through the snow and chirping as happily as though it were a day in June, and say they despise it? They give cheer to many and brighten the lives of the disheartened and the ill, and afford amusement and inspiration to countless children.

(Butler, 1907:48)

Around the same time, a midwestern ornithologist dramatically declared in *The Birds of Ohio* that

[w]ithout question the most deplorable event in the history of American ornithology was the introduction of the English sparrow. The ... passing of the Wild [passenger] Pigeon ... [was a trifle] compared to the wholesale reduction of our smaller birds, which is due to the invasion of that wretched foreigner, the English Sparrow.

(Dawson, 1903:40)

This rhetoric was typical of many of the sparrow's critics, who, like most Americans in the nineteenth century, characterized birds as good or bad and assigned human qualities to them accordingly. Negative feelings toward the English sparrow were tied to American antagonism toward Great Britain ("that all-conquering nation") and urban immigrants (Italians and eastern European Jews) at a time when opposition to immigration was strong in the USA (Coates, 2007:41). Though ornithologists claimed to base their opinions entirely on fact, most shared the prevailing biases of their time. Because the English sparrow was not native and was abundant in cities, this "rowdy," "lazy," "bird of the street and gutter," was a "noisy little foreigner" that occupied "tenements" in "alarming numbers" "at the expense of our own avian favorites" (Coates, 2007:38–40, 52). Scientists and the general public alike used the sparrows to draw moral lessons. The English sparrow was said to set a poor example for its human neighbors, especially working-class boys. Since people fed the birds, they "set the unwholesome example of consuming what they do not earn" and "thought they had been sent to run the town." "[F]ond of low society and full of fight, stealing, and love-making," these "ignoble," "disgusting exotics" lacked "domestic ethics." In contrast, native birds were portrayed as clean, tuneful, "industrious ... good Citizens" (Coates, 2007:40,44). Attacks on the English sparrow, which "would never make a good citizen" and, according to Hornaday, displaced native birds by crowding out "its betters," evoked fears of human immigrants (Hornaday, 1913:334). In her guide to bird study, the nature writer Neltje Blanchan explicitly aroused fear of Asian immigrants ("As the 'yellow peril' is to human immigration, so is this sparrow to other birds") and went on to rue the fact that there was no sparrow exclusion act comparable to the Chinese Exclusion Act of 1882 (Blanchan, 1913:208–209; Coates, 2007:45,47;).

4.3.3 Government Predator Control

In addition to the bounty system, during the early days of formal wildlife management, government predator control agents in the USA set out to exterminate predatory birds and mammals in the name of protecting domestic stock and game. In 1917, the Chief of the US Bureau of Biological Survey boasted that "in five years we can destroy most of the gray wolves and greatly reduce the numbers of other predatory animals. In New Mexico we have destroyed more than fifty percent of the gray wolves and expect to get the other fifty percent in the next two or three years" (quoted in Trefethen, 1975:165). In 1931, the US Animal Damage Control Act authorized federal efforts to control animals thought to be harmful to crops and livestock. From the 1930s to mid-century, there was little opposition to this federal effort to control animal damage. But gradually some concerns arose.

In 1944, Aldo Leopold (Section 1.5.2.2) wrote a now-famous essay entitled *Thinking Like a Mountain*, where he described a hunting trip on which he and his compatriots killed a group of wolves:

We were eating lunch on a high rimrock at the foot of which a turbulent river elbowed its way. We saw ... a wolf [and a] half dozen others, evidently grown pups ....

In those days we had never heard of passing up a chance to kill a wolf. In a second, we were pumping lead into the pack, but with more excitement than accuracy: how to aim a steep downhill shot is always confusing ....

We reached the old wolf in time to watch a fierce green fire dying in her eyes. I realized then, and have known ever since, that there was something new to me in those eyes – something known only to her and to the mountain. I was young then, and full of trigger itch; I thought that because fewer wolves meant more deer, that no wolves would mean hunters' paradise. But after seeing the green fire die, I sensed that neither the wolf nor the mountain agreed with such a view.

I now suspect that just as a deer herd lives in mortal fear of its wolves, so does a mountain live in mortal fear of its deer. And perhaps with better cause, for while a buck pulled down by wolves can be replaced in two or three years, a range pulled down by too many deer may fail of replacement in as many decades.

(Leopold, 1966:130)

Recent archival evidence has revealed that the hunt took place in 1909; however, Leopold continued to support wolf control for many years after that. Christian Diehm, a professor of philosophy and environmental ethics, suggests that rather than experiencing an instant revelation when he saw "something new ... in those eyes" in 1909, Leopold had a gradual change in his attitude during decades of research and field work (Diehm, 2013).

The essay is admired as a powerful statement about the interconnections among elements in an ecosystem. It also reflects Leopold's humility and honesty in his willingness to revise his ideas.

Wolves continued to remain controversial after the publication of Leopold's essay (and controversy over wolves shows no signs of abating). However, beginning in the 1960s, several developments caused the controversy to intensify. Wolf populations in parts of the USA were listed as endangered under the federal Endangered Species Act (Section 7.3.2), conflicts with livestock proliferated as wolves recovered because of their newly acquired protected status, and research provided evidence suggesting that wolves and other large carnivores might have far-reaching beneficial effects on ecosystems (Section 8.7).

4.4 Integrated Pest Management

At around the time that Leopold's essay appeared, entomologists at the University of California published articles in agricultural journals proposing "supervised control," which emphasized measures to make pest control as efficient as possible by (1) integrating biological control with chemical control, and (2) applying insecticides only when economic damage is imminent (Smith and Smith, 1949; Stern et al., 1959). Subsequently, they expanded that idea to include diverse methods and disciplines in *integrated pest management*. This approach combines biological and chemical control with habitat manipulation, modified agricultural practices, and education aimed at maximizing the effectiveness, efficiency, and safety of pest control while minimizing environmental harm.

4.4.1 Using Media Technology to Control an Agricultural Pest in Rural Africa

The invasive fall armyworm threatens food security and livelihoods throughout much of Africa. Like an army, this "worm" (actually the larva of a moth) has marched across sub-Saharan Africa (Figure 4.2) since it was first recorded in Africa in 2016. It attacks maize (corn), a staple food crop in the region, as well as rice, cabbage, tomatoes, and other crops in over 40 countries. Several synthetic and organic options are available for integrated pest management of this agricultural pest. But because of its recent arrival and its prevalence in rural areas, education about how to control fall armyworm is challenging, and access to expertise is often limited.

Fortunately, advances in communications technology have created options for addressing these problems. CornBot is a mobile audiovisual Internet app (see Figure 4.3) that interacts with farmers in their local languages, guiding them through identification, control, and management of this pest and linking them to additional sources of information (Anon., No date). After a farmer using the app confirms that they have fall armyworm, they can see information on options for control. These include directions for safe use of synthetic pesticides as well as alternatives derived from naturally occurring plants.

As environmental awareness increased, concern about pest control mounted. In addition to ecological issues, concerns arose about environmental, social, and economic effects. The examples discussed below illustrate some challenges of addressing these issues.

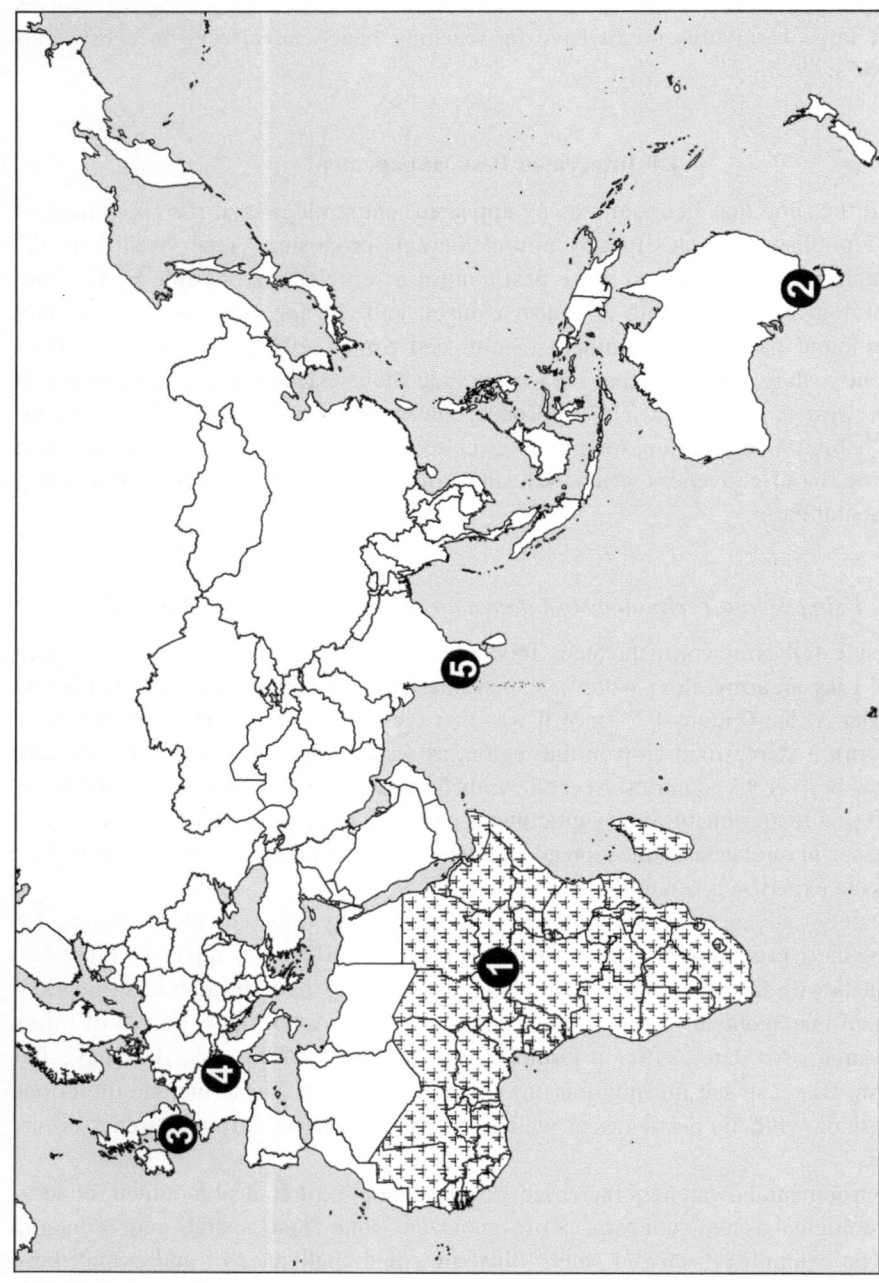

Figure 4.2 Locations of 1, sub-Saharan Africa (cross-hatched area); 2, the first release of rabbits on the island of Australia; 3, southwest England; 4, Lake Geneva and Basel, Switzerland; 5, Bhadra Sanctuary and Tiger Reserve, India. Map created by Eva Strand using Esri, DeLorme World Countries Generalized Data & Maps for ArcGIS 2013, with permission.

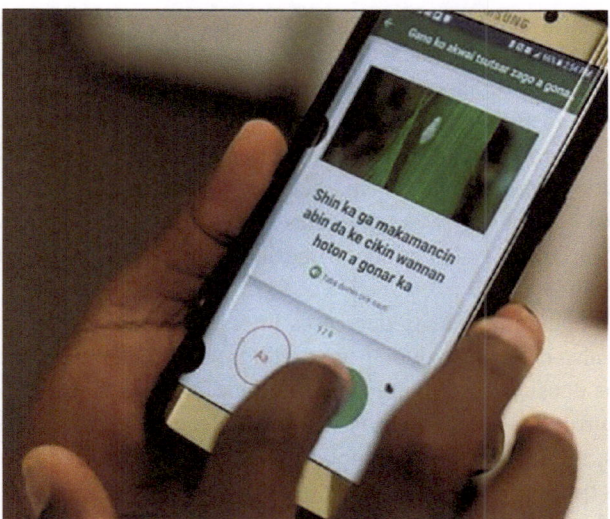

Figure 4.3   An example of integrated pest management. Screen shot from the CornBot app, for fall armyworm control in sub-Saharan Africa. Users are guided through identification of the pest and presented with options for control and sources of additional assistance. From Adewale (2018), with permission.

4.5 Pests That Compete with People or Valued Species

In the next two examples, species were targeted for control because they were associated with declines in valuable resources, and it was assumed that the declines were caused by competition from the pests. In Section 10.1.2, we will consider how experimental studies can provide more rigorous evidence of competition.

4.5.1 Competition for Livestock: Coyote Control on Rangelands in the Western USA

Like most pests, coyotes have positive and negative values. Many Native American tribes revere Coyote as a trickster, a hero, and an important character in legends. In rural areas, livestock producers target coyotes because they prey on sheep, hunters kill them because they kill game, and trappers harvest them for their fur. Urban and suburban dwellers may value the opportunity for wildlife viewing that coyotes provide, but also fear them because they occasionally attack people, especially children, and pets.

Prior to Euro-American settlement, coyotes in North America probably inhabited only the deserts and prairies of the West. But beginning in the nineteenth century, they expanded their geographic range. Coyotes now live in Alaska, the lower 48 states, all of Mexico, parts of Central America, and throughout Canada except for the far northeast. They thrive in environments modified by humans, which allows them to occupy suburbs and the urban/wildland interface.

With the disappearance of most large predators, especially the gray wolf and the grizzly, from most of North America, concerns about predators focused on coyotes, which kill substantial numbers of sheep on western ranges, as well as game animals, particularly fawns (Gese and Bekoff, 2004). The North American sheep ranching industry has declined in recent decades because of economic and social factors, including the availability of synthetic fibers and falling markets for lamb. In this context, ranchers argue that the additional burden of losses to coyotes is hard to bear, but critics suggest that the decline of the sheep industry is not due primarily to predation and should not be used to justify killing wildlife.

When coyote population density decreases in response to control, litter sizes increase and coyotes may also begin reproducing as yearlings instead of as two-year olds (Knowlton et al., 1972). This density-dependent compensation can make coyote populations resilient in the face of control. It is another reason why coyote control is controversial. Where compensation occurs, killing coyotes accomplishes little because reproduction increases, and other individuals quickly replace those that die.

Some livestock producers use lethal control methods, such as trapping, den hunting (the destruction of litters in the den, often by asphyxiation with carbon monoxide), shooting, and poisoning. These methods are inefficient because they do not selectively target breeding males, which are most likely to kill livestock. Nonlethal methods such as the use of guard dogs, llamas, and deterrents – for example electric fences, plastic flags, sirens, or flashing lights – have met with mixed success.

4.5.2 Competition for Range Vegetation: Rabbit Control in Australia

Various segments of the community see rabbits either as appealing
characters from cartoons and literature, a commercial resource, a
subsistence food source, an animal welfare concern or a major pest ....
Scientists and state and territory land management agencies must
communicate to legislators, land managers, landholders and the public
the damage caused by rabbits and ensure these people have sufficient
information to make appropriate rabbit management decisions.
(Williams et al., 1995:4)

On Christmas day in 1859, Thomas Austin, a hunter and member of an acclimatization society (Box 4.2), brought 24 European rabbits to the Australian mainland, where he put them in enclosures. Some soon escaped or were released. Although they spread slowly at first, once the rabbits became established, they rapidly colonized most parts of Australia, where they can live anywhere except at the highest altitudes. In Australian rangelands, rabbits attained the fastest rate of spread recorded for any mammal in the world.

Although they can live in shallow depressions on the soil surface if cover is available, rabbits are especially abundant in *warrens*. These complex networks of burrows provide shelter from extreme temperatures and protection from predators. Warrens are key to rabbits' success, but they also serve as foci for control because they harbor dense concentrations of rabbits that are easily targeted.

Because the flora and fauna of Australia evolved in isolation from other large land masses, many types of animals and plants that occur in other inhabited continents were not present in Australia when Europeans first arrived. There were relatively few species of wild predators. Ferrets and stoats, two species of small, slender weasels that can enter warrens and kill young rabbits, were absent. Some kinds of parasites and diseases that could have limited rabbit numbers also were not present in Australia.

Rabbits affect native flora throughout Australia by browsing, grazing, and stripping bark from around the bases of trees and shrubs. They keep vegetation from recovering by eating seeds and seedlings. These impacts in turn affect many wild and domestic herbivores that depend on the vegetation. Degradation of pastures and croplands by rabbits has had serious economic repercussions. Feral rabbits are also believed to have contributed to the extinction of several native mammals.

Initially bounties were offered for dead rabbits, with the usual lack of results. Efforts to implement biological control of rabbits began with the introduction of the virus myxomatosis.

This microorganism occurs naturally in two species of cottontails of the Americas, where it does not normally cause them to become seriously ill. It is transmitted from one rabbit to others by mosquitoes. Within a few years of its introduction to Australia in 1950, myxomatosis spread throughout the country wherever there were rabbits. The strain that was initially introduced caused 99.8% mortality in rabbits. By the next year, however, only 90% of the rabbits died, and mortality continued to decline. The initial high rate of mortality had selected against the severe strain of the virus that was introduced, because most rabbits died before that strain could be taken up by mosquitoes and transmitted to other rabbits. Consequently, less severe strains of the virus became more prevalent, "in a way analogous to the application of chemical insecticides" (Main, 1987:142). During this process, natural selection also favored rabbits that had some resistance to the virus. As a result of this coevolution myxomatosis no longer provided effective rabbit control.

As the promise of control through myxomatosis faded, new biological controls emerged. One of these involves a genetically modified virus with potential to inhibit rabbit reproduction. Although this initially seemed promising, the sterile rabbits lived longer than those that reproduced, which led to an increase in population that compensated for the decrease that resulted from the treatment.

Rabbit control methods such as shooting, poisoning, and *ripping* (in which a bulldozer dragging sharp tines is driven over occupied warrens) reduce population only temporarily because rabbits quickly move in from nearby areas. Rabbit-proof fencing can be effective, but this method is expensive because fences need to be both deep and tall in order to exclude rabbits and must be checked often for breaches.

More than a century and a half after they arrived in Australia, feral rabbits remain a persistent problem despite enormous amounts of money and effort expended on their control.

4.5.3 Competition for Water: Shrub and Tree Control in Rangelands

Rangeland vegetation usually consists of a layer of grasses and forbs plus an overstory of medium-sized shrubs, along with trees near water courses or in floodplains or other moist habitats. To utilitarian range managers, the precipitation that falls on rangeland shrubs and trees is "lost" because it is unavailable for forage production or to recharge streams. To improve water availability, they may remove the encroaching vegetation. Control involves fire, herbicides, or mechanical means such as bulldozing, chopping, or mowing.

In dry climates, utilitarian range managers have also targeted *riparian vegetation* (vegetation growing along streams and rivers) for eradication because they felt that *evapotranspiration* (evaporation plus the movement of water vapor from plant tissues into the atmosphere) competed with valuable forage plants for water:

Evapotranspiration from the flood plain of a major river depletes the water contributed from upstream areas of the watershed. This depletion can be significant in arid regions where water supplies are inadequate .... [C]ontrol [of deep-rooted woody vegetation] offers a method of reducing the evapotranspiration and thus increases the water available to downstream users. This control is achieved by removing the [deep-rooted woody plants] and replacing them with grasses having a lower consumptive use.

(Culler, 1970:684)

A strictly utilitarian approach that controls shrubs and other vegetation in *riparian zones* (places adjacent to flowing fresh water) in order to promote livestock production conflicts with management for other objectives because that vegetation has other values. The roots of riparian plants and rangeland shrubs stabilize soil and reduce erosion. The removal of those shrubs and trees converts structurally complex vegetation to simpler, more homogeneous configurations. Their leaves, branches, flowers, and seeds provide critical food and cover for many kinds of amphibians, bats, ungulates, and birds, especially migratory songbirds.

4.6 Pests That Transmit Diseases to People or Domestic Animals

Transmission of diseases by wild animals can potentially be controlled either by reducing contact between wildlife and people and domestic animals (Section 4.6.1) or by treating the affected animals (Section 4.6.2). Most people approve of controlling pests that transmit diseases in some situations, but once again approval depends on the value that people put on the species which are targeted for control and the species that are harmed.

The bison and elk in Yellowstone National Park carry brucellosis, or undulant fever, a disease with the potential to cause spontaneous abortions in cattle. Management of the roaming bison has involved both lethal and nonlethal control to keep them from coming in contact with cattle. Ranchers in the area are likely to support lethal control, but public opposition to killing bison is strong.

The determination of whether a wild animal transmits a disease to people or domestic organisms is often made on the basis of anecdotal evidence. However, that kind of evidence should not be the basis for control, because it can lead to excessive kills of animals that are inaccurately assumed to be causing disease, and it may have unintended consequences. In order to definitively establish transmission from a wild to a domestic animal, it is necessary to show (1) that the pathogen exists in the wild population, (2) that the wild population comes into contact with the domestic or human population of concern, and (3) that enough of the disease-causing organism is transferred to cause disease. This is difficult, time-consuming, and expensive, as the case of the link between Eurasian badgers and tuberculosis in cattle illustrates.

4.6.1 Controlling Badgers That Potentially Transmit Tuberculosis to Cattle in England

Badgers are members of the weasel family, a group of small to medium-sized carnivores that includes mink, ferrets, stoats, otters, and wolverines. The European or Eurasian badger (Figure 4.4) is widespread throughout Europe and adjacent parts of Asia. Eurasian badgers live in large, communal burrow systems that they keep clean by depositing feces in latrines outside their dens. They play several games outside their burrows, including leapfrog (Nowak and Paradiso, 1983). They are common in a variety of habitats including farmlands, where they may carry the bacterium that causes tuberculosis in cattle.

Figure 4.4   Eurasian badger. Credit: DamianKuzdak / E+ / Getty Images.

Bovine tuberculosis, the form of tuberculosis that affects cattle, is under control in most of Europe, but in the British Isles except for Scotland, it remains problematic. The persistence of the disease in British and Irish cattle has been linked to badgers. To minimize transmission, badgers have been removed for many years. However, when an interdisciplinary team of scientists in medicine, conservation, and statistical modeling from the UK and the USA conducted a controlled experiment on the relationship between this *culling* (removal) and bovine tuberculosis in southwest England (Figure 4.2), their results were surprising. In their large, randomized, replicated field experiment, they found that, contrary to their expectations, the incidence of tuberculosis in cattle was higher in areas where badgers had been culled than in nearby control areas where there had been no culling (Donnelly et al., 2003).

Subsequently, another controlled experiment was conducted to test the hypothesis that culling altered badger behavior in ways that could increase the amount of contact between badgers and cattle. Investigators studied badger activity in 13 study areas with different levels of culling (the treatments). To do this, they devised an inexpensive and minimally invasive (and clever) method of measuring badger movements. Near badger burrows, they placed baits consisting of a mixture of peanuts and treacle plus indigestible colored beads. When latrines in the baited study areas were surveyed, the locations of feces with color markers were used as an index of the distance badgers moved in areas with and without culling. Analysis of the number and distribution of these bait returns showed that home ranges were significantly larger in areas with culling, indicating that removal of badgers influenced badger movements. They concluded that the resulting changes might affect tuberculosis transmission between badgers or from badgers to cattle and that their findings "may help to design more effective management policies and should be taken into account in deciding what role badger culling should play in controlling cattle" (Woodroffe et al., 2006:9).

4.6.2 Controlling Foxes That Potentially Transmit Rabies to People in Western Europe

Self-administered wildlife vaccines offer a possibility for reducing disease transmission from wildlife to people. In Western Europe, a combination of medical technology and wildlife management has made it possible to eliminate or reduce the live rabies virus in wild foxes.

Rabies kills tens of thousands of people annually. Routine vaccination of domestic dogs and cats can limit transmission of rabies from pets to humans, but the virus is also present in some wildlife. Because the rabies virus is shed in saliva, this fatal disease is transmitted by bites from infected animals. Most of its victims live in impoverished countries, where dogs are often unvaccinated and people who are bitten lack access to medical treatment or cannot afford it. In North America, bats, skunks, foxes, and raccoons harbor the virus, but in Western Europe, foxes are the only significant wild source of rabies.

In the 1950s, health workers in Europe tried to control rabies by reducing the density of the fox population to the point where an infected animal would be unlikely to transmit the disease to another individual. This was ineffective, expensive, and controversial. Biologists concluded that more than 60% of the target population would have to be removed to achieve the desired objective, and killing that many foxes was not acceptable to the public.

Scientists next turned their attention to methods of vaccinating free-ranging animals. Since it would be impractical to trap animals, vaccinate them, and release them, interest in self-administered baits mounted. American researchers adapted the *coyote getter*, a device originally developed for poisoning coyotes, to this purpose. The apparatus consists of a small pipe stuck in the ground and baited with a tuft of scented wool. When an animal tugs on the bait, a cartridge fires a jet into the animal's mouth. Coyote getters fired the lethal poison cyanide. It was a simple matter to substitute a dose of vaccine for the poison, but unfortunately, this device damaged the target animal's mouth. While this was not a problem with the coyote getter, which quickly killed the animal, it was a serious drawback for the vaccine program because it made the vaccinated animal unable to eat.

Next, investigators developed a device consisting of a buried trigger pan that fired a vaccine-loaded syringe into the target animal's side when it stepped on the buried pan. This worked, but it was not economical, and it was hazardous to nontarget animals.

These difficulties suggested that baits might be the best way to administer the rabies vaccine. This, too, presented technical challenges. Researchers had to develop a concentrated vaccine that would penetrate the mucous membranes of the mouth and throat, and they had to find a bait that would appeal to the target species but not to others. By the mid-1970s, these obstacles had been overcome, and field trials began in Switzerland. When rabies spread among foxes on the eastern shore of Lake Geneva in 1978, scientists were ready to use baits to vaccinate the affected population (Figure 4.2). They succeeded in containing the outbreak by distributing chicken heads laced with live rabies virus in the affected area (Winkler and Bögel, 1992).

Self-administered oral rabies vaccines are a nonlethal, relatively economical method of rabies control that is acceptable to the public. As of 2013, this approach had been successful in eliminating wild reservoirs of rabies from much of Western Europe and Central Europe (Freuling et al., 2013).

The good thing about successful vaccination of wild foxes is that it changes the status of a valued species so that it is no longer a pest in most situations. Henhouses, of course, are an exception.

4.7 Pests with Recreational Value

The control of wildlife pests in urban settings poses special challenges. Cemeteries, parks, college campuses, backyards, and even building ledges provide habitat for wildlife in and around residential and business areas. But wild animals in cities are prone to becoming pests for two reasons. First, wild organisms that are found in urban areas tend to be capable of reaching high population densities around people. Pigeons, English sparrows, rats, and house mice are familiar examples. Second, in cities and suburbs, wild animals become established in places where they are not wanted and damage buildings, make noise, and spread diseases to people or their pets. This is true even with native species that are not usually regarded as pests. For instance, Canada geese become a nuisance when large flocks congregate in parks and on golf courses, leaving copious amounts of droppings. In some instances, deer or other wild animals in cities or suburbs become unpopular because they damage landscaping, or they face starvation because of inadequate resources.

On the other hand, wildlife in metropolitan areas has positive recreational and educational values. People who live in cities typically value contact with wildlife and have few opportunities for such contact. Thus, although it is often desirable for economic, esthetic, or humanitarian reasons to reduce wildlife populations in or near urban areas, city residents may react negatively toward control that kills wildlife. As a consequence, urban wildlife control becomes a matter of public relations as well as biology, and managers must strive to find acceptable methods of controlling unwanted organisms. This is certainly the case in cities where people feed pigeons.

4.7.1 Managing Human Behavior to Control Pigeons in Switzerland

High densities of street pigeons (rock doves) occur in many cities around the world, where they often have both negative and positive values. They provide city dwellers with an opportunity to see and interact with wild animals, but they transmit diseases and parasites to people and domestic animals, and their droppings damage buildings and statues. The result is a paradoxical situation in which city governments kill pigeons on the one hand, while city residents feed them on the other.

Efforts to control pigeons by lethal means are expensive, controversial, and often ineffective. Between 1961 and 1985, game inspectors in the city of Basel, Switzerland (Figure 4.2), trapped and shot about 100,000 street pigeons, yet these measures had no lasting effect on the population. Like coyotes, street pigeons are resilient in the face of population reductions. When adult mortality increases, there is a compensatory decline in juvenile mortality, and young birds quickly replace individuals that disappear from a population.

Faced with the ineffectiveness of lethal control by itself, city officials decided to add another component to their strategy. Besides removing individuals from the population, they reduced the birds' food supply. Since pigeon feeding was a popular activity that provided substantial food to the birds, it was important to educate the public about the effects of high pigeon populations. Pamphlets and posters that showed young pigeons suffering from density-dependent diseases and parasitic infections were distributed, with text explaining that excessive pigeon densities were bad for pigeons as well as people.

But pigeon feeding can fulfill an important social function. It often provides people who have no one to care for with emotional ties and a sense of being useful. Recognizing this, the organizers of the Pigeon Action Project at the University of Basel came up with a creative way to provide opportunities for pigeon feeding by maintaining a small number of flocks in supervised lofts where population density was carefully controlled.

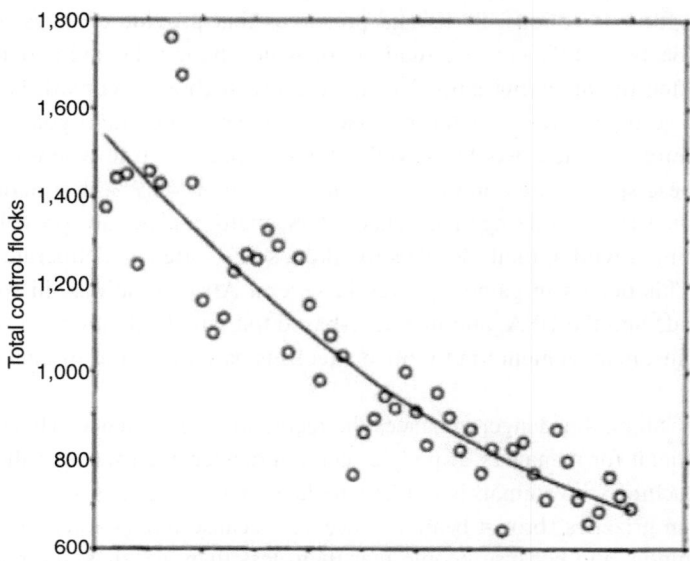

Figure 4.5 Decline in the total number of pigeons in 13 feeding flocks counted in Basel, Switzerland in response to reduction in pigeon population and restriction of pigeon food from 1988 through 1992. Used with permission of John Wiley & Sons Ltd., from Haag-Wackernagle (1995).

The project met with mixed success. Unregulated pigeon feeding declined markedly, as did the number of pigeons in the city (Figure 4.5) and the amount of damage they caused, but the designated pigeon encounter areas were less successful. Few people used them, in part because the project had been so successful in getting its anti-feeding message across that it created social pressure against all pigeon feeding. This example illustrates the complexities of controlling animal damage while providing opportunities for positive interactions between people and wildlife.

4.8 Pests That Are Both Rare and Dangerous

Control of problem species usually involves population reduction, but this is complicated when the target species is rare. Tigers, elephants, and rhinos are rare but also potentially dangerous to people, crops, and livestock and the source products that bring exorbitant prices in the (often illegal) international wildlife trade. People have long interacted with these charismatic species. In different contexts, they are revered, feared, exploited, and avoided. Besides weighing the needs of such animals and people, wildlife agents need to distinguish between poaching and legitimate cases where the removal of problem animals is warranted to protect people.

Wherever people and large animals overlap, there is potential for conflict. This occurs where people and wild animals live close to each other, and in situations where people unaccustomed to interacting with wildlife seek contact by visiting places that offer opportunities for interactions. Parks and preserves that provide chances for people to see and approach wildlife create situations in which human behavior may contribute to attacks. Often the attacking animal is a carnivore such as a leopard, bear, shark, or dingo, but large herbivores – including bison, elephants, mountain goats, hippos, and rhinos – sometimes attack people as well. This is especially true where encounters by people and these species are common, as in *nature-based tourism* destinations. Despite education aimed at minimizing risky encounters, many attacks are provoked by tourists approaching a wild animal, perhaps to take a selfie, often in deliberate violation of regulations. This occurs in game reserves in several African nations, in some national parks in Canada and the USA, and in nature-based tourism destinations in Australia and elsewhere. Thus, management to minimize attacks is partly a matter of managing human behavior.

For some visitors, this danger enhances the recreational experience. This makes things especially difficult for managers. If people seek out danger, then warning them about the risks of approaching wild animals isn't likely to deter them. Black bears are considered less dangerous than grizzlies (brown bears). However, because interactions with black bears are more common, and because people fear them less than grizzlies, the rate of injuries from black bear encounters in national parks in the lower 48 states of the USA exceeds the rate of injuries from grizzlies.

When wild animals attack people, those species are sometimes culled in retaliation or to prevent future attacks. This is true regardless of whether or not the attacks are provoked. Sometimes, the individual animals that attacked are culled, but sometimes the killing is not selective. Therefore, minimizing wildlife–human conflict is important both to protect people's safety and to minimize losses of the offending species. Since species that attack people are often large carnivores, which tend to be rare, balancing safety and conservation is crucial (but tricky).

4.8.1 Minimizing Conflicts between Bears and People in the USA and Canada

4.8.1.1 Where Visitors Visit Parks with Bears

In 1967, two young women were unexpectedly killed by grizzlies in separate, unprovoked attacks during the same night in Glacier National Park, in Montana (Figure 4.1). In response to this situation, the US National Park Service sought to identify factors that increased the likelihood of bear attacks. The resulting policies addressed changing circumstances affecting interactions between people and bears. An increase in the number of backcountry visitors had set the stage for the attacks because it increased the likelihood of encounters. In addition, grizzlies were attracted to food in the campgrounds and at garbage dumps, which made them dependent on artificial sources of food. In 1969, the Park Service closed the dumps and implemented strict regulations for disposing of all food and dishwater in campgrounds and in the back country. Films, signs, and pamphlets educated people about the importance of garbage control and about what to do if they saw a bear. Park bears were monitored, and areas where the probability of an attack was considered high were closed to camping or hiking. If problem bears were identified, those individuals were sometimes moved to other locations. If that was not successful, problem bears were sometimes killed (Wright, 1992).

4.8.1.2 Where Bears and People Are Neighbors

Bears and people come in contact in rural areas where substantial populations of bears inhabit areas near human communities. In 1994, electric fencing was installed around the landfill in Revelstoke, British Columbia, Canada (Figure 4.1), a city of about 8,000 people. This kept bears from accessing a significant food source. When bears turned to garbage within the city, problems developed, and 62 bears were either killed or relocated. Then when the berry crop failed the following year and bears lost access to another important food source, 25 additional bears were relocated, and 23 more were killed.

In response, community members – alarmed about the bear hazard and concerned about the number of bears that were being killed – formed the Revelstoke Bear Awareness Society. The group focuses on the same issues that park management agencies address: reducing bear access to food sources near people (in this case, garbage, fruit trees, and pet food) and educating people about reducing bear attractants. These measures led to a marked decline in bear killings. In 2019, the Conservation Officer Service of British Columbia destroyed only three black bears (which had become conditioned to garages). There were 11 instances in which bears broke into houses and destroyed property, and one case in which a bear injured a person. As part of its efforts to decrease bear-garbage interactions, the Society provided financial assistance for residents to purchase bear-resistant bins (Spizzirri, 2019).

These government and community programs have reduced conflicts between bears and people. However, since the densities of large animal populations are low, culling even a few individuals may put carnivore populations at risk. In these cases, the tradeoff between public safety and conservation is stark.

In Revelstoke, community support for reducing human–wildlife conflict while minimizing wildlife mortality is high. People value proximity to the bears and are motivated to take actions that promote coexistence.

In other situations, interactions between people and dangerous wildlife are more problematic, especially where local communities have little role in deciding how to manage dangerous animals.

The costs to people of conflict with tigers are high (including substantial risk of death) and so are the costs of conservation (relocation). Furthermore, the opportunities for community participation in conservation are limited.

4.8.2 Managing Human–Tiger Conflict in India

Tigers, like other large carnivores, require large home ranges that support substantial amounts of prey, and they have a long history as targets of exploitation and predator control. Currently, illegal trade for high-value tiger products – including bones (for use in traditional Asian medicine), skins, teeth, claws, and meat – continues to put pressure on tiger populations.

The tiger is a globally endangered species. It is widely distributed in South and Southeast Asia, and an isolated population persists in eastern China. But its distribution has contracted, and its numbers have declined dramatically.

Tigers are an iconic species. They feature prominently in literature, art, legends, songs, and dances. In the developed world, tiger conservation generates sympathy and funds in many quarters. The World Wildlife Fund offers symbolic tiger adoption kits to donors, who receive a plush toy tiger, photo, and "adoption" certificate.

Where tigers and people overlap, tiger attacks on people and livestock occur. Sometimes this mortality is substantial. In the Sundarbans, a mangrove forest of coastal India and Bangladesh, tigers killed about 40 people from 2000 to 2010. In some parts of India, the families of tiger victims get financial compensation, but the widows of men killed by tigers face social stigma.

Tiger conservation focuses on establishing an international network of large *Tiger Protection Areas* where human activities are restricted and tiger mortality is minimized. It is assumed that the removal of people from areas in and around the protected areas will result in an improvement of tiger habitat.

When attacks occur, tigers are typically killed or removed. To minimize human–tiger conflict between people and tigers, people in and around many tiger reserves have been relocated. Where people remain near tiger reserves, traditional uses of resources – such as grazing and collection of forest products – are prohibited. In 2005, a report by the Tiger Task Force estimated that 80 villages, with 2,904 families and 46,341 people, have been relocated for tiger reserves (Lasgorceix and Kothari, 2009).

The earliest relocations from Tiger Protection Areas were forced, but in a few areas, such as Bhadra Sanctuary and Tiger Reserve in southwestern India (Figure 4.2), there has been a trend toward voluntary relocations. Initially, relocations were characterized by a lack of transparency and poor communication between the villagers and the Forest Department (which implemented the removals), and villagers resisted resettlement by setting fires, demonstrating, and signing petitions. Around 1990, the situation improved. Villagers were offered better terms; voluntary participation became an option; communication and coordination between the Forest Department, local NGOs, and villagers got better; villager participation in the process increased; and fewer conflicts occurred.

The situation with regard to relocations is changing. As awareness of the problems caused by forcible evictions grew, the conditions offered to those who relocated improved, and communities began to accept or ask for relocation. This was partly due to recognitions that staying was not an attractive option because "life within the PA [Protected Area] is very difficult" (Lasgorceix and Kothari, 2009:38).

4.9 Values and Tradeoffs

Although attempts to remove species deemed undesirable may have unfortunate consequences, the control of invasive species may be an important part of ecosystem protection or restoration. If people had been able to eliminate the chestnut blight before it devastated American chestnuts, would that have been a good thing?

Insects that compromise food security by attacking crops, like the fall armyworm in Africa (Section 4.4.1), or that transmit serious diseases, like the mosquitoes that spread the microorganism which causes malaria (Box 4.3), don't have a lot of champions. Nevertheless, efforts to eliminate them entirely should not be taken lightly.

Whether and how to minimize conflicts with organisms that cause economic harm, ecological damage, suffering in animals, or threats to human health or safety are issues that are subject to debate. Pest control always involves tradeoffs, and decisions about tradeoffs are always linked to values. Managers should weigh potential impacts of control on human well-being, interspecific interactions, environmental quality, and animal welfare.

The choices are difficult. If one interest group wants a species to be reduced or eliminated while another group wants it to be abundant, how much control is appropriate, who should decide, and who should pay for it? If society wants to coexist with a species that

---

Box 4.3
Should We Try to Eliminate the Main Mosquito Species That Carries Malaria?

It might be worth losing one species … because the burden of human suffering is pretty high.

(Steven Juliano, mosquito-ecology researcher, quoted in Zhang, 2018)

Malaria is a major health concern throughout the world, especially in poor countries. A person develops malaria after the single-celled parasite enters their blood during a mosquito bite. High fevers, chills, and flu-like symptoms follow. The disease is sometimes fatal; however, most often a person with malaria survives but experiences repeated, debilitating relapses. Most cases of malaria occur in warm climates, but it also occurs in temperate regions, especially where stagnant water abounds, because that is where mosquito larvae mature.

Hanging mosquito nets over beds is somewhat effective in reducing the transmission of malaria, especially if the nets are treated with insecticide. Typically, nets are impregnated with chemicals that are not very toxic to people. Spraying the inside walls of homes with insecticides also lowers transmission. Several African nations permit the use of DDT for this purpose, which is controversial because of DDT's toxicity to people and wildlife (Section 5.2.1.1). In 2006, after weighing the possible benefits to human health from improved malaria control against the potential adverse human health and environmental effects, the World Health Organization approved the use of DDT for indoor spraying.

Many species of mosquitoes transmit malaria, but the worst forms of the disease are caused by bites from a group of species known as *Anopheles gambiae*. As part of efforts by an international research consortium that develops and shares technology for malaria control, researchers investigated the ecological effects of *A. gambiae*. Realizing that pest control has a long history of leading to unintended consequences, scientists from Ghana and England began studying how *A. gambiae* interacts with other insects, as well as flowers, fish, and bats. They set up small pools of water to serve as artificial breeding sites and then studied what happened when they took out the *A. gambiae* larvae. Would a different species of mosquitoes take over the functions previously performed by *A. gambiae*? Would species that prey on or are pollinated by *A. gambiae* be seriously affected if it disappeared? The scientists concluded that *A. gambiae* does not play a key role in the diets of animals that eat it and is not an important pollinator for plants with which it interacts. But research to test these hypotheses further continues (Zhang, 2018).

causes hardship for some groups, should the negatively affected group be compensated? What is a fair distribution of these benefits and losses? How do we weigh the value of human life, health, and livelihood compared to the life of a tiger? Or of an ecosystem? There are no objective scales for measuring these things.

4.10 Evaluating Pest Management

In spite of a long history of management to reduce pest species, data on the long-term effectiveness of such management is nevertheless hard to come by. Assessments often draw conclusions from poorly designed studies or rely on anecdotal or descriptive accounts rather than scientific evaluation.

To address the need for evidence on the effectiveness of predator control, a team of three scientists (one each from the USA, Slovenia, and South Africa) searched peer-reviewed scientific literature for experimental tests of the effectiveness of efforts to control carnivore predation on livestock in North America and Europe. In a paper entitled *Predator control should not be a shot in the dark*, they reported that they found only 12 tests (five that used nonlethal methods and seven studies of lethal methods) that met the accepted standard of scientific inference without bias. Of those twelve, prevention of livestock predation was demonstrated in only half the examples. In two cases of lethal control, predation actually increased. The scarcity of well-designed studies and the results from the few rigorous experimental tests this group discovered led them to "recommend that policy makers suspend predator control efforts that lack evidence for functional effectiveness and that scientists focus on stringent standards of evidence in tests of predator control" (Treves et al., 2016).

The lesson from this work is relevant not just to studies of predator control but to all pest control (and in fact to scientific research in general). Control programs should identify objectives in terms of specific desired outcomes, and the performance of those policies in meeting their stipulated objectives should be objectively evaluated on the basis of scientific criteria that are specified ahead of time. Did control result in a reduction in the density of the target population? If so, was that effect lasting? Were associated outcomes such as declines in economic losses, disease transmission, or accidents achieved? How were those things measured?

If objectives are not stated clearly at the outset, control can become an end in itself regardless of whether it effectively addresses the original problem. The effectiveness of control should not be judged by the number of coyotes or badgers or rabbits killed or the extent of sagebrush eradication, but rather in terms of whether desired results of those interventions occurred.

In addition, the ecological, social, economic, and political contexts of pest control should be considered as part of evaluation. Did the results justify the expenses of the operation? Were the environmental and public health, animal welfare, and public relations consequences acceptable? Did different groups have different opinions about whether the outcomes were acceptable? Can a control program be improved so as to do a better job of meeting its objectives?

The efforts of early utilitarian managers to decrease or eliminate predators and other unwanted species were often wasteful and ecologically harmful. More recently, the emphasis shifted to developing methods of control that are more efficient and that specifically target problems while minimizing harmful impacts. Nevertheless, pest control often has unforeseen effects. If a program is not carefully designed and monitored, it may waste money and labor and accomplish little or, worse yet, cause harm. Synthetic chemicals that are administered to control pests may poison nontarget species and accumulate in the food chain (Box 5.2), and successful control may reduce or eliminate species that perform significant ecological functions. Unintended evolutionary consequences might also result, because control creates selective pressures for certain characteristics (Box 6.2).

In this and the preceding chapter we have seen how the strategies of utilitarian conservation strive to address the needs of some kinds of plants and animals by maximizing the productivity of valued species while minimizing problems from others. The methods developed for those utilitarian goals can also be applied successfully in different contexts. In the second half of the twentieth century, preservationist conservationists began to apply the same methods to new goals. They modify habitats to benefit rare and endangered species, apply insights about the effects of edge to the design of nature reserves, and try to manage pests with methods that minimize negative ecological and social impacts.

Historically, harvest management, habitat management, and pest control tended to overlook complex interactions and environmental impacts. This set the stage for new developments in conservation.

Chapters 5 through 8 consider management aimed primarily at preserving species and habitats from human impacts. The roots of this approach go back as far as those of utilitarian management, and the two approaches developed simultaneously. However, the preservationist approach reached its heyday as problems stemming from intense resource utilization became apparent in the wake of World War II. We will consider those next.

Part II

Protecting and Restoring Populations and Habitats: A Preservationist Approach to Conservation

5

Historical Context

Rising Concerns about Human Impacts

All human societies use natural resources and have impacts on their environment. Those impacts may be deliberate or unintended. They may lead to environmental collapse, long-term use, or a range of outcomes in between. In Part I, we saw how conservation of living natural resources developed in response to unregulated exploitation. This has been successful in sustaining the take of many harvested species, controlling unwanted organisms, and manipulating habitats for the benefit of certain organisms. But it also had unintended consequences due in part to the simplification of managed habitats.

Furthermore, changing conditions in the middle of the twentieth century created new pressures on species and habitats, and changing values dictated a broader focus, one that considered the needs of many species. This chapter describes the development of an approach to conservation that seeks to preserve living things regardless of their economic value.

Influential writers in industrial societies have been sounding alarms about the impacts of people on the Earth's resources since the end of the eighteenth century (Malthus, 1798; Marsh, 1874). In the middle of the twentieth century, a constellation of interrelated crises triggered more widespread awareness of and concern about those impacts. Toxins in the environment, habitat degradation, accelerated rates of species' extinctions, and widespread soil erosion contributed to a mounting sense of alarm.

5.1 Economic, Demographic, and Technological Changes after World War II

In the wake of World War II, fundamental changes took place in both rich and poor countries. These changes resulted in new pressures on habitats and species. Industrialized nations experienced simultaneous population growth and rising standards of living, both of which raised levels of resource consumption. In wealthy countries, agriculture became increasingly mechanized. Because mechanized farming required less labor, much of the rural population was displaced and migrated to urban centers. At the same time, people began moving out of cities. Instead of returning to rural areas, they settled in suburban bedroom communities and commuted to urban jobs. These demographic changes profoundly altered the landscape. Fields planted to a single crop replaced small, diverse farms, and discrete cities surrounded by farms were replaced by suburban sprawl.

In developing countries, a different set of forces came into play in the late twentieth century. In different ways, foreign investors and international aid programs put pressure on living natural resources. Because large-scale commercial exploitation of resources requires capital and sophisticated technology, developing nations came to depend on foreign investments and expertise for economic development. Large-scale, international timber harvest, in particular, requires large amounts of capital; this is usually provided by multinational corporations backed by financial muscle in developed countries (Myers, 1979). This dependence tended to make host countries timid about imposing environmental regulations that might scare off foreign investors. Because of this unbalanced relationship, foreign investors faced few restraints on how they managed resources. At the same time, they were "driven to apparently reckless forms of exploitation" by interest rates on their substantial investments (Myers, 1979:193). Those conditions favored short-term profits rather than ecological stewardship.

Simultaneously, rising living standards in the developed world created a demand for resources from the developing world, such as timber, paper pulp, and beef.

In addition, international development agencies began projects that put novel pressures on resources in Asia, Africa, and Latin America. These projects often encouraged people to abandon traditional modes of resource use, and in some cases to replace them with management strategies based on Western models that were not appropriate for conditions in the developing world (Homewood and Rodgers, 1987). The agencies also built roads, dams, and other forms of infrastructure that had unforeseen environmental and social impacts.

During and after the World War II, technological developments led to the production of new and more powerful weapons, synthetic chemicals, and machinery. By the 1950s, concern about widespread environmental changes had mounted. Public awareness of connections between environmental quality, the well-being of wild organisms, and human health and welfare grew. This increase in environmental awareness was accompanied by heightened worry about species and habitats that do not provide commodities. As the pace of resource extraction quickened, people began to notice more and more undesirable effects of resource consumption, even if it was regulated.

5.2 Awareness of Environmental Impacts

5.2.1 Harmful Substances in Air and Water

Because toxic substances are often introduced in small amounts and appear to be rapidly diluted in air, water, and soil, people initially thought that pollutants such as radioactivity and pesticides would not reach harmful concentrations in the environment. In the 1950s and 1960s, however, it became apparent that complacency about the vastness of the Earth was not justified.

5.2.1.1 Pesticides

During World War II, a powdered form of the synthetic insecticide DDT was used to control lice – which transmit the deadly disease typhus – among soldiers, prisoners, and refugees. (Anne Frank, the young Jewish prisoner whose famous diary chronicled her life in hiding during the German occupation of the Netherlands, probably died of typhus in a German concentration camp.) Because powdered DDT is not readily absorbed through the skin, people thought it was safe. After the war, the production of pesticides expanded. Large amounts of insecticides, including DDT, rodenticides, herbicides, and fungicides were used to control agricultural pests, and pesticides were also developed for use in homes, offices, and factories.

In his influential book on introduced species (Section 2.1.2.5), Charles Elton expressed alarm about an "astonishing rain of death upon … much of the world's surface," due to the "incredibly massive use of insecticides now carried on in every part of the crop-growing world" (Elton, 1958:137–38,142). The book attracted the attention of a marine biologist and prize-winning nature writer named Rachel Carson. At the time, Carson was working on a book about the effects of synthetic insecticides on wildlife and people. In *Silent Spring*, the book that resulted from this work, Carson summarized the known effects of synthetic pesticides on the environment, people, and wildlife. Although the book was often misinterpreted as an indictment of all use of synthetic pesticides, Carson stated explicitly that she was not opposed to all pesticide use but rather was concerned with the widespread use of insecticides without adequate understanding of their potential effects or consideration of the rights of people who were exposed.

It is not my contention that chemical insecticides must never be used. I do contend that we have put poisonous and biologically potent chemicals indiscriminately into the hands of persons largely or wholly ignorant of their potentials for harm. We have subjected enormous numbers of people to contact with these poisons, without their consent and often without their knowledge. If the Bill of Rights contains no guarantee that a citizen shall be secure against lethal poisons distributed either by private individuals or by public officials, it is surely only because our forefathers … could conceive of no such problem.

I contend, furthermore, that we have allowed these chemicals to be used with little or no advance investigation of their effect on soil, water, wildlife, and man himself. Future generations are unlikely to condone our lack of prudent concern for the integrity of the natural world that supports all life.

(Carson, 1962:12–13, emphasis added)

Even before its publication in 1962, it was clear that *Silent Spring* would reach a wide popular audience and have a major influence. The book was by turns praised as a work of artistic and scientific merit by a respected professional and criticized as an alarmist tirade by a woman who was a sentimental fanatic (Box 5.1). Shortly before its publication, a lawyer for Velsicol, a chemical company that produced some of the pesticides discussed in *Silent Spring*, wrote to her publisher implying that the company would sue if the book were released. Using language typical of Cold War conspiracy theories, he suggested that the book reflected "sinister influences" interested in serving the interests of communist nations in Eastern Europe by creating "the false impression that all business is grasping and immoral, and [reducing] the use of agricultural chemicals in this and the countries of western Europe so that the supply of food will be reduced to east-curtain parity" (quoted in Lear, 1997:417) *(see Q5.2)*.

Box 5.1

Some Responses to the Publication of *Silent Spring*

Miss Carson writes with passion and with beauty, but with very little scientific detachment. Dispassionate scientific evidence and passionate propaganda are two buckets of water that simply can't be carried on one person's shoulders.

The fact that the public does not recognize the enormous debt it owes the scientist can have disastrous consequences many times more serious than the alleged risks Miss Carson fears from life-saving chemistry.

(Frederick J. Stare, Chair, Department of Nutrition, Harvard University in Stare, 1963:242)

There is a very real danger that the general public, knowing this book to be written by a biologist, will accept it as a genuine scientific version of this problem. This it certainly is not and was probably never intended to be. In a highly emotional and over-emphatic way, *Silent Spring* will draw the attention of the public to acute problems that biologists and chemists have been striving, and with considerable success, to solve for two decades or more.

(Ieuan Thomas, United Kingdom, Ministry of Agriculture in Groshong, 2007:369)

Her book is replete with examples of the damage which she believes to have been caused by various programs for the extermination of agricultural parasites. There is only one paragraph where the fact that these programs increase agricultural output is discussed .... To solve difficult problems we need accurate information and serious thought. Dr. Carson's new book provides neither; it is, instead, an obscurantist appeal to the emotions.

(Gordon Tullock, economist, in Groshong, 2007:370)

Her evidence, including a bountiful list of source material, appears to be overwhelming. Of course, in this kind of a book, the crusader takes pains to present only her side of the argument. Coupled with that, she writes extremely well. It is doubtful if anyone could do a better job of describing some of the biochemical effects of such toxic substances as DDT, chlordane, parathion, malathion and arsenic.

(Graham Berry, journalist, in Groshong, 2007:372)

Perhaps the most important service to be rendered by Miss Carson's book will consist in the enlightenment it brings the public regarding the high complexity and interrelatedness of the web of life in which we have our being.

(Nobel Prize-winning geneticist Herman Muller, in Groshong, 2007:375)

Miss Carson is not a faddist, she is not an anti-science "nut" .... She is instead a biologist, retired from government service, technically competent and qualified to discuss such matters as are discussed in *Silent Spring*. That she feels regulation ... is needed to guard against misuse of chemical insecticides is a logical result of her training and of her experience. I found nothing I could disbelieve, little with which I could not agree in her review of how man uses and abuses chemicals as pesticides.

(Mike Baker, chemist, in Groshong, 2007:374–75)

DDT attacks the central nervous systems of insects and affects liver function in birds. This can lead to reproductive failure, because the liver controls calcium transport in the oviduct and therefore eggshell formation. As a result, the eggs laid by affected birds do not have enough calcium in their shells. This means the shell is too thin and soft to support the weight of the incubating parent, and the chick dies before it hatches. Pelicans, grebes, terns, and several species of *raptors* (birds of prey), including ospreys and bald eagles, experienced pronounced declines in reproduction after exposure to DDT.

DDT dissolves in fat. At times when a bird's need for energy is high and its fat is used to meet that need, the concentration of DDT in the remaining fatty tissue increases. This threat is especially serious for migratory birds because of the high energy demands of long-distance flight.

Toxins can be passed from the tissues of one organism to those that feed on it. This phenomenon, termed *secondary poisoning*, occurs when an animal that has not consumed a toxin is poisoned indirectly by feeding on a poisoned animal. DDT and related compounds persist in the environment for a long time. In addition to being passed from one organism to another through secondary poisoning, stable toxins can become increasingly concentrated with each step in the food chain, a process termed *biomagnification* or *bioconcentration*. In this way, they can accumulate in animal tissues, reaching concentrations that are several times higher than their original concentrations in the environment (Box 5.2). This occurs because each trophic level has less biomass than the one below it (Figure 2.6). Herbivorous animals must eat many grams of plant matter to gain one gram of biomass. The same is true for primary consumers feeding on herbivores, and for secondary consumers feeding on primary consumers. As biomass is transferred from one trophic level to another up the food chain, pesticides and other toxins become increasingly concentrated, even if they are present at low concentrations initially. Because of this bioconcentration, carnivores and scavengers can accumulate high concentrations of persistent toxins in their tissues (Woodwell, 1967).

Box 5.2
**Passing Toxins up the Food Chain: Secondary Poisoning
and Biomagnification**

Clear Lake in northern California (Figure 5.1) is a shallow, warm body of water that historically produced good crops of sport fish, making it a favorite spot for anglers. Unfortunately, however, it was also favored by the Clear Lake gnat, whose larvae develop in mud at the bottom of the lake. Although this insect is related to mosquitoes, it does not bite people. But because of its sheer numbers, it was considered it a threat to tourism at Clear Lake (Hunt and Bischoff, 1960; Carson, 1962).

Efforts to control gnats at Clear Lake began as early as 1916, but the gnats continued to be a problem. When synthetic pesticides were developed, studies of DDT and DDE, a closely related compound, demonstrated that these insecticides were effective in killing gnat larvae. Of the two, DDE appeared to be less harmful to fish. Preliminary studies indicated that DDE caused relatively low fish mortality when applied at a concentration of 0.014 parts DDE to 1 million parts water (expressed as 0.014 parts per million, ppm). The lake bottom was surveyed, lake volume was calculated, and in 1949 DDE was applied to the lake at that rate.

Initially, control seemed to be successful. Follow-up studies showed that 99% of the gnat larvae were killed, and very few gnats were observed for two years following the treatment. But in 1951, larvae showed up again, and by 1952 they had reached problem levels. DDE was reapplied in 1954, this time at the higher rate of 0.02 ppm. Again, the treatment appeared to be effective. But again, the gnat population rebounded, this time even more quickly. That led to a third treatment in 1957.

The first sign of problems from gnat control came in December 1954, when 100 dead western grebes (a species of fish-eating bird) were found on the lake. Investigators who examined the birds'

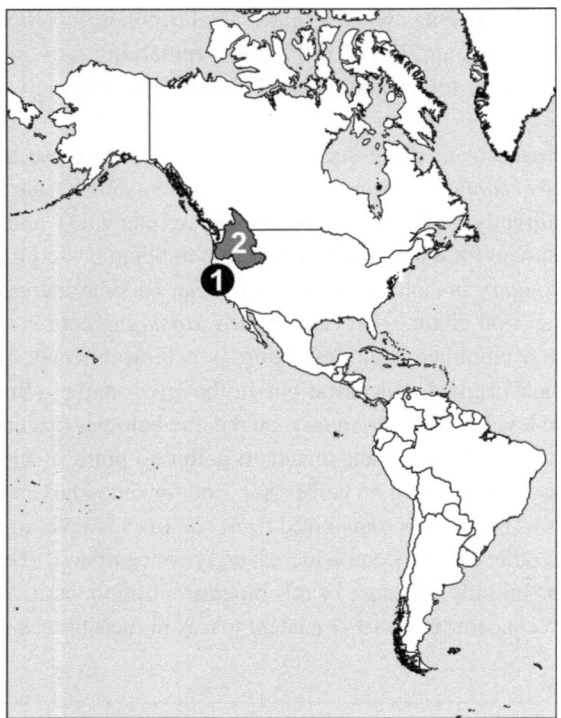

Figure 5.1   Locations of: 1, Clear Lake, California; 2, Columbia River watershed, Canada and USA. Map created by Eva Strand using Esri, DeLorme World Countries Generalized Data & Maps for ArcGIS 2013, with permission.

carcasses found no signs of infectious disease. More dead grebes were found in March of 1955 and in December of 1957. Once again, none of the birds showed any evidence of disease. However, chemical analyses of the birds' fat tissue revealed very high concentrations of DDE, up to 1,600 ppm. This was 80,000 times higher than 0.02 ppm, the rate of the highest application (1,600/0.02 = 80,000).

What brought about this extraordinary increase? Additional tissue studies of fish, frogs, and birds revealed that DDE was present in all animals that were sampled. Concentrations increased with trophic level, according to the following progression:

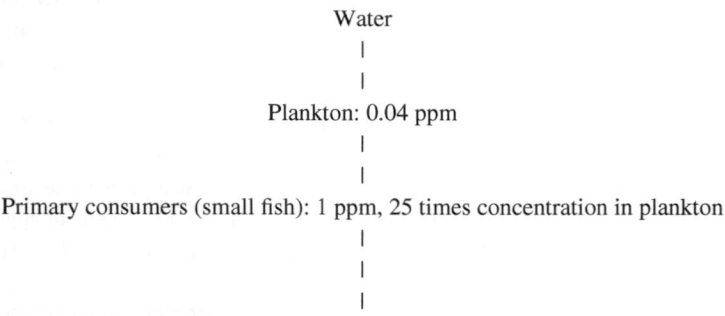

Water
|

|

Plankton: 0.04 ppm
|

|

Primary consumers (small fish): 1 ppm, 25 times concentration in plankton
|

|

|

Secondary consumer (ring-billed gull): 75 ppm, 1,875 times concentration in plankton

This is illustrated in Figure 2.6. Note that the dots, which represent molecules of a substance that does not break down, become more concentrated as they are passed to each successive trophic level. Therefore, the amount of toxin stays virtually the same at each trophic level. However, biomass decreases at each successive level, so that the toxin becomes more concentrated at the top of the biomass pyramid.

A comparison of DDE levels in fishes with different diets provided further evidence that feeding patterns were responsible for the observed tissue concentrations. Fish that fed on *plankton* (minute aquatic organisms that drift passively in the ocean), algae, or plants had lower levels of DDE in their tissues than fish that ate fish, which of course were higher on the food chain. Young fish also had lower levels of pesticide than older fish, indicating that the effects of ingesting DDE-laden food were cumulative.

Although about 1,000 pairs of western grebes had nested on the lake prior to the first treatment, less than 25 pairs were found during nesting surveys in 1958 and 1959. Grebes were not found in areas where they had previously nested, and pairs that did attempt to nest were apparently unsuccessful. Even after all traces of DDE had vanished from the water itself, DDE had remained part of the "fabric of life" of Clear Lake (Carson, 1962:48), continuing to affect the lake's organisms.

In addition to pesticides, several other environmental changes with long-lasting and far-reaching consequences began to worry scientists in the 1960s. This category included radioactive fallout and alterations in the composition of the Earth's atmosphere.

5.2.1.2 Radioactive Fallout

Nuclear explosions propel radioactive material into the atmosphere. When it falls to Earth, this is known as *fallout*. When aboveground testing of nuclear weapons began in the 1950s, scientists knew that fallout could damage cells in ways that cause genetic damage and disrupt cellular growth and reproduction. But they predicted that people would not be exposed to fallout in dangerous doses. They reasoned that that as wind and water currents moved around the globe, fallout would be diluted to harmless levels. That was not the case. In *Science and Survival*, a book for the general public, the American plant physiologist Barry Commoner pointed out that scientists seriously underestimated the effects of fallout because of faulty assumptions about how it would spread around the globe, the rate at which it would fall, the amount that would be deposited in soil, the way it would move through food chains, and its effects on the human body (Commoner, 1966).

Because the amount of fallout that reaches the ground near the poles is low, scientists thought that not much fallout would end up in the food of Indigenous peoples living in the Arctic. However, since fallout accumulates in lichens – which are abundant in the Arctic and absorb nutrients directly from the air – it moved up the food chain to lichen-eating caribou. Then it accumulated in even higher concentrations when native peoples ate the caribou. Because they did not take this into account, scientists were wrong about the amount of radioactivity that would accumulate in the diets of arctic peoples. (In my own life, the potential risks from radioactive contamination of food were impressed on me in a dramatic way in my childhood in the early sixties. My nutrition-conscious mother suddenly switched from urging me to drink three glasses of milk a day to telling me to cut back on milk after she learned about the risks from the radioactive form of the element strontium, which, like calcium, accumulates in bones.)

Like Carson and Elton, Commoner criticized scientists for promoting technological advances without understanding the effects of those discoveries.

Unwittingly we have loaded the air with chemicals that damage the lungs, and the water with substances that interfere with the functioning of the blood. Because we wanted to build nuclear bombs and kill mosquitoes, we have burdened our bodies with strontium-90 and DDT, with consequences that no one can now predict. We have been massively intervening in the environment without being aware of many of the harmful consequences of our acts until they have been performed and the effects—which are difficult to understand and sometimes irreversible—are upon us .... We are, in effect, conducting a huge experiment upon *ourselves*.

(Commoner, 1966:27–28, emphasis in original)

5.2.1.3 Ozone-Depleting Substances

Ozone is a form of oxygen in which each molecule consists of three atoms (O_3) instead of the usual two (O_2). It occurs in two parts of the atmosphere. In the *stratosphere* (the upper level of the atmosphere), ozone forms a protective shield that screens out most of the sun's harmful ultraviolet radiation, which can damage eyes and cause skin cancer (Introduction). At ground level, on the other hand, ozone is a component of smog and a human health irritant that can aggravate lung disease.

In the 1970s, scientists discovered that the ozone-rich layer in the stratosphere was becoming thinner in some places and that increasing levels of *chlorofluorocarbons* (CFCs) – gases that were found in aerosol spray cans, air conditioners, and refrigerants at that time – were responsible for this destruction of ozone (Commoner, 1971). The loss was especially pronounced over Antarctica during September and October, a phenomenon that became known as the *ozone hole*. At the extremely low temperatures found during the south polar winter in the stratosphere above the poles, CFCs attach to ice particles in clouds. In the polar spring (September and October), the ice particles melt, releasing the ozone-depleting CFCs, which then destroy ozone and cause the ozone layer to thin. Because a CFC molecule can break apart a molecule of ozone without itself being used up, a single CFC molecule can destroy many molecules of ozone.

5.2.1.4 Greenhouse Gases

In the nineteenth century, several European scientists realized that carbon dioxide, CO_2, in the atmosphere absorbs infrared radiation, which then warms the Earth. Since the amount of carbon dioxide in the atmosphere rose after the Industrial Revolution, when the production of CO_2 due to fossil fuel combustion also rose, scientists became concerned that further increases in CO_2 could warm the planet, with disastrous effects. "Carbon dioxide makes a huge greenhouse of the earth," wrote Commoner, "allowing sunlight to reach the earth's surface but limiting radiation of the resulting heat into space" (Commoner, 1966:11). This could lead to a rise in the Earth's temperature, he reasoned, melting the Antarctic ice cap, and causing catastrophic flooding (Section 10.3).

5.2.1.5 Acid Rain

In the 1950s, scientists in Europe and the Soviet Union began collecting information on the chemistry of air and water. Within about a decade, some unexpected patterns became evident. Swedish scientist Svante Odén pointed out in 1968 that precipitation in Europe had become more acidic since record keeping began. He noted that this effect was related to increased burning of sulfur-containing fuels (Odén, 1976).

In 1963, the US Forest Service set up the Hubbard Brook Ecosystem Study, a long-term experimental study of a large hardwood forest in New England. An interdisciplinary team initiated comprehensive monitoring of water chemistry to compare paired watersheds within the study area. From the outset, the data showed that rain collected by the study was very acidic. Coincidentally, a few years later, a member of the team traveled to Sweden, where he learned from Odén about a similar trend in Scandinavia.

This phenomenon became known as *acid rain* or *acid precipitation*. Acidity is measured by the pH scale. A pH of 7 indicates a neutral solution, a pH below 7 indicates an acid, and pH values greater than 7 indicate basic solutions. The scale is exponential (Section 2.1.1.1). The further below 7 the pH is, the stronger the acid. Thus, a liquid with a pH of 5 is 10 times more acidic than a liquid with a pH of 6, and 100 times more acidic than a solution with a pH of 7. In the mid-1970s, monitoring stations in Western Europe detected precipitation with pH values lower than 4.0 (and even as low as 2.4 in one instance in Scotland). American scientists reported similar findings.

Precipitation can become acidic when oxides of nitrogen or sulfur are released into the atmosphere. Scientists reasoned that acid precipitation might result from the burning of the *fossil fuels* coal and natural gas, which releases oxides of sulfur and nitrogen that form acids when they dissolve in precipitation. Testing this hypothesis required complex analyses of changes in the acidity of precipitation over time and across space. It was not possible, however, to go back in time to before the Industrial Revolution. Instead, to search for reference sites where human-caused pollution was minimal, Eugene Likens – the aquatic ecologist from Hubbard Brook – and some of his colleagues went to a few of the most remote parts of the Earth and investigated the chemistry of water samples. They found background pH values of about 5.1. To find the source of nitrogen and sulfur oxides that might cause precipitation which was as acidic as what they found in Europe and eastern North America, scientists analyzed data on regional movements of air masses. These data showed that acid rain could be caused by the transport of sulfur and nitrogen oxides across hundreds and even thousands of miles.

The story is more complicated, however, because a region's geology and soils influence how pollution affects acidity. In regions where the underlying bedrock contains limestone, a form of calcium carbonate, acid precipitation may be neutralized. But in areas where the bedrock is high in granite, this buffering does not take place. In those regions, soils and surface waters readily become acidified (Likens et al., 1979).

Acid precipitation has complex effects on forests. Tree growth and survival decline, with implications for animals and plants of forest habitats. Aquatic invertebrates, fishes, and amphibians are directly affected when levels of acidity in their environment exceed their physiological tolerance. Acidity interferes with calcium metabolism in fishes, producing deformed skeletons and death. Carnivores and scavengers are affected indirectly by declines in their food supplies. The rate at which organic matter decomposes also slows with acidification – perhaps because of changes in the microbial communities of acidified waters – which means that less organic matter is incorporated into soil. In addition, acid precipitation damages buildings and statues.

These problems led researchers to warn that "[T]he extent to which the SO_2 [sulfuric acid], concentration may ultimately rise is a matter for serious concern, for increased concentrations may promote still more serious and widespread acid rain with all that this outcome implies for the structure and function of both natural and man-manipulated ecosystems" (Likens and Bormann, 1974:1179).

Fallout and greenhouse gases affect all parts of the globe. Therefore, no place on Earth is truly pristine.

A common thread in these discussions of specific environmental concerns was a more general worry. Like Charles Elton, Barry Commoner, and Rachel Carson, other scientists were concerned about the scale of the changes wrought by modern technology. The changes brought about by synthetic pesticides, fallout, ozone depletion, greenhouse gases, and acid rain were unprecedented in their effects. "The new hazards are neither local nor brief," wrote Commoner,

Air pollution covers vast areas. Fallout is worldwide. Synthetic chemicals may remain in the soil for years. Radioactive pollutants now on the earth's surface will be found there for generations .... Excess carbon dioxide from fuel combustion eventually might cause floods that could cover much of the present land's surface for centuries .... Modern science and technology are simply too powerful to permit a trial-and-error approach.

(Commoner, 1966:28–29)

These concerns went along with a sense that human-caused changes were altering environmental conditions so rapidly that evolving life-forms could not adapt.

It took hundreds of millions of years to produce the life that now inhabits the earth.... For time is the essential ingredient; but in the modern world there is no time....

The rapidity of change and the speed with which new situations are created follow the impetuous and heedless pace of man rather than the deliberate pace of nature. Radiation is no longer merely the background radiation of rocks, the bombardment of cosmic rays, the ultraviolet of the sun that have existed before there was any life on earth; radiation is now the unnatural creation of man's tampering with the atom. The chemicals to which life is asked to make its adjustment are no longer merely the calcium and silica and all the rest of the minerals washed out of the rocks and carried in rivers to the sea; they are the synthetic creations of man's inventive mind, brewed in his laboratories, and having no counterparts in nature.

To adjust to these chemicals would require time on the scale that is nature's; it would require not merely the years of a man's life but the life of generations.

(Carson, 1962:6–7)

5.2.2 Habitat Modification

As scientists raised alarms about harmful substances in the environment, concerns about the effects of resource use on habitats mounted. Scientists and the public alike realized that forestry, farming, ranching, urban development, and many other activities had unintended consequences for wild organisms and their habitats.

The most obvious forms of habitat modification involve landscapes that are paved over for development. This clearly has serious impacts for conservation: not many kinds of wild organisms survive in habitats dominated by concrete and asphalt. But most habitat modification involves more subtle transformations. Farms, pastures, and tree plantations may look natural, in the sense that they are green and open, but they are not necessarily friendly for many wild things.

5.2.2.1 Forest Use

Harvesting for timber, fuel, pulp, lumber, or nontimber forest products alters forest communities. The types of impacts depend upon the amount of biomass removed, the methods used, the scale of the disturbance that is created, and the type of forest community.

In addition to the direct removal of trees, logging compacts soil, enhances erosion, alters microclimates, and changes disturbance regimes. By creating openings, enhancing edges, and promoting pioneer communities, timber harvest improves habitat for some species, but it also degrades habitat for species that require late-successional communities.

In the 1960s and 1970s, scientists studying tropical ecosystems raised concerns about rapid rates of deforestation in the tropics (Croat, 1972), especially in moist tropical forests. Tropical forests are slow to return to pre-disturbance conditions after large disturbances (Section A.1.8). In some instances, the removal of vegetation can also cause irreversible changes in their soils. This sensitivity is cause for concern for several reasons. First, tropical forests contain a great many different kinds of organisms. Although these forests contain many different species, the geographic ranges of those species are often quite small. Thus, when a tropical forest is cleared, many species may be completely wiped out. Second, many birds that summer in North America or Eurasia migrate to the tropics in winter. Loss of wintering habitat could mean the extinction of these species. Third, because of their great biomass, tropical forests fulfill important ecosystem functions including controlling floods, moderating climate, and storing carbon. Fourth, the pace of tropical forest destruction escalated in the last half of the twentieth century, for the reasons mentioned at the beginning of this chapter.

5.2.2.2 Water Management

People alter water on a landscape by diverting it or blocking it with dams or other structures. *Wetlands* – habitats that are flooded or saturated for long enough to have distinctive vegetation and soils – absorb floodwaters, treat pollution, provide habitat for many species of animals and plants, and support diverse cultures. For millennia people have drained wetlands to create farmland and building sites, but like many other land uses, rates of wetland drainage accelerated in the twentieth century (Box 5.3).

Box 5.3

How Much Wetland Did the World Lose between 1900 and 1981?

By the early 1980s, many wetlands in the contiguous USA had been drained for agriculture and development. In 1985, two scientists at the University of Wisconsin–Madison published a paper in which they stated that "the biggest changes in land-use since 1900 have been a 50% decrease in wetlands globally" (Winkler and DeWitt, 1985:325). To support this statement, they cited two papers from the proceedings of a conference on wetland values and management. Both papers dealt with wetland losses from only a few types of inland wetland habitats in the midwestern United States. However, the extrapolation of that assertion to global wetland loss was often repeated in the scientific literature.

Other scientists, however, noticed that this received wisdom was unsubstantiated. To get a better idea of the extent of historical wetland losses, Nick Davidson, a scientist in New South Wales, Australia, searched scientific journals and reports for data on changes in wetland extent. From the 189 cases he found, he concluded that the figure of 50% wetland loss was actually low. By the early twenty-first century, up to 70% of natural wetlands that existed before 1900 were gone. Losses were larger and faster for wetlands that were located inland than for coastal wetlands and greater in Asia than elsewhere (Davidson, 2014).

This story illustrates a couple of important points. First, it is a sobering reminder of the importance of critically evaluating sources. It seems likely that many or perhaps most of the writers who repeated the 50% loss figure had not read the original papers. If they had, perhaps someone would have noticed the problem with extrapolating from the USA to the entire world and lumping all types of wetlands together. But this example also illustrates the self-correcting potential of science. Although a conclusion based on a limited sample in a specific region was applied to the entire world, eventually someone noticed the problem and designed a more rigorous study. The result contributed to a more nuanced understanding of different kinds of wetland losses in different regions at different times during the twentieth and early twenty-first centuries. This was possible because information from hundreds of studies representing many geographic contexts was available.

Hydroelectric power is a renewable resource that lacks many of the drawbacks associated with other methods of power generation in industrialized societies. Since water does work as it moves downhill, dams provide power without generating nuclear waste, acid rain, or greenhouse gases. In the words of the American folksinger Woody Guthrie "river while you're rambling you can do some work for me" (*see Q5.3*). Impoundments behind dams also allow for flood control and provide reservoirs for transportation and some kinds of recreation. In the USA in the early 1940s, many people viewed federal dam construction on the "wild and wasted" Columbia River in the Pacific Northwest as good for "the farmer and the factory and all of you and me." It provided power for rural electrification and for the manufacture of military aircraft used in World War II: "to fight for Uncle Sam, spawned upon the King Columbia [River] by the big Grand Coulee Dam" (Woody Guthrie, No date).

But dams are not a win for everyone. The Columbia River watershed includes parts of seven states in the USA and one Canadian province (Figure 5.1). Dams altered Indigenous fishing grounds and prevented salmon from reaching their spawning gravels in more than a thousand miles of that upstream habitat (Hunn, 1990).

When the availability of water in time and space is altered, some people benefit and others lose out. Major diversions such as large dams and canals have far-reaching effects. By the second half of the twentieth century, concerns about the potentially harmful environmental and social impacts from dams were mounting. In 1968, the French physicist J. R. Rothé warned of the potential for large dams to cause earthquakes, evoking the fable of a sorcery student who causes havoc by using magical powers he cannot control: "When he builds these [dams], Man plays the role of the Sorcerer's Apprentice: in trying to control the energy of rivers, he brings about stresses whose energy can be suddenly and disastrously released" (Rothé, 1968:434) (*see Q5.4*).

In 1979, the Australian zoologist K. F. Walker noted that the regulation of flows in the Murray-Darling River System had far-reaching ecological ramifications (Walker, 1979). Dams alter the environment of fish and other river organisms as well as associated riparian plant communities and the cultural systems they sustain. When water is prevented from moving downstream by a dam, its velocity, temperature, and chemistry change. Relatively warm, slow, deep waters behind a dam replace the cool, fast-moving waters in a free-flowing river. In addition to interrupting water flow, dams interfere with the movements of salmon and other migratory fishes.

By changing the amount and timing of flowing water, dams modify the disturbance regimes of rivers. Some organisms do well in dry years while others prosper in years with high flows s (Poff et al., 1997). By decreasing variability in flow dynamics, dams decrease environmental heterogeneity and associated species diversity.

Dikes and levees have often been built to straighten river channels and control floodwaters. The diverse and productive ecosystems of natural floodplains adjacent to lowland rivers depend on floods, however (Box 11.5). Flood control structures cut off inputs of sediment, water, and nutrients to the adjacent landscape.

Most temperate-zone floodplains have been extensively modified by water control structures because floodplains attract agricultural and residential development where flood control is desired. Ironically, by confining flowing waters to a straight channel and not allowing flood waters to spread out, those measures often make flooding worse.

In deserts and drier portions of *steppe* (grassland dominated by perennial plants (Section A.1.5)), agriculture depends upon irrigation. But when irrigation waters are withdrawn from rivers and streams, the resulting drop in water levels negatively impacts organisms that depend upon habitats associated with rivers. Irrigation can also degrade soil quality by causing salts to build up as water evaporates.

5.2.2.3 Agriculture

Agricultural systems affect much of the Earth. They have the potential for positive or negative impacts on soil conservation, wildlife habitat, water quality, livelihoods, food security, climate, culture, and animal welfare. In the process of producing food and fiber crops, soil is disturbed, vegetation is removed, and nutrients are consumed. Farm animals provide meat, milk, leather, wool, and other products as well as labor and fertilizer. They can also damage vegetation and contribute to erosion and pollution. Livestock also produce greenhouse gases.

Some regions have been farmed for centuries. Examples of long-term agricultural systems that maintained or enhanced fertile soils include rice/mulberry production in China; rice in Bali, Indonesia; wheat cropping in the Nile Delta; terrace farming of potatoes and other crops in Peru; and corn-based systems in the southwestern USA.

However, under some circumstances, industrial agriculture in wealthy societies and subsistence agriculture in poor societies both contribute to unsustainable environmental impacts. Farming often decreases the quantity and quality of vegetation covering soil and leads to increased erosion from either clearing followed by burning or from tillage.

Shifting cultivation: In tropical moist forests, a family cuts down the trees on a few hectares of forest and burns them before planting crops. This small-scale cultivation is known as *slash-and-burn agriculture, milpa, swidden*, or *shifting cultivation*. Ashes, weeds, and human excreta supply nutrients to the soil. Nevertheless, after a few years, fertility declines, the site is abandoned, and the process is repeated elsewhere. If population density is low in relation to the amount of available habitat and if rates of nutrient cycling are high, this type of farming has sometimes been sustained for long periods of time.

But the situation is different under contemporary conditions. Because of a combination of human population growth and conversion of tropical forest to other uses, the land available for shifting cultivation has shrunk. As a result, fallow periods are shorter. Because the time between cycles of clearing has decreased, there is not enough time for fertility to recover. In addition, government policies often discourage shifting cultivation (in some cases, criminalizing it), and forest departments and conservation organizations oppose it (sometimes resettling cultivators from lands under their control). A review of 157 studies of changes in shifting cultivation in the early 2000s found that transitions from shifting cultivation to other land uses often contributed to deforestation, decreased species diversity, increased weeds, and accelerated soil erosion and declining soil fertility (van Vliet et al., 2012).

Industrial agriculture: In industrialized countries, a different set of conditions puts pressure on soils and cultivated land. A fundamental agricultural transition took place in wealthy nations early in the twentieth century, when steam engines began to substitute for horses. Yet another agricultural revolution happened after World War II, when agriculture in wealthy nations became dependent on fossil fuels, machinery, synthetic chemicals, and high-yielding crop varieties. As dependence on external inputs increased, farming became more capital intensive.

To facilitate the operation of the big machines that were required for mechanized agriculture after World War II farmers cultivated larger fields and removed vegetation along field borders. This large-scale tillage left large expanses of soil exposed to wind and water. Habitat diversity declined, and populations of insect pests built up. At the same time, agricultural land was taken out of production and used for urban or suburban development, which created pressure to intensify agricultural production by using synthetic fertilizers, herbicides, and insecticides. Paradoxically, agricultural production increased, yet soil declined in fertility and was lost through erosion (Pimentel, 1993).

In the mid-twentieth century, concern about environmental, ecological, social, and ethical problems from modern agriculture mounted. These included contamination of groundwater with agricultural runoff, soil erosion, unstable farm incomes, poor working conditions among farm laborers, health risks to workers and the public from exposure to agricultural chemicals, declining soil quality, greenhouse gas emissions, and animal suffering. One approach to addressing some of these problems is covered in Section 11.5.1.

5.2.2.4 Rangeland Use

Large animal grazing is one of the principal land uses in *arid* climates (where available moisture is too low to support agriculture without irrigation) and in cold climates where the growing season is too short for crop production. In such environments, pastoralists traditionally undertake seasonal migrations with their livestock. Where a dry or cold climate limits plant growth, the most efficient systems for converting plant material into products that support human populations involve large grazing mammals: reindeer in the Arctic, wildebeest, cattle, and other ungulates in Africa's Serengeti-Mara Ecosystem, kangaroos in Australia's rangelands, camels and horses across the Eurasian steppe, llamas and alpacas in South America's mountains and plains, yaks on the Tibetan Plateau, bison in the tall-grass prairies of midwestern North America, and livestock throughout much of the world. In some parts of Africa, India, Central Asia, and the Arctic, nomadic pastoral tribes have tended herds of grazing animals for centuries. Although some development professionals have suggested that traditional nomadic grazing led to overgrazing, ecological disaster, and famine in parts of Africa (Picardi and Seifert, 1977), some ecologists disagree. They suggest that the reverse was true. According to that view, the disruption of traditional grazing strategies that resulted when nomadic pastoralists became sedentary was a principal cause of overgrazing (Sinclair and Fryxell, 1985) (Section 10.1.1.2).

Nevertheless, grazing can cause serious habitat degradation in some situations. This is more likely where the effects of grazing are not well understood, where vegetation is grazed continuously, where non-native grazers have been introduced, and where grazing systems that are not well suited to local environmental conditions are imposed. Many plant species are able to cope with some grazing, but excessive grazing will exceed their ability to regenerate lost tissue. Grazing promotes the growth of increaser species at the expense of decreasers (Section 2.1.2.3). Large grazing animals also remove or kill vegetation by trampling and wallowing, which compact soil. As a result, the capacity of soil to hold water decreases, causing runoff and erosion to increase.

5.2.2.5 Habitat Fragmentation

Many of the activities discussed in this chapter divide habitats into smaller fragments. Even if these patches are themselves undisturbed, their small size and isolation compromise their usefulness as habitat for many kinds of wild things. Some species – such as the giant panda, the northern spotted owl, and many other large mammals – require vast expanses of continuous habitat. Others may inhabit fragmented habitats but fail to thrive there. Small patches are windier, hotter, drier, sunnier, and weedier than large ones.

By the middle of the twentieth century, scientists studying in the tropics worried about the high rates at which species-rich tropical forests were being cut down and converted to cattle pasture. Because this process left behind landscapes dotted with patchy forest remnants surrounded by a matrix of unsuitable habitat, it became important to understand the effects of habitat fragmentation on the biota of tropical forests. In 1979, a scientific organization based in Brazil collaborated with an international non-governmental conservation organization to establish a long-term, large-scale, controlled experiment designed to address that question. Within a study area in the in the Amazon Basin, different-sized fragments of tropical moist forest were left standing when a large area was cleared, resulting in forest remnants of different sizes and distances from uncut forest. In the decades that followed, researchers collected standardized data on populations of trees, birds, mammals, amphibians, butterflies, ants, termites, beetles, and other organisms before and after experimental isolation of the forest fragments (Laurance, 2007). Many species of large mammals, birds, and insects disappeared after the fragments were created. On the other hand, for some groups of organisms, the number of species stayed the same or even increased after fragmentation because disturbance-adapted species from the matrix moved into the small fragments. For example, the number of frog species increased after non-rainforest frog species moved into the small patches.

Many songbirds that spend their winters in Central and South America migrate north to breeding grounds in North America. These birds, known as *Neotropical migrants*, are doubly vulnerable because loss or fragmentation of their habitats on either their breeding or their wintering grounds can put them at risk. The fragmentation of tropical forests on migrants' wintering grounds in Latin America has contributed to their decline, but reduction and fragmentation of nesting habitat in North American forests also played a role. Although small woodlots persisted into the middle of the twentieth century in the eastern forests, many of those patches were too small to support breeding populations of the Neotropical migrants, which require large patches of late-successional vegetation (Galli et al., 1976; Whitcomb et al., 1981).Neotropical migrants have several characteristics that make them especially vulnerable when their breeding habitat is fragmented. First, each spring these birds return to the site where they bred the previous year. If that site no longer contains suitable habitat, they do not settle and breed elsewhere. Paradoxically, they move long distances during their migrations but are inflexible about their breeding site after they arrive.

Second, Neotropical migrants typically require habitat found in forest interiors. Forest edges are usually unsuitable for them. Many permanent residents of temperate forests – like chickadees and woodpeckers – lay their eggs in cavities, where they are fairly safe from predators. Migrants, on the other hand, often construct open, cuplike nests on the ground. Since small patches of forest have a large proportion of edge, they receive many visits from humans, pets, raccoons, and skunks. Ground nests near edges are often destroyed, either by predation or trampling. *Brood parasitism* (a type of parasitism in which some birds – including some cuckoos and cowbirds – lay their eggs in the nest of a host species (Figure 5.2), which then feeds and rears the young of the brood parasite) (Box 7.4) also increases in fragmented forests (Brittingham and Temple, 1983).

Finally, Neotropical migrants have low biotic potential. Because of the energy demands of migration, they may have only enough fat stores left over to lay a single clutch of eggs. If that attempt to nest is unsuccessful, they will not get another chance to reproduce until the following year.

1597.—Young Cuckoo in Hedge-Sparrow's Nest.

Figure 5.2   Brood parasitism. A young cuckoo in a sparrow nest is fed by the sparrow even though the cuckoo is much larger than the sparrow's young. From engraving by D. Walker, 1844. Credit: duncan1890 / DigitalVision Vectors / Getty Images.

Because of Neotropical migrants' dependence on large patches of mature forest, scientists got a surprise when they attached tiny radio-transmitters to birds and followed them after they left their breeding grounds. Some of the classic forest-interior-breeders turned up in forests that had been clear-cut. This pattern was especially true for fledglings. It suggests that resources in the shrub thickets and other early successional vegetation in clear-cuts may benefit the young birds as they build up fat stores for their first migratory journey (Vitz and Rodewald, 2006).

5.2.3 Disappearing Species

By the middle of the twentieth century, populations of the bald eagle, Bengal tiger, peregrine falcon, giant panda, all large whales, and many other striking species or subspecies of wildlife had declined to alarming levels. "Millions of wild animals have already disappeared from Africa this century …. What, if anything, can be done to safeguard it?" These words from one of three articles by the British evolutionary biologist Julian Huxley appeared in a British newspaper in 1960. "If wildlife is destroyed," wrote Huxley, "it is gone forever, and if it is seriously reduced, its restoration will be a lengthy and expensive business" (Kellaway, 2010). Along with photographs of impressive wildlife – a rhinoceros in Kenya, a giraffe in South Africa, a fish eagle in Uganda – Huxley's words were influential in raising concern among scientists and the public about species on the verge of extinction. In addition, popular accounts of dwindling populations of large, appealing birds and mammals, which became known as *charismatic megafauna*, generated public sympathy and funds for conservation.

It has been estimated that the extinction rate of birds and mammals between 1600 and 1975 was five to 50 times higher than it had been throughout most of the history of life on Earth (Ehrlich et al., 1977). There is evidence that at least some of the extinctions after 1600 were caused by exploitation, habitat loss, or introduced species. In other cases, too little information is available to know what caused extinction.

Because of this accelerated loss of species, scientists such as the entomologist E. O. Wilson and others argued that we should be concerned about the loss of *biological diversity*, or *biodiversity*. Initially, the term biological diversity meant the total number of species in a particular context. (A more precise term for this is *species richness*.) It has been expanded, however, to encompass different levels of organization. Most biologists now consider biodiversity to denote the diversity of life at all levels of organization, including genetic material, species, and communities. However, often the term biodiversity refers just to species richness. (This is partly because surveying species is easier than measuring some other aspects of biodiversity.)

No one can predict with certainty the effects of accumulated extinctions, but many scientists and non-scientists alike contend that it would be prudent to try to minimize extinctions. Paul and Anne Ehrlich used the metaphor of an airline mechanic removing rivets from a plane wing to illustrate the potential dangers of species loss.

As you walk from the terminal of your airliner, you notice a man on a ladder busily prying rivets out of its wing. Somewhat concerned, you saunter over to the rivet popper and ask him just what the hell he's doing.

"I work for the airline—Growthmania Intercontinental," the man informs you, "and the airline has discovered that it can sell these rivets for two dollars apiece."

"But how do you know you won't fatally weaken the wing doing that?" you inquire.

"Don't worry," he assures you. "I'm certain the manufacturer made this plane much stronger than it needs to be, so no harm's done. Besides, I've taken lots of rivets from this wing and it hasn't fallen off yet...."

You never *have* to fly on an airliner. But unfortunately, all of us are passengers on a very large spacecraft—one on which we have no option but to fly. And, frighteningly, it is swarming with rivet poppers behaving in ways analogous to that just described.

(Ehrlich and Ehrlich, 1985:xi–xii, emphasis in original)

5.2.3.1 How Much Is a Species Worth?

Western societies have increasingly asked the monetary value of items
and qualities formerly regarded as priceless .... Behind this search has
been the hope that, by weighing the benefits to society of nature in the

undeveloped state against the benefits of resource development, an
objective basis for decision-making will be achieved.

(Westman, 1977:960)

The heightened awareness of extinctions went along with a reappraisal of the ways we
value species. It had been clear for some time that the utilitarian classification of species
as good or bad solely on the basis of what they ate was too narrow. Species make valuable
contributions to our lives in many other ways. They have the genetic potential from which
new crops, medicines, and other products are developed; they contribute to ecosystem
services such as flood control, pollination, climate regulation; and so on. To bring attention
to this issue, many scientists turned their attention to the economic benefits from the goods
and services that wild species and ecosystems provide and the high costs to society of
extinctions. However, assigning value to species – and ecosystems – gets tricky.

In 1977, the ecologist Walter Westman began an essay on the value of nature by quoting
the poet William Wordsworth about the intangible value of flowers: *"To me the meanest
flower that blows can give/Thoughts do often lie too deep for tears."* Westman then went
on to ask "How much was this mean flower worth to a poet like Wordsworth?"

There is a downside to the emphasis on economic value. If we could definitively show
that that a species was of no use to us, that it had no potential economic value, would that
mean we should not be concerned about its fate? Do all forms of life on Earth deserve protection, regardless of their economic value or emotional appeal?

5.3 Diagnosing the Problem

As concerns about the fate of endangered species mounted, scientists echoed a theme that
Carson and Commoner had raised: people are bringing about changes at unprecedented
rates with potentially dangerous results. These trends and other criticisms of post-industrial
society generated a search for underlying causes. Public discourse about the roots of these
environmental problems involved questions of religion, ethics, human behavior, economic
organization, and political power.

5.3.1 Population Growth

Extending the views of the Reverend Thomas Malthus (1798) (Section 2.1.1.1), Paul Ehrlich
(1968) and the human ecologist Garrett Hardin (1968) argued that the Earth's population
had reached a level which put excessive pressure on the Earth's resources. Hardin suggested
that unregulated population growth was the inevitable result of people acting in their own
short-term self-interest. He took as his starting point an argument articulated by the amateur
nineteenth century mathematician William Lloyd. In 1833, the question of whether there
should be public relief for the poor was hotly debated in Britain. Lloyd published a pamphlet
suggesting that people will add additional calves to a common pasture or bear additional
children, since it is in their short-term interest to do so and there are no immediate negative
incentives for restraint. (This is the problem of externalities discussed below.)

In a famous essay on *The Tragedy of the Commons*, Hardin elaborated on this idea:

The tragedy of the commons develops in this way. Picture a pasture open to all. It is to be expected that each herdman will try to keep as many cattle as possible on the commons. Such an arrangement may work reasonably satisfactorily for centuries because tribal wars, poaching, and disease keep the numbers of both man and beast well below the carrying capacity of the land. Finally, however, comes the day of reckoning, that is, the day when the long-desired goal of social stability becomes a reality. At this point the inherent logic of the commons remorselessly generates tragedy.

As a rational being each herdsman seeks to maximize his gain. Explicitly or implicitly, more or less consciously, he asks. "What is the utility *to me* of adding one more animal to my herd?" This utility has one negative and one positive component.

1) The positive component is a function of the increment of one animal. Since the herdsman receives all the proceeds from the sale of the additional animal, the positive utility is nearly +1.
2) The negative component is a function of the additional overgrazing created by one more animal. Since, however, the effects of overgrazing are shared by all the herdsmen, the negative utility for any particular decision-making herdsman is only a fraction of −1.

Adding together the component partial utilities, the rational herdsman concludes that the only sensible course for him to pursue is to add another animal to his herd. And another; and another .... But this is the conclusion reached by each and every rational herdsman sharing a commons. Therein is the tragedy.

(Hardin, 1968:1244; emphasis and ellipsis in original)

The only alternatives to the tragedy of the commons, Hardin concluded, are privatization or state regulation. Privatization, he thought, increases individual responsibility because the individual bears the costs and reaps the benefits of their actions. On the other hand, Hardin considered government control a necessary evil. In his view, "injustice is preferable to total ruin" (Hardin, 1968:1247). Action motivated by a sense of social responsibility was not a possibility in his view because those who were socially responsible would lose out to those who were not.

Hardin linked his concern about population control with ideas about whose populations should be controlled. He believed that society should not rely on persuading people to limit their reproduction, because only educated people (whom he believed were genetically superior) would respond to such arguments. Hence, Hardin supported coercion, including involuntary sterilization, to limit the fertility of the "feeble minded" and people in categories he considered inferior. He supported *eugenics (eu,* true; *gene,* born*),* a movement that advocates improving the genetic quality of human populations through differential reproduction. This was to be achieved by promoting reproduction among people with superior traits and preventing others from reproducing. Hardin also opposed both food aid for poor countries and immigration into rich countries, measures that he believed would benefit inferior groups (Hardin, 1974; Locher, 2013).

When Hardin described the consequences of grazing the *commons,* he meant a communally owned "pasture open to all." He assumed that small communities are incapable of regulating the use of resources they own in common, and thus they cannot restrict use of their commons. However, economists recognize that under some circumstances, communities do regulate their communally owned resources. We shall return to this point in Section 10.1.3. Although Hardin's presentation of the tragedy of the commons was oversimplified, it is still widely cited and taught in its original form.

Like Hardin, Paul Ehrlich held that dealing with population growth would require difficult decisions. In *The Population Bomb*, he compared population growth to a malignancy.

A cancer is an uncontrolled multiplication of cells; the population explosion is an uncontrolled multiplication of people. Treating only the symptoms of cancer may make the victim more comfortable at first, but eventually he dies—often horribly. A similar fate awaits a world with a population explosion if only the symptoms are treated. We must shift our efforts from treatment of the symptoms to the cutting out of the cancer. The operation will demand many apparently brutal and heartless decisions. The pain may be intense. But the disease is so far advanced that only with radical surgery does the patient have a chance of survival.

(Ehrlich, 1968:166–67)

5.3.2 Disparities in Resource Consumption

Some writers focused on inequities in resource distribution as much as, or in addition to, problems of population growth. As early as 1953, Judge Samuel Ordway suggested that Americans' high levels of consumption were straining resources (Ordway, 1953, 1956). The economist John Kenneth Galbraith (1958a, b) picked up this theme, pointing out that since World War II American "consumption of most materials has exceeded that of all mankind through all history before that conflict" and arguing that such levels of resource use were not sustainable (Galbraith, 1958b:90).

Paul Ehrlich agreed that the enormous disparity in consumption patterns in rich and poor nations was problematic. "At the moment the United States uses well over half of all the raw materials consumed each year," he wrote in 1968. "Less than 1/15th of the population of the world requires more than all the rest to maintain its inflated position" (Ehrlich, 1968:133). A few years later, Ehrlich published *The End of Affluence*, in which he articulated this critique more forcefully.

Overdeveloped countries [ODCs] are those in which population levels and per capita resource demands are so high that it will be impossible to maintain their present living standards without making exorbitant demands on global resources and ecosystems....

People in ODCs, especially the United States, aren't eating more *food*; they are eating more meat, poultry, and dairy products. Americans (6 percent of the world's population) not only consume about 30 percent of the world's natural resources, they also consume 30 percent of the world's meat.

The protein-rich, highly varied diet of the average American requires nearly *five times* the agricultural resources (such as land, water, fertilizers, and pesticides) that are needed to feed a citizen in a UDC [underdeveloped country].

(Ehrlich, 1974:21,23; emphasis in original)

5.3.3 Misuse of Science and Technology

Barry Commoner argued that neither population growth nor increased affluence was sufficient to account for the level of environmental degradation in developed countries (Commoner et al., 1971:61). Instead, the problem was the misuse of science and technology, particularly the "erosion of the principles which have long given science its remarkable capability to understand nature." This occurred in the face of pressure from military and industrial interests during and after the second world war, so that "even *basic* scientific work" often came to be "controlled by military and profit incentives that impose secrecy on the dissemination of fundamental results" (Commoner, 1966:48,61; emphasis in original).

5.3.4 Market Factors

In spite of Commoner's insistence that affluence was not the culprit, by the late 1970s, ecologists began to scrutinize Western import patterns and their influence on resource use in the developing world, particular the tropics. "Species disappear because of the way we prefer to live, all of us," wrote Norman Myers in his preface to *The Sinking Ark.* "For example, the expanding appetites of affluent nations for beef at 'reasonable,' that is non-inflationary prices, encourages [sic] the conversion of tropical moist forests into cattle ranches" (Myers, 1979:xi). Likewise, the "booming demand on the part of the developed world for tropical hardwoods" resulted in increased pressure on tropical forests (Myers, 1979:158).

The American demographer Ansley Coale addressed the question of why the workings of the marketplace had not prevented environmental degradation. He concluded that pollution and other forms of environmental degradation proliferate because free markets do not require people to be responsible for the environmental consequences of production. Markets pay people for the "goods" they produce, but people do not have to "pay for the bads."

The way our economy is organized is an essential cause, if not *the* essential cause, of air and water pollution, and of the ugly and sometimes destructive accumulation of trash. I believe it is also an important element in such dangerous human ecological interventions as changes in the biosphere resulting from … the accumulation in various dangerous places such as the fatty tissue of fish and birds and mammals of incredibly stable insecticides ….

The economist would say that harmful practices have occurred because of a disregard of what he would call *externalities.* An externality is defined as a consequence (good or bad) that does not enter the calculations of gain or loss by the person who undertakes the economic activity. It is typically a cost (or a benefit) of an activity that accrues to someone else …. Air pollution created by an industrial plant is a classic case of an externality; the operator of a factory producing noxious smoke imposes costs on everyone downwind, and pays none of these costs himself—they do not affect his balance sheet at all. This, I believe, is the basic economic factor that has a degrading effect on the environment: we have in general permitted economic activities without assessing the operator for their adverse effects. There has been no attempt to evaluate—and to charge for—externalities.

(Coale, 1970:132; emphasis in original)

The writings of scientists like Ehrlich, Hardin, Commoner, Myers, and Coale outline specific policy recommendations involving regulations and incentives aimed at changing the behavior of people and businesses. To some people, this did not go far enough. What was needed, they suggested, was a basic philosophical shift.

5.3.5 Anthropocentrism

According to the historian Lynn White, the root cause of the daunting environmental problems that became obvious in the twentieth century was an anthropocentric philosophical tradition that encourages the exploitation of nature. In an influential article published in the journal *Science* in 1967, White suggested that the ethic of Western Christianity encourages environmental domination. "Especially in its Western form," he wrote,

Christianity is the most anthropocentric religion the world has ever seen .... Christianity ... not only established a dualism of man and nature but also insisted that it is God's will that man exploit nature for his proper ends ....

The present increasing disruption of the global environment ... cannot be understood historically apart from distinctive attitudes toward nature which are deeply grounded in Christian dogma. The fact that most people do not think of these attitudes as Christian is irrelevant .... Hence, we shall continue to have a worsening ecologic crisis until we reject the Christian axiom that nature has no reason for existence save to serve man.

(White, 1967:1205,1207)

White saw a remedy for this situation in an alternative interpretation of Christianity. According to White, Saint Francis of Assisi rebelled against the idea that people should dominate nature. Instead, he believed in "the virtue of humility ... for man as a species. Francis tried to depose man from his monarchy over creation and set up a democracy of all God's creatures" (White, 1967:1206, 1207).

Around the same time, the Norwegian philosopher Arne Naess also wrote about the need to go beyond an anthropocentric worldview. He advocated a change to "a deep-seated respect, or even veneration, for ways and forms of life," which have an *"equal right to live and blossom"* (Naess, 1973:95,96, emphasis in original). Naess coined the term *deep ecology* as an alternative to what he termed a shallow ecology concerned solely with "improving the health and affluence of people in the developed countries" (Naess, 1973:95). In this view, which Naess called *ecological egalitarianism* and others have referred to as biocentrism (in contrast to anthropocentrism) (Preface), the root of our ecological crisis is arrogance toward the natural world. This attitude should be replaced by "a keen, steady perception of the profound *human ignorance* of biospherical relationships" (Naess, 1973:97, emphasis in original).

Although Naess agreed with Lynn White that the fundamental problem was philosophical arrogance, he went further than White in outlining social and economic remedies. "Ecologically inspired attitudes," wrote Naess, "favour diversity of human ways of life, of cultures, of occupations, of economies. They support the fight against economic and cultural, as much as military, invasion and domination, and they are opposed to the annihilation of seals and whales as much as to that of human tribes or cultures" (Naess, 1973:96). He concluded that the Earth's environmental problems could be solved only by a radical shift to an economic system characterized by decentralization and local autonomy that would foster ecological and *cultural diversity* (the variety of different cultures) and minimize class distinctions.

5.4 Response: Protection of Species and Habitats

These concerns and insights led to a critical re-examination of utilitarian conservation. The utilitarian approach had failed to protect many species of wild organisms from threats posed by excessive exploitation, habitat alteration, pollution, and alien species. The seriousness of the environmental problems described in this chapter led scientists to search for new approaches to conserving species and their habitats. In particular, they turned their attention to the threats to biodiversity, especially in the tropics.

In response, a preservationist approach to the conservation of living resources flowered. This approach seeks to maintain or prevent the loss of biodiversity by preserving and restoring species and habitats threatened by the activities of people. In connection with this trend, an academic discipline termed *conservation biology* developed (Soulé and Wilcox, 1980). Conservation biologists apply scientific knowledge about how species arise, persist, and disappear to the task of designing policies that protect biodiversity.

This emerging awareness had important consequences for how scientists and managers think about and implement conservation strategies. Chapter 6 explores some of the scientific concepts underlying this approach, and Chapters 7 and 8 examine strategies for preserving habitats and species.

6

Central Concepts

Evolution, Adaptation, and Extinction

We saw in Chapter 5 how the events that followed World War II directed attention toward escalating threats to biodiversity and human well-being. Because of this situation, it became crucial for conservation scientists and resource managers to understand the processes that give rise to species, that allow them to persist or cause them to disappear, and that govern how species interact with each other. Evolution provides a unified explanation for those processes and their practical applications. This pivotal concept provides insights into how species become resistant to pesticides, why some populations are more likely to develop resistance than others, and why some species are more vulnerable to immediate threats like exploitation and habitat loss than others. These insights can guide managers and conservationists as they prioritize their efforts.

In this chapter, we begin with the concept of natural selection and the role it plays in adaptation, coevolution, and the formation of species. We then consider the mechanisms of heredity, the process of extinction, and the risks of extinction that contemporary species face.

6.1 Natural Selection

In Section 2.1.1.1, we saw that under optimal conditions all organisms can produce more offspring than can survive. Not all individuals are equally likely to survive, however. In the mid-nineteenth century two British naturalists, Charles Darwin and Alfred Russel Wallace, working independently in different parts of the world, pondered these points and proposed the theory of *natural selection*, which states that (1) individuals that are better adapted to their environment will have a greater probability of surviving and reproducing than those that are not as well adapted and (2) those reproducers will pass on their characteristics to their offspring. Darwin summarized this in the introduction to *The Origin of Species*:

As many more individuals of each species are born than can possibly survive; and as, consequently, there is a frequently recurring struggle for existence, it follows that any being, if it vary however slightly in any manner profitable to itself, under the complex and sometimes varying conditions of life, will have a better chance of surviving, and thus be *naturally selected*. From the strong principle of inheritance, any selected variety will tend to propagate its new and modified form.

(Darwin, 1958:29; emphasis in original)

141

Darwin and Wallace held that natural selection produces *adaptation*, the process by which organisms become more able to survive in their environment, and *speciation*, the development of new species. We consider these next.

6.1.1 Adaptation

Adaptation is the process by which populations of organisms evolve in a way that results in their becoming better suited to their environment. This happens because natural selection favors traits that are advantageous in a particular environment and the genetic basis for those traits is passed on to successive generations. Eventually, the favorable traits become predominant in that population.

Adaptation is a passive process in which natural selection acts on genetic variations that favor survival and reproduction. This is neither conscious nor voluntary. Organisms do not decide to adapt, and they do not develop adaptations because they try to or want to. Genetic changes that produce favorable traits occur at random. They do not spring forth because they are needed (Gregory, 2009:156).

The more surviving offspring an individual produces, the higher its *fitness*. In evolutionary biology, the word "fitness" has a different meaning from everyday usage. An individual's evolutionary fitness is defined in terms of its success in passing its genetic material to succeeding generations. This need not have anything to do with physical fitness in the sense the term is used in athletics. Evolutionary fitness is not necessarily related to endurance, strength, or any other characteristic of physical condition. Nor does evolutionary fitness refer to social status or success, although the term has often been misinterpreted that way.

6.1.1.1 Does the Theory of Natural Selection Have Implications for Social Policy?

Social Darwinism is the idea that status in human society is due to biological superiority in a competitive struggle.

The idea that natural selection no longer operated in modern society worried Darwin. He feared that laws that benefitted the poor, medical care, and mental asylums might prevent natural selection from eliminating those who were physically or mentally "inferior." Coupled with high birth rates among the poor, this was thought to cause those members of society to ultimately replace their betters: "If the prudent avoid marriage, whilst the reckless marry," he wrote in *The Descent of Man*, "the inferior members tend to supplant the better members of society" (Darwin, 1871:403).

Despite these concerns, Darwin recognized that factors unique to human society moderated the influence of natural selection on human evolution. "The moral qualities" he wrote, "are advanced, either directly or indirectly, much more through the effects of habit, the reasoning powers, instruction, religion, &c., than through natural selection" (Darwin, 1871:404).

Darwin did not favor ending charity or interfering with human reproduction, but some of his contemporaries interpreted his theory as a justification for social engineering. Darwin's cousin Francis Galton coined the term eugenics (Section 5.3.1), which developed from an initial emphasis on positive incentives for reproduction among some groups at first to coercive measures aimed at preventing persons deemed unfit from breeding. In the mid-twentieth century, laws based on those ideas were enacted in many places. In 1933, Nazi Germany adopted a law that allowed for compulsory sterilization of anyone judged by a genetic health court to have a supposed genetic disorder. By 1940, laws allowing sterilization of "defectives" without their consent had also been passed by many American states and three Canadian provinces, as well as Turkey, Japan, Cuba, and many European nations (Paul, 2003).

Social Darwinism conflates evolutionary fitness with political, economic, and social success. Even if inherited ability contributes to success, other things undoubtedly play a role. Inherited position, education, health care, stress, trauma, luck, social interactions, or other potential influences are not considered in Social Darwinism.

There is an ironic contradiction in Social Darwinism. The theory of natural selection holds that individuals that leave more surviving offspring are more fit than those with fewer offspring. Since the wealthy and powerful have lower birth rates than the poor and powerless, they are less successful at passing on their genes than the poor and powerless. Would that mean that the wealthy and powerful are less fit (at least in the evolutionary sense of the term) than the poor and powerless?

Although social Darwinists argued that Darwin and Wallace's theory supported their agenda, social critics with quite different agendas, including socialism, feminism, or pacifism, argued that the same theory justified their programs. In short, the theory of natural selection drew attention to fundamental questions about human biology and progress, and it stimulated discussions about the relevance of that theory for society. The fact that people found support for divergent ideas highlights the importance of examining the broader implications of science, and the importance of making clear distinctions between what the scientific facts of a topic tell us and the interpretations that we layer onto them.

6.1.2 Coevolution

Darwin recognized that interactions with other species affect survival and reproduction and that, therefore, interacting species influence each other's evolution. This results in coevolution (Section 2.1.2.4) (Thompson, 1982). For example, natural selection in a plant species subject to herbivory will favor *mutations* (randomly occurring genetic changes) that allow the production of chemicals which defend against animals that eat or parasitize that plant. But then natural selection will also favor mutations that allow those herbivores to break down the plant's chemical defenses when they eat that plant. The sequences of mutations will be repeated as natural selection favors mutations that bring about improved survival and reproduction in each of the antagonistic species – the eater and the eaten. This process has been compared to an arms race, with each side evolving an escalating array of defenses against its evolutionary enemies in a stepwise fashion. The plant evolves physical or chemical defenses against its enemies, such as plant-eating insects. The herbivore develops adaptations that allow it to get around the plant's defenses, and so on. A similar metaphorical arms race takes place between predators and prey.

Coevolution doesn't have to involve enemy species. Sometimes it involves mutualistic interactions (Box 6.1). Unlike interactions between antagonists, mutualistic interactions are beneficial to both the interacting species. The benefit may be in the form of a material resource, such as food, or one species may aid the other by providing a service, such as pollination, seed dispersal, or defense against enemies.

In popular media, the term coevolution is used loosely in a way that implies voluntary cooperation among different species, or even ecosystems. This is very different from what the term means in evolutionary biology. Like other forms of evolution, coevolution is not due to purposeful action. Organisms do not choose to help each other, nor do they develop defenses against their enemies on purpose. Coevolution occurs between species that are responding to selective pressures that favor their own survival in their dealings with other species that harm them (antagonists) or help them (mutualists).

Natural selection also drives the evolution of resistance to chemicals that we use to control unwanted species (Box 6.2).

Box 6.1
"What Insect Can Suck It?" – A Striking Example of Coevolution

In a population of a flowering plant that is pollinated by insects, individual plants with traits that increase the amount of pollinator visits to their flowers have more offspring than plants lacking such adaptations. Similarly, individuals in a population of the pollinating species are favored by natural selection if they have adaptations that allow them to harvest more resources from the flowers they visit. In this way, natural selection promotes traits in both species that lead to a mutualistic interaction.

Some flowering plants are pollinated by hawkmoths in a mutualistic relationship that allows the moths to obtain nectar from the plant's flowers and the plant to be pollinated by the moth. When a moth reaches into a flower to obtain nectar, pollen becomes attached to its mouthparts. It then carries the pollen to the next flower it visits, which results in cross pollination.

When Darwin examined a Star-of-Bethlehem orchid from Madagascar that a colleague sent to him in London, he realized that it must have an unusual mutualist. The orchid specimen had a foot-long tubular extension of its flower, with nectar only at the very bottom. Darwin knew that this species of orchid could not have persisted unless it interacted with an insect that would accidentally pick up some pollen while it was feeding on nectar pooled at the bottom of the floral tube and then transport the pollen to the next flower it visited.

Because the tube was too narrow for a pollinator to crawl down to the nectar, Darwin was sure that there had to be a pollinator with a tongue that was long enough to reach the nectar from above. Since the only way such an insect would be able to do that would be with an exceptionally long tongue, this seemed remarkable.

"Good Heavens," Darwin wrote, "what insect can suck it?"

He predicted that the orchid's coevolutionary partner – an insect with sucking mouthparts that were a foot long – would eventually be discovered.

Darwin was right. Two decades later the giant hawkmoth of Madagascar was discovered. Its sucking mouthparts were a foot long (California Academy of Sciences, No date).

Box 6.2
Evolution of Resistance to Pesticides and Antibiotics

When an insecticide is sprayed on a population of insects, some individuals may have genetic traits that give them superior resistance to it. They will survive and reproduce after being exposed to the pesticide, and they will therefore leave more offspring than susceptible individuals will leave. In the next generation, the resistant individuals will be more common than they were initially. The same thing will happen in each successive generation. Eventually, all or most of the individuals that remain will be descendants of the original resistant insects. At that point, natural selection will have resulted in the evolution of resistance to the pesticide in the affected population. The application of the pesticide will have created a strong selective pressure for the evolution of resistance to it.

A similar process allows disease-causing organisms such as bacteria or fungi to evolve resistance to antibiotics. Since their discovery, antibiotics have been widely used in human medicine and have saved millions of lives. They are also used in industrial production of beef, pork, and poultry. As a result, human resistance to antibiotics is now a serious health threat. Since resistance to antibiotics developed after they began to be used in medicine, it might be tempting to conclude that the antibiotics caused the resistance directly. But that is not the case, except in the sense that the use of antibiotics created a selective pressure for resistance.

Insects occupying a variety of trophic levels can be found in an untreated field. Some are herbivores that feed on the crop; others are carnivorous insects that feed on the crop-eating, herbivorous insects. For example, ladybird beetles, or ladybugs, prey on aphids and other insects that eat agricultural crops. From an economic standpoint, the herbivorous insects are harmful, while carnivorous insects are beneficial because they attack the crop eaters. To achieve optimum pest control, it would be best if the herbivorous insects could be eliminated while leaving the beneficial carnivorous ones to assist in pest control by eating any remaining plant-eating insects. But unfortunately, pesticide resistance is more likely to evolve in herbivorous insects than in carnivores. This is because of coevolution.

Throughout their evolutionary history, plants have evolved defenses against herbivorous insects. The plants' defenses might involve physical structures like thorns, or they might be in the form of toxic chemicals. As plant-eating insects (the herbivores) encountered these plant defenses in the plant species they fed on, they evolved chemical defenses against some of the plant toxins. Natural selection favored insects that possessed mutations allowing them to break down naturally occurring plant chemical defenses in their plant foods.

When herbivorous insects are first exposed to a new pesticide, they might not be able to digest the toxins it contains. But they might evolve the adaptations for doing so relatively rapidly because the chemical pathway for breaking down those novel toxins might be similar to an adaptation that they had as a result of previous selection. It might take only a few changes in a metabolic pathway that the herbivorous insects already possess. In other words, because of their long history of coevolution in response to plant toxins, herbivorous insects generally evolve the ability to break down toxins in synthetic insecticides faster than carnivorous insects do.

Predatory insects have little evolutionary experience with breaking down toxins. For them, the evolution of chemical pathways that can break down insecticides typically requires the development of complex new adaptations starting from scratch. This will require many mutations and many generations.

So the application of pesticides is likely to eliminate beneficial predaceous insects such as la-dybugs while favoring resistant strains of the most harmful crop-destroying insects. Clearly, this is just the opposite of what we would like pesticides to accomplish.

The take-home lessons from these experiences are that (1) it is important to understand the evolutionary consequences of our actions and (2) antibiotics and other pesticides should be used only selectively because indiscriminate use increases selection for resistance (*see Q6.2*).

6.2 Speciation: The Formation of Species

6.2.1 The Development of Darwin's Ideas

As a young man, Darwin spent nearly five years aboard the *HMS Beagle*. While the other men aboard this ship surveyed and mapped the lands and waters they visited, Darwin and his servant collected animals and plants, including a group of songbirds from the Galápagos Islands off the coast of Ecuador (Figure 6.1). When he returned to England, he took up pigeon breeding because he was interested in how *artificial selection*, or *selective breed-ing*, in which a breeder creates new breeds by selecting and breeding animals with desired

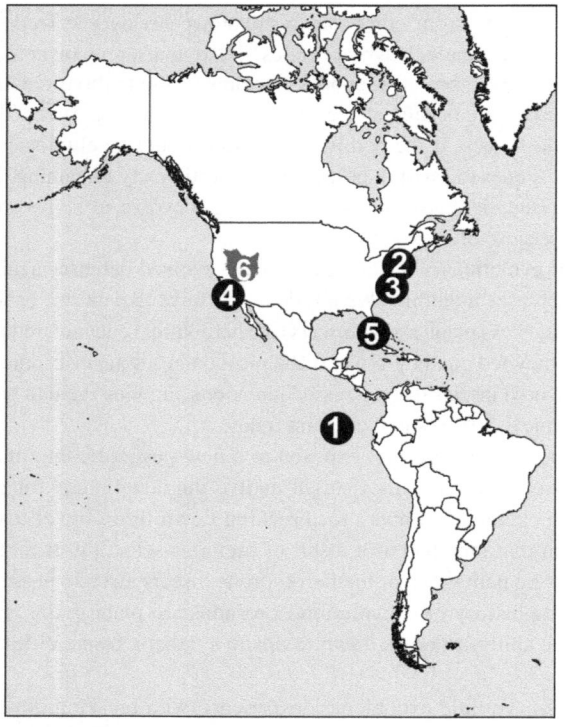

Figure 6.1 Locations of: 1, Galápagos Islands; 2, New York City; 3, Alligator River National Wildlife Refuge, North Carolina; 4, Channel Islands, California; 5, Florida Bay; 6, Great Basin. Map created by Eva Strand using Esri, DeLorme World Countries Generalized Data & Maps for ArcGIS 2013, with permission.

traits that differ markedly from their ancestors. He reasoned that natural selection operating in a similar manner over long periods of time would give rise to new species.

Darwin also pondered his specimens from the Galápagos Islands. Although in many ways the birds on different islands were similar to each other and to birds on the mainland, the birds from different islands also differed consistently from each other in some ways. Darwin concluded that all the birds in this group (which later became known as *Darwin's finches* although they are no longer classified as finches), were descended from an ancestral species that colonized the different islands. After they reached the islands, they eventually developed into more than a dozen new species. In other words, speciation, the evolution of new species, had occurred.

Darwin's observations of the Galápagos birds, along with his observations of diversity in pigeon breeds and the diverse forms of plants and animals that he encountered on his travels convinced him that species resulted from the action of natural selection on inherited variations.

6.2.2 Reproductive Isolation of Different Species

A *species* is generally regarded as a kind of organism that is reproductively isolated from other groups in nature, that is, it is unable to reproduce with members of other species. Because of *reproductive isolation*, members of a species share a common evolutionary history. *Gene flow*, the exchange of genetic information, can occur among members of a species but usually not between members of different species. A modern-day group of frogs illustrates how reproductive isolation can lead to a new species (Box 6.3).

Box 6.3

A "Weird" Croak Leads to the Discovery of a Previously Unknown Species of Frog

Leopard frogs are common and widespread. Many students dissect these amphibians in biology classes or encounter them in ponds in the wild. Scientists at first thought that all leopard frogs belonged to a single species. However, genetic research subsequently showed that frogs in this group belong to several different species even though they look very much alike. The explanation for this apparent paradox lies in their mating calls, which are complex and species-specific. Some species have a long, snore-type call, whereas others have a rapid, chuckle-call (at least to human ears). Females mate only with males that give the appropriate call for their species. By preventing the different species from breeding with each other, the different calls cause reproductive isolation between the different species.

A team of scientists from California, New Jersey, and Alabama studying frogs in the vicinity of New York City made a surprising discovery in 2012 (Figure 6.1). In an area where the ranges of two species of leopard frogs overlap, some intermediate frogs with a "weird," short, repetitive croak turned up. Genetic analysis revealed that those frogs did not belong to either of the species known to occur there. In fact, they were a species previously unknown to science.

Additional searching revealed that the center of its current range is Yankee Stadium. "I was really surprised and excited once I started getting data back strongly suggesting it was a new species. It's fascinating in such a heavily urbanized area," said Jeremy Feinberg of Rutgers University in New Jersey (Anon., 2012).

Because of reproductive isolation, each species is a repository of unique genetic material. Once a species becomes extinct, its genetic information is lost forever. This is one (of many) of the reasons why the loss of species is of such concern.

Hybridization (crosses between different species or subspecies) usually produces offspring that are sterile. This is because even if mating does occur, usually the genetic material from the parents of two different species is so different that it cannot pair up when the sperm and eggs of the different parent cells meet during fertilization. The mule, the offspring of a cross between a female horse and a male donkey, is a familiar example. Because mules are sterile (with a few rare exceptions), horses and donkeys remain reproductively isolated from each other.

In some cases, however, different species of animals hybridize after abrupt changes in habitat bring together two closely related species that would not normally breed with each other. We will see an example of this below.

New species of plants can sometimes arise when an entire set of chromosomes is duplicated. Because many plant species can reproduce through *self-fertilization*, an individual with an extra set of chromosomes may be able to fertilize itself. In such cases, finding a mate with matching chromosomes is obviously not a problem.

6.3 Classifying Species

The science of classifying things is termed *taxonomy*. Taxonomists classify life forms according to a hierarchy of ranked categories that organizes different species into a framework that expresses the relationships between them. Western science uses a system developed in the eighteenth century by the Swedish biologist and physician Carolus Linnaeus, in which each species known to science is designated by a two-part Latin name consisting of a species and a *genus* (plural: *genera*). Indigenous cultures have their own taxonomic systems.

The Linnaean system organizes species according to degrees of relatedness. Species that are closely related are grouped together in a genus; genera that are closely related are placed in a family; related families are grouped together in an order; related orders are grouped in a class, and so on (Table 6.1). The gray wolf (*Canis lupus*), the coyote (*Canis latrans*), and the domestic dog (*Canis familiaris*) are all placed in the genus *Canis* within the dog family (Canidae). At any level in the hierarchy, an organism can belong to only one rank.

Linnaeus and classical taxonomists who followed him decided how closely related organisms were on the basis of their external appearance, anatomy, and, for living forms but obviously not extinct ones, their behavior. The amount of information available for judging relatedness increased astronomically in the late twentieth century when scientists developed techniques for comparing different organisms' genetic material. The information provided by this technology is more accurate than that provided by classical methods, because, as we saw in Box 6.3, species that appear to be very similar can have different genetic material. The reverse is also true; species that look very different can be quite closely related.

Table 6.1 *Classification of the fox squirrel,* Sciurus niger

Category	Example	Description
Kingdom	Animalia	multicellular heterotrophs
Phylum	Chordata	vertebrates and close relatives
Class	Mammalia	vertebrates with mammary glands
Order	Rodentia	rodents (mammals with front teeth that grow indefinitely)
Family	Sciuridae	squirrels
Genus	*Sciurus*	bushy-tailed squirrels
Species	*Sciurus niger*	fox squirrel

6.3.1 Genetic Differentiation within Species

Species consist of one or more populations. Each population possesses adaptations to its local environment. As a consequence, populations of the same species differ genetically to some extent, but not enough to prevent them from interbreeding. This becomes an important consideration when managing declining populations.

When a species includes populations that inhabit a distinct part of the species' geographic range and differ in some consistent respects, scientists sometimes designate *subspecies*, *races,* or *varieties.* The category of race or subspecies is different from other taxonomic categories for three reasons. First, although subspecies are different enough to be distinguished, they are not reproductively isolated from each other. By definition, members of a subspecies are capable of breeding with other members of the same species. Thus, gene flow is possible between subspecies. Second, the designation of subspecies or races is subjective. There is no objective standard for how different from each other subspecies need to be. Third, the category of subspecies need not be used at all. Every organism must belong to a species; every species is part of a genus; every genus belongs to a family; and so on, but an organism does not have to belong to a subspecies (Gould, 1977). Nevertheless, the designation of races or subspecies can have important implications for conservation (Box 6.4).

Box 6.4
What Is a Red Wolf, and Why Does It Matter?

The red wolf (usually classified as *Canis rufus*), is a small, reddish member of the dog family that once occurred throughout much of eastern North America. After European settlement, red wolves declined in abundance as their geographic range shrank. At the same time, the range of the closely related coyote (*Canis latrans*), which does well in settled landscapes, expanded.

By the turn of the twentieth century, red wolves were gone from most of their former range, but coyotes were common. Red wolves became so scarce that they probably had trouble finding mates of their own species. Mating with coyotes occurred, and the hybrid offspring sometimes mated with individuals from one of the parent species. This created headaches for taxonomists later on.

By the 1970s, when the red wolf had been reduced to a remnant population of fewer than 100 animals, the US Fish and Wildlife Service captured the remaining red wolves and began breeding them. After the captive population increased sufficiently, biologists began reintroducing the animals into portions of their natural habitat in North Carolina.

The red wolf is intermediate between the coyote and the gray wolf in many respects. This has raised questions about whether it is really a separate species. One hypothesis is that the red wolf is a distinct species that recently hybridized with coyotes. In that interpretation, hybridization was due to human-caused habitat changes and is a threat to the genetic integrity of the red wolf. A competing hypothesis is that the red wolf is a hybrid between the gray wolf and the coyote. In either case, hybridization occurred in the ancestors of modern red wolves.

When biologists Robert Wayne and Susan Jenks analyzed genetic material from study skins in museums and the blood of captive animals, they did not find any genetic material that was unique to red wolves. All the material that they examined was also found in either gray wolves or coyotes (Wayne and Jencks, 1991). They concluded that the red wolf is not a true species. That interpretation is controversial, however (Nowak, 2002; Hinton et al., 2013).

The unusual genetic history of red wolves complicates their management. If the red wolf is actually a hybrid it could lose its protected status as an endangered species (O'Brien and Mayr, 1991). This scientific controversy has important political and practical ramifications. Captive breeding and reintroduction of red wolves are costly and time-consuming. The American Sheep Industry Association unsuccessfully petitioned the federal government to remove the red wolf's listing as an endangered species on the grounds that it is not a separate species. But the red wolf also has great public appeal. The US Fish and Wildlife Service sponsors popular "Howling Safaris" that offer visitors to Alligator River National Wildlife Refuge a chance to hear reintroduced red wolves (Figure 6.1).

This example illustrates some of the challenges of reconstructing evolutionary history and historical ecology. Studies of the mating behavior and reproductive success of wild red wolves in areas where coyotes are present might eventually help to unravel this evolutionary puzzle.

The red wolf is not the only species or subspecies whose status as a subspecies (race) has political implications. Controversy about the scientific validity of human races is closely tied to ideas about the causes of racial inequality, a topic that is influenced by underlying cultural assumptions (Box 6.5).

Box 6.5
What Can We Learn from Morton's Measurements of Human Skulls?

Late in the eighteenth century, a German anthropologist named Johann Friedrich Blumenbach classified people into five races, which he called Caucasian (white), Mongolian, Ethiopian, American, and Malay. Blumenbach based his classification in part on his subjective judgment that a single skull specimen from the Caucasus, an area between the Black and the Caspian Seas (Figure 6.2) produced "the most beautiful race of men" (quoted in Gould, 2011:357).

Today, Blumenbach's pro-European bias seems obvious. We no longer believe that there is a universal standard of beauty (and one that can be objectively determined from skulls, at that). But in the nineteenth century, scientific arguments for ranking human races, with whites at the top, prevailed in Western science.

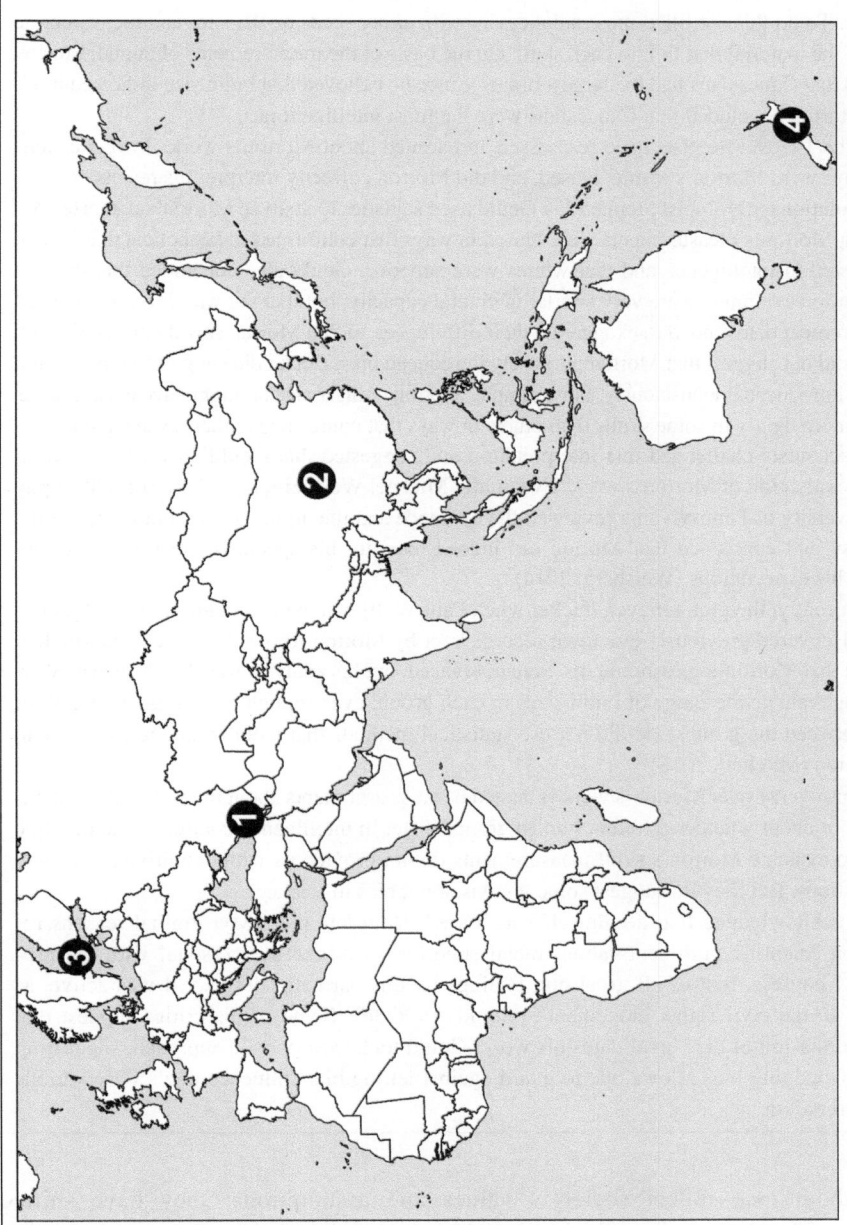

Figure 6.2 Locations of: 1, Caucasus region, Western Europe; 2, center of giant panda distribution, China; 3, Åland Islands, Finland; 4, North Brother Island, New Zealand. Map created by Eva Strand using Esri, DeLorme World Countries Generalized Data & Maps for ArcGIS 2013, with permission.

From 1830 to 1851, Samuel George Morton, a prominent American physician and physical anthropologist, amassed a large collection of human skulls. He measured the volume of the *cranium* (the part of the skull that encloses the brain) of 623 skulls from Europe, North America, and Egypt. To do this, he filled the cranial cavity with either seeds or BB shot and measured the volume of the material that fit into each skull. On the basis of the measurements obtained, Morton concluded that Caucasians had the largest brains. Since he believed that brain size indicated intelligence, Morton concluded that Caucasians were the most intelligent race.

Since the 1970s, scientists have reassessed and argued about Morton's work. Two questions are relevant: were Morton's results biased, and did Morton correctly interpret his results?

The evolutionary biologist Stephen Jay Gould used statistics to analyze Morton's data. He concluded that Morton's measurements were biased in ways that conformed to his beliefs that human races differed in intelligence and that whites were superior. Gould also concluded that the five groups Morton examined were very similar in cranial capacity, but that the way Morton analyzed his measurement data tended to exaggerate their differences so that Morton saw them as different.

Gould did not suggest that Morton purposely introduced bias. But he did suggest that bias could have been introduced unconsciously. For example, Morton or his assistant might have packed seeds or pellets more tightly in some skulls than others in ways that conformed to their expectations.

Other scientists challenged this interpretation and suggested that Gould's own biases influenced his evaluation of Morton's work. Eventually Michael Weisberg, a professor of philosophy at the University of Pennsylvania reviewed the evidence and the arguments on both sides of the controversy and concluded that Morton did indeed measure his specimens in ways that conformed to his expectations (Weisberg, 2014).

More recently, the plot got even thicker when Paul Wolff Mitchell, an anthropology graduate student, discovered previously unknown records kept by Morton. From these archives Mitchell concluded that Morton's measurements were not biased, but his analysis was. For instance, Morton did not evaluate the range of brain sizes in each group, even though there was quite a bit of overlap between the groups. He did not use statistical methods that would take the overlap into consideration (Mitchell, 2018).

The controversy over Morton's work is heated because arguments about the designation of human races or about whether different human groups differ in intelligence continue. The multiple attempts to evaluate Morton's skull measurements do not provide us with definitive answers to those questions. But they do suggest some lessons about bias in science.

Gould acknowledged that he himself was biased. He wrote that "My original reasons for writing [on scientific arguments about racial superiority] mixed the personal with the professional. I confess, first of all, to strong feelings on this particular issue.... I was active, as a student, in the civil rights movement" (Gould, 1996:36). Some of his critics suggest that Gould's admission of bias invalidates his work. Gould took the opposite approach, suggesting that acknowledging bias allows one to guard against letting bias influence one's observations and interpretations.

Ideas about race reflect society's values and assumptions; they have shifted many times. The designation of races is subjective (Saini, 2019). The scientific consensus is that the differences between groups of people are minor compared to the differences between races of other species. In the words of the American geneticist Alan Templeton: "human evolution has been and is characterized by many locally differentiated populations coexisting at any given time, but with sufficient genetic contact to make all of humanity a single lineage sharing a common evolutionary fate" (Templeton, 1998:632).

6.3.2 Diagramming the History of Life

As the story of the red wolf illustrates, classification is not just an academic exercise. In order to conserve life forms, scientists need to know what they are conserving. In addition to being a useful tool for naming and keeping track of the different kinds of living and extinct organisms, and for prioritizing conservation efforts, classification provides a powerful conceptual framework that summarizes scientists' understanding of the relatedness of the Earth's life forms. As new species are described and new information about their relationships is uncovered, some aspects of the scientific classification system are revised and updated, new hypotheses are formulated, and our understanding of evolutionary relatedness is broadened.

Over 1.5 million species had been cataloged as of 2012. This total includes plants, animals, fungi, algae, and some microorganisms (but not bacteria and archaea (Section 2.3.2.2)). The animals that we are usually most aware of are not the most diverse groups. Most animals are invertebrates; most invertebrates are insects; and most insects are beetles, which are the most species-rich group of animals on the planet. Mammals account for only a small proportion of the known species diversity of animals. By contrast, the 400,000 beetle species make up about 25% of the total number of known animal species.

The relationships between different species have often been represented with a branching diagram, analogous to a genealogical tree. Both existing (or *extant*) species and extinct species are included on such diagrams. The powerful image of a tree of life appears in many cultures, but recently scientists have moved to using a circular diagram, or *supertree*, with time radiating outward. The center represents the origin of life on Earth, and evolutionary groups are represented by lines arrayed around the circle. Lines that start close to the center represent species that arose relatively early. The lines branch, and the branches represent evolutionary relationships, with closely related forms close together. A supertree for just a small portion of the animal domain, members of the Class Mammalia, Order *Carnivora* (mammals specialized for eating flesh), is shown in Figure 6.3.

Because there are so many species alive on Earth today, a complete supertree would be hard to show on paper. However, the data about each group can be stored in a digital version of the circle. Researchers from around the world add more information to the supertree using interactive features available on the Internet, thereby improving our understanding of the interrelatedness of Earth's biodiversity. That happened in a dramatic way in 2016 (Box 6.6).

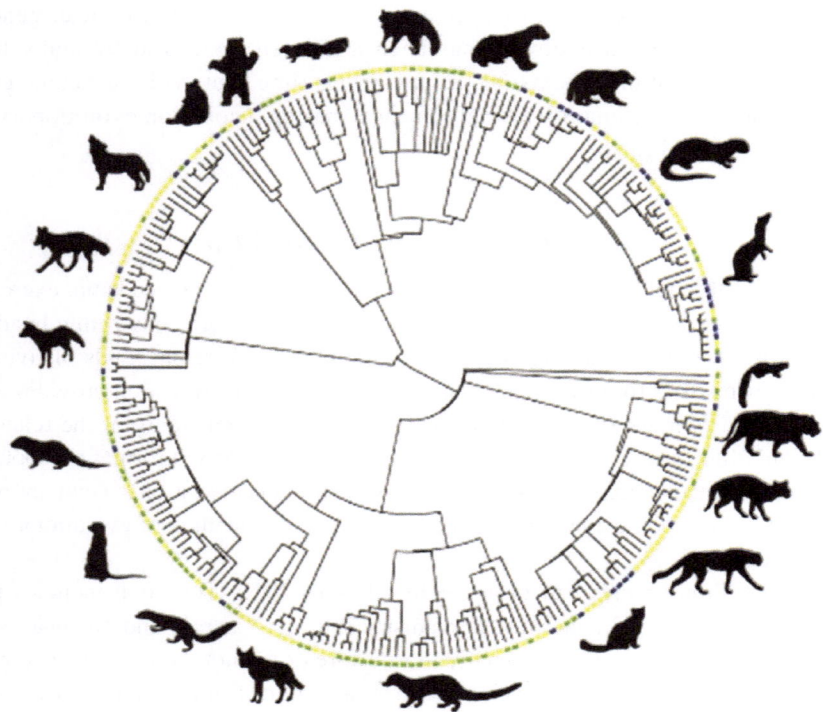

Figure 6.3 Diagram of supertree showing the evolutionary relationships of carnivores (Order Carnivora, Class Mammalia). Distances between branches indicate degree of relatedness of different groups, with the most recently evolved forms farthest out on the circle. Used with permission of John Wiley & Sons Ltd., from Rolland et al. (2015).

Box 6.6
A New Branch on the Supertree of Life

An international team of scientists led by Laura Hug at the University of California, Berkeley, added an entire branch of previously unknown bacteria to the supertree. Until this research was done, scientists knew only those kinds of bacteria that they could grow in a laboratory. The team discovered so many unknown forms because they began studying *DNA* – the molecule that is the main carrier of genetic information – isolated from previously unexamined environments (Hug et al., 2016). According to Hélène Morlon, an evolutionary ecologist with the French *Centre National de la Recherche Scientifique*, there is a lesson to be learned from this dramatic discovery.

We didn't spot [this group of bacteria] because it doesn't look like what we thought life should look like.... Until now, we thought all bacteria shared a universal [genetic] fingerprint .... In fact, the corresponding gene sequence of bacteria from the 'new' bacterial group is quite different ....

That this group of bacteria is so dominant means that it's probably one of the main players in our ecosystems – in terms of nutrient cycling, for example. Yet we know close to nothing about it! It is very distinct from the rest of the bacteria: half of the genes are unlike other known genes. We might even find new biological functions that could be useful for designing new drugs or degrading pollutants. As we learn more about this huge group of organisms, we will certainly encounter many new discoveries and surprises.

[This discovery] forces us to stay humble about what we do know. The big lesson: we missed an enormous chunk of [biodiversity] because it looked just a bit different from what we assumed life should look like. It's a beautiful reminder to always question what knowledge we take for granted.

(Eng, 2016)

6.4 Extinction: The Disappearance of Species

When all representatives of a group have died, leaving no living descendants, the group is *extinct*. Extinction can affect a population, a species, or a subspecies. If every population of a species goes extinct, then that species is of course extinct. Sometimes some but not all subspecies of a species may become extinct. When that happens, the genetic material of some subspecies is gone, but other subspecies of that species persist. If all populations of a species in a local area die out, the species is said to be locally extinct, although other populations might exist elsewhere. For example, although two subspecies of reindeer or caribou – the East Greenland caribou and the Queen Charlotte Islands caribou –are extinct, 12 caribou subspecies still exist in suitable habitats of North America and Eurasia.

Extinction occurs when losses from a population consistently exceed gains, that is, when death and emigration are greater than birth plus immigration. This may happen because a species is not adapted to its environment (perhaps because the environment has changed due to succession, climate change, development, geological change, catastrophes, or a combination of factors) or because of excessive exploitation or pollution.

6.4.1 Past Extinctions

There have always been extinctions. Darwin recognized that the number of species on Earth had not increased markedly despite ongoing speciation. He pointed out that for this to be the case some species must have become extinct as new ones arose. In addition to isolated extinctions, the geological record is marked by a few *mass extinctions*, in which many species disappeared at about the same time. Perhaps the most familiar example is the dramatic disappearance of the dinosaurs around 66 million years ago. The extinction of these reptiles was accompanied by the disappearance of many other groups of terrestrial and marine organisms as well.

Four other mass extinctions took place prior to the extinction of the dinosaurs. The most extensive of those occurred about 252 million years ago, when at least 90% of the world's species of marine invertebrates disappeared. (Scientific understanding of the speciation that followed this cataclysmic event got a boost in the early 2000s when a teenager named L. J. Krumenacker went fossil hunting near his home in southeastern Idaho and found several fossilized shark teeth in the first rock he picked up. Subsequently, he and 17 paleontologists from seven nations published a scientific paper on their insights from those specimens plus additional discoveries at this previously unknown site (Brayard et al., 2017).)

6.4.1.1 The Prehistoric Overkill Hypothesis

The relatively sudden extinction of so many forms suggests a drastic, far-reaching event such as widespread environmental change, volcanic activity, or collision with a meteor. Between 50,000 and 10,000 years ago, many large mammals in North and South America, Eurasia, and Australia – including mammoths, giant beaver, and giant ground sloths – became extinct (Figure 6.4). These disappearances are referred to as the *Pleistocene extinctions* after the geological epoch when the ice ages ended.

Because several things changed at around the same time, it is difficult to sort out causal relationships for these extinctions. The roughly simultaneous disappearance of many species, the changes in climate as the ice sheets waned, and the arrival in North America of people from *Beringia* (the land connection between Alaska and Siberia that was present when the large amount of water locked up in glaciers caused sea level to fall and a land bridge to be exposed) present a complicated puzzle. Earlier mass extinctions were characterized by the disappearance of many different groups, but this instance was unique in that mainly large animals were affected.

In 1967, a Soviet climate scientist named Mikhail Budyko proposed that prehistoric hunting tribes in Europe had played a decisive role in bringing about these extinctions. He based his reasoning on mathematical models of predator–prey dynamics in which people were the predator and mammoths were the prey (Budyko, 1967). At about the same time, Paul Martin, a geoscientist at the University of Arizona, developed a model of interactions between prehistoric hunters and their prey in North and South America. His model used assumptions about the rate at which people migrating into North America moved south, the growth rate of their populations, and the effects of their hunting on large mammals (Martin and Wright, 1967; Martin, 1973). He concluded that "by analogy with other successful animal invasions," the movements of big game hunters into North and subsequently South America triggered a population explosion and unsustainable hunting of large prey (Martin, 1973:973). Martin also proposed that prey animals were easily wiped out because they had not evolved adaptations to human predators, and he used the emotionally laden military term *blitzkreig* (a misspelling of the German word *blitzkrieg*, referring to a sudden violent offensive) for this hypothesis (Mosimann and Martin, 1975).

The selective disappearance of the prehistoric megafauna is striking. On the other hand, because prehistoric hunters lacked the kinds of weapons that allowed modern market hunters to drive innumerable species to extinction, some scientists question whether prehistoric hunting could cause the extinction of large numbers of species. It is also possible that the warming trend that occurred as the ice sheets waxed and waned caused, or contributed to, the wave of extinctions. Some evidence for and against the prehistoric overkill hypothesis for Eurasia and North America is summarized in Table 6.2.

Fortunately, in the nearly seven decades since this controversy began, many new tools for testing hypotheses about the extinctions have been developed. Ancient DNA sequences from teeth, bones, or other tissues, combined with statistical techniques and computer modeling, allow scientists to assess the genetic diversity of extinct populations and to test hypotheses about past population dynamics and climate change. Of course, some hypotheses about the past cannot be tested directly. We cannot measure extinct animals' behavioral

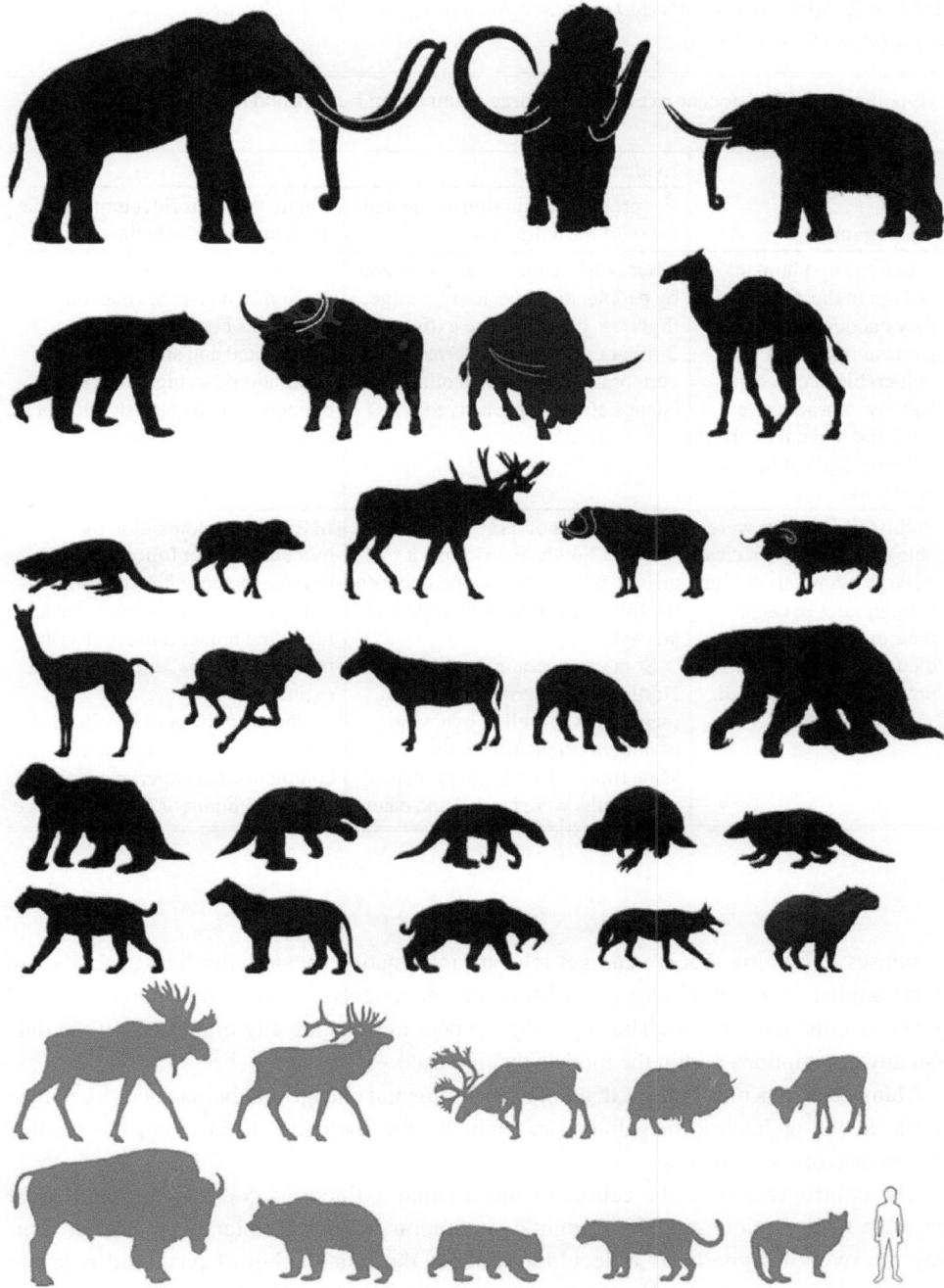

Figure 6.4 Silhouettes of selected extinct species (black) from the Pleistocene and living species (gray) of mammals of northern Eurasia. Outline of modern human being at lower right gives approximate scale. Used with permission of John Wiley & Sons Ltd., from Stuart (2015).

Table 6.2 *Some arguments and evidence for and against the Prehistoric overkill hypothesis (Stuart, 2015)*

Hypothesis: the Pleistocene extinctions of large mammals in Eurasia and the Americas were caused by overhunting.

The argument	Evidence	
	For prehistoric hunting as the main cause of the extinctions	Against prehistoric hunting as the main cause of the extinctions
When the first humans arrived in the Americas, they encountered big game animals that were vulnerable because they had low reproductive rates and had not evolved defenses against human predators.	After New Zealand was colonized by people, nine species of a large, flightless birds became extinct. Similar extinctions of terrestrial vertebrates occurred on other islands after people arrived.	In general, extinction rates are higher on islands than on continents because islands are isolated and support small populations; so high rates of extinction on islands should not be extrapolated to continents.
Prehistoric hunters were able to kill many species of large mammals at high enough rates to cause their extinction. Predators died out too when their prey based disappeared.	Many species of large mammals in North or South America prior to the arrival of humans disappeared from the fossil record soon after people arrived. Some archaeological sites in North America contain bones of mammoths as well as evidence of human occupation at the same time, which suggests that mammoths were hunted and eaten.	In Eurasia, colonization by humans was not followed by a marked increase in extinctions. Many species that were hunted, including horses and reindeer in northern Eurasia, did not become extinct. The proportion of species that went extinct was greatest on continents that experienced the greatest amount of climate change.

responses to hunting, but we can get relevant information by studying living relatives of those animals and combining this with computer models of what might have occurred given specific assumptions. The key things to bear in mind are any limitations of the data and any assumptions within the models that are used.

Many scientists now believe that both environmental change and human activity played a role in the prehistoric megafaunal extinctions. But they still differ about the relative importance of those factors.

The controversy over the causes of megafaunal extinctions comes with ideological baggage related to emotionally fraught contemporary issues. The language used to convey the overkill hypothesis reflects ideas about the relationship of past, and possibly present, people to nature. If people were culprits in a catastrophic onslaught on biodiversity, does that tell us anything about contemporary hunter-gatherers? If hunting was not the sole or even the principal cause for the megafaunal extinctions, and climate change played a major role, does that mean that the contemporary combination of human activities and a changing climate will be disastrous? What does the extirpation of many species in a relatively short period of time tell us about the consequences of the high rates of extinction that we currently face? If people were responsible for the prehistoric extinctions, does that mean we have a moral obligation to bring back megafauna where we can? Should we try to compensate for some past extirpations of large species by reintroducing animals with similar ecologies into places where they are now missing (Section 8.7)?

These issues illustrate the potential for science to both reflect and influence our values and assumptions.

6.5 Contemporary Extinctions

6.5.1 Why Are Some Populations More Vulnerable Than Others?

Many biologists believe that a sixth mass extinction is underway. If mass extinctions have occurred repeatedly in the past, why is there so much concern about contemporary extinctions? One reason is that under current conditions of development and habitat fragmentation it is unlikely that the Earth's new species would develop fast enough to replace lost ones. While the Earth's extinction rate has escalated, the rate of speciation has not shown a corresponding increase. In fact, species are probably arising less often today than in the past. Paul and Anne Ehrlich compare this situation to a sink in which species are disappearing down the drain, but the flow of new species through the faucet is not increasing (Ehrlich and Ehrlich, 1985).

6.5.1.1 Ultimate and Proximate Risk Factors

Why do cockroaches, dandelions, pigeons, and house mice proliferate as land development progresses, while many other organisms disappear? In the next section, we will consider what, according to contemporary ecological theory, affects a population's risk of extinction. Chapters 7 and 8 will cover practical applications of those insights.

Factors that predispose a species to decline and eventual extinction can be divided into two categories: *ultimate* (or *intrinsic*), and *proximate* (or *extrinsic*) *factors*. The vulnerability of a species ultimately depends on characteristics of its reproduction, ecology, anatomy, and behavior (Wolfheim, 1976). These *ultimate risk factors are* genetically determined and result from a species' evolutionary history. We cannot change them. Species with those risk factors recover relatively slowly after they decline. Superimposed on those characteristics are short-term, or *proximate, factors* – such as exploitation, persecution (for example, predator control or unregulated commercial exploitation), habitat change, pollution, introductions of non-native species, climate change, or other environmental catastrophes – that affect the balance between births and deaths. These factors may be considered the immediate drivers of extinction when they trigger the population changes that eventually lead to extinction.

Although managers cannot alter ultimate factors that affect vulnerability to extinction, an understanding of those factors is useful for identifying species that are at risk. Species which are especially likely to become extinct typically have one or more characteristics that make them especially vulnerable to immediate factors that put them at risk. Late sexual maturity, long intervals between breeding, and small numbers of young, all of which are associated with low biotic potential, contribute to vulnerability. Specialized habitat or food requirements; restricted and/or isolated geographic range; limited ability to disperse; high trophic level; and limited exposure to recently encountered evolutionary antagonists such as predators, competitors, or disease-causing organisms, are also associated with high risk of extinction.

Combinations of these ultimate and proximate causes of decline contributed to the extinctions of the species featured in Sections 1.3.3 (Table 6.3).

Large animals with large home ranges are often at risk because they come in conflict with people and are negatively affected by habitat change, exploitation, and persecution. On the other hand, many extinct species had small geographic ranges. The destruction of their habitat contributed to their demise. Many of those species required specialized habitats or food resources, and often they had limited potential to disperse. Animals or plants that exist only in small areas and cannot colonize elsewhere can easily be wiped out. Specialization and limited geographic range increase the likelihood that habitat alteration will make it impossible for a species to meet its requirements. The larvae of many herbivorous insects feed on only one or a few species of plants. The fates of these insect specialists (such as the Xerces blue butterfly) depend on the fate of their food plants, even if the insects themselves have high reproductive rates.

Ultimate risk factors lead to decline and possible extinction when proximate factors come into play. A highly specialized species may be able to obtain the resources it needs until changes in proximate risk factors put it at risk (Box 6.7).

Risk of extinction is especially acute for organisms that evolved without exposure to antagonistic species which eat them or cause diseases. Because most mammals are incapable of crossing large bodies of salt water on their own, plants on oceanic islands typically evolved without selective pressure from large herbivores. Consequently, they never evolved defenses against introduced herbivores such as deer, goats, or cows. Similarly, island biota are vulnerable to introduced predators, competitors, parasites, and diseases.

For animals, high trophic level is an ultimate risk factor. In comparison to herbivores, populations of carnivores and scavengers are small, and their reproductive rates are low. In addition, toxic substances accumulate in their tissues (Box 5.2). Animals that are high on the food chain typically have large home ranges because they need large areas within which to acquire enough food. (Note that either a small geographic range or a large home range can contribute to vulnerability.) Animals with large home ranges often come into conflict with people. They compete with people for prey such as livestock, they may attack people, and their habitats are likely to be negatively affected by anthropogenic (human-influenced) activities. For these reasons, predatory birds and mammals often become rare. Many large, carnivorous birds and mammals, such as the grizzly, gray wolf, tiger, and California condor, are threatened or endangered in many places due to combinations of these factors.

Table 6.3 *Factors that contributed to the decline of organisms featured in Section 1.3.3. "?" indicates possible risk factor but insufficient information available to conclude whether this factor was a significant factor in extinction risk.*

	Passenger pigeon	Chilean sandalwood	Extinct giant tortoises	Tasmanian wolf	Caribbean monk seal	Xerces blue butterfly
Ultimate factors						
Conspicuous coloration or behavior	X		X			
Formation of large groups	X				?	
High trophic level				X		
Limited ability to disperse						X
Limited experience with evolutionary antagonists or human predators			X	X		
Restricted geographic range		X	X	X	X	X
Low biotic potential			X			
Restriction to islands		X	X	X		
Specialized habitat or food requirements						X
Proximate factors						
Exploitation	X	X	X	X	X	
Habitat change				?		X
Introduced diseases				?		
Introduced species			X	X		
Persecution (predator control)				X		

Box 6.7
Extreme Dietary Specialization in a Plant-Eating Species of Bear

The giant panda is an unusual bear. Its chubby body and distinctive facial markings, which give the impression of enormous eyes, evoke affection rather than fear. The public knows individual zoo pandas by their first names and celebrates the rare occasions when zoo pandas give birth. One of the most prominent conservation NGOs in the world uses the giant panda as its logo.

This charismatic species differs from its bear relatives in its biology as well as its appeal. Like other bears, the giant panda is a member of the Order Carnivora. But it does not eat meat. Unique among bears, it is specialized for eating bamboo, a group of large, evergreen grasses native to Asia. Compared to the foods of other bears, bamboo has little nutritional value. It is low in protein and high in fiber.

Because of its evolutionary history, the giant panda's digestive tract is similar to the digestive tract of its carnivorous ancestors. Consequently, pandas do not have the adaptations for digesting plants that horses or deer have. A panda's digestive tract is short, and its gut does not contain microorganisms that can break down cellulose and other plant fibers. This means that a lot of the biomass that a panda eats passes through its gut without being digested, so pandas do not get much energy from their diet. To meet its daily energy requirement, a panda must spend about 14 hours foraging and consume about 12.5 kg bamboo per day. And because pandas cannot digest most of the bamboo they eat, pandas defecate more than 100 times in 24 hours.

In the wild, giant pandas inhabit remote mountain ranges of central China (Figure 6.2). Their extreme dietary specialization is an ultimate factor that makes them vulnerable. In addition, bamboo has a characteristic that contributes further to the panda's vulnerability. Bamboo flowers rarely and then dies. These episodes of *mass flowering* happen unpredictably at intervals ranging from 15 years to a century. Mass flowering prevents populations of species that eat bamboo, including pandas, from building up. Sometimes pandas starve because of mass die-offs of bamboo.

Despite its inefficient dietary system and dependence on an unpredictable resource base, the giant panda has persisted for thousands of years. However, the interaction of those ultimate risk factors with proximate threats from relatively recent loss and fragmentation of bamboo forests, along with poaching for panda skins, caused the species to decline to the point where it was in danger of becoming extinct.

Fortunately, giant panda populations have increased as a result of intensive and high-profile conservation efforts including monitoring, research, anti-poaching efforts, and the establishment of a network of reserves, and the species is no longer considered endangered (Swaisgood et al., 2016) (*see Q6.5*).

Finally, traits that increase the likelihood of detection by predators (including people) pose additional risk if exploitation is intense. These include large size, conspicuous coloration, and the formation of dense breeding colonies, flocks, herds, or *arribadas* (Box 3.2). Breeding aggregations of some invertebrates, amphibians, birds, seals, and bats are easily located and exploited or targeted for control. Because they occurred in large groups, both the passenger pigeon and the bison were overexploited, even though they were initially present in enormous numbers. Like the birds that were collected during the plume trade, butterflies and some plants such as cacti are threatened by collection. Defensive behaviors also can increase vulnerability. For example, when muskox are threatened they collectively defend themselves by forming a line or circle. This adaptation allows them to defend against wild predators without expending the energy they would use if they fled, but it does not protect them against human predators with guns.

6.5.2 Genetic Diversity and Risk

Genetic diversity affects potential vulnerability to extinction. Unlike ultimate risk factors, which are characteristic of each species, genetic diversity differs between populations. To understand how genetic diversity influences extinction risk, we need to understand the genetic basis of inheritance (Box 6.8).

Box 6.8
Inheritance and Genetic Change: Genes, Alleles, and Chromosomes

Although Wallace and Darwin recognized that natural selection acting on genetic variation drove adaptation and speciation, they knew nothing about how organisms pass variation on to their offspring. When these naturalists were writing, the mechanisms of inheritance were not well understood.

We now know that genetic material is located on *chromosomes* and that chromosomes are composed of *deoxyribonucleic acid* (*DNA*) and protein. In organisms with complex cells, DNA occurs in the cell nucleus, which contains most of a cell's genetic material, and in parts of cells that control energy flows. A molecule of DNA consists of two, long coiled strands held together in a *double helix* (two-stranded spiral) by bonds between subunits of the DNA. The arrangement of those subunits is the *genetic code*. The functional unit of genetic information is the *gene*, a segment of DNA. Genes occur in definite positions called *loci* [singular: *locus*]) on chromosomes. Genes control hereditary traits by regulating the cell's manufacture of proteins and by turning other genes on or off.

Most animals are *diploid* (*di*, two; *ploid*, sets of chromosomes) throughout most of their life cycle; that is, their cells normally possess two sets of chromosomes, one from each parent. (There are a few exceptions, such as the social insects – ants, bees, and wasps. In this group, females, except the queen, possess only one set of chromosomes.) In most diploid organisms, *gametes* (eggs and sperm) normally contain one set of chromosomes; they are *haploid*. When a haploid egg and a haploid sperm are united in the process of fertilization, they form a diploid *zygote*, a fertilized egg that normally has two sets of chromosomes.

Different forms of a gene are termed *alleles*. If both alleles at a given locus are the same, an individual is said to be *homozygous* (*homo*, same; *zygous*, zygote) for the trait controlled by that allele. If two different alleles are present at a locus, the individual is *heterozygous* (*hetero*, different; *zygous*, zygote) for the trait in question. Normally, a diploid individual has two copies of each allele. An exception occurs with the X chromosome in mammals, where females normally have two X chromosomes (one from the male parent and one from the female parent) and males have one X chromosome (from their mother) and a shorter Y chromosome (from their father). Because the Y is shorter than the X chromosome, in males some alleles on the X chromosome do not have a corresponding allele on the Y chromosome. The X and Y chromosomes are termed *sex chromosomes* because they influence the sex of offspring in mammals. A similar mechanism with different sex chromosomes exists in birds.

Within a population, many alleles for a given locus may be present, but normally a diploid individual cannot have more than two of those alleles at that locus. An exception to this occurs in certain conditions in which an individual has an extra chromosome or part of a chromosome. Down Syndrome is an example.

If an individual is heterozygous at a given locus on a chromosome, one allele but not the other might be expressed. The one that is expressed is termed *dominant*, and the one that is not expressed is *recessive*. For a recessive allele to be expressed in a diploid situation, an individual must have two copies of that allele. In other words, that individual must be homozygous at that locus, unless the trait is on a part of the X chromosome that has no counterpart on the Y chromosome. A recessive allele on that part of the X chromosome will be expressed because it is not paired with an allele on the Y chromosome that could dominate it. The most common form of color-blindness in people is such a trait. In this *sex-linked* trait, a person who inherits a recessive allele for colorblindness on their maternal X chromosome will be colorblind.

A mutation (change in genetic material), can involve a change in the structure of a DNA molecule (a gene), a rearrangement of genes on a chromosome, or a change in the number of chromosomes or parts of chromosomes. The genetic variants introduced by mutations provide the raw material for natural selection.

Natural selection drives evolution through differential survival of individuals with favorable mutations, which occur at random. Natural selection does not create those mutations. (Note that the indirect manipulation of genetic material through artificial selection is fundamentally different from *genetic engineering*, also termed *genetic modification*) (Box 6.9).

In the late twentieth century, research on the regulation of gene expression proliferated, a field of study known as *epigenetics*. Research into the molecular basis of heredity revealed mechanisms that can affect the expression of a gene without changing its structure, by influencing how genes are turned on or off. In some cases, research has shown that these epigenetic changes are inherited and, furthermore, that they can be influenced by age, disease state, the environment, and lifestyle. These discoveries have prompted a great deal of research, as well as interest from the media and some remarkable, unsubstantiated claims. Although epigenetics has shed light on how gene expression is controlled, it does not contradict existing knowledge about the genetic basis of heredity (Powledge, 2011; Deichmann, 2016).

Box 6.9
Genetically Modified Organisms (GMOs)

In selective breeding, offspring with new combinations of existing genetic material result from crosses of parents with desirable traits. Genetic engineering, however, produces *genetically modified organisms* (GMOs) by directly altering the genetic material of an organism. Specifically, it alters the structure of DNA by introducing foreign DNA. For example, genetic material from *Bt*, a species of bacteria that produces a substance that is toxic to some insects, has been added to corn and a few other crops to produce genetically modified corn, a GMO. Although the genetic material for producing the *Bt* toxin occurs in nature, it does not naturally occur in corn or any other crops, so *Bt*-modified corn is a GMO.

In the case of *Bt*, the foreign genetic material that is inserted is from another species (the bacterium), but that is not the case in all GMOs. Some GMOs are constructed using DNA that is synthesized in a lab.

GMOs have many applications in agriculture, medicine, and other fields, but because genetic modification creates organisms that do not occur in nature it raises concerns about ethics, human health, and environmental impacts (FAO, 2001).

6.5.2.1 Causes of Low Genetic Diversity in Small Populations

The *genetic diversity* of a population is the amount of variation in the genetic material of the individuals in that population. Small populations or even populations that were small at some time in the past generally have low genetic diversity. They are likely to have fewer alleles than populations that remained large throughout their evolutionary history because of *genetic drift*, the chance loss of alleles that occurs from one generation to the next.

A finite number of alleles occurs in the zygotes that give rise to a new generation. Some alleles in the population's gene pool are in this sample that is passed on to progeny, and some are not. The gene pool is analogous to a jar containing a population of many beans of many colors. If a handful of beans (representing alleles) is removed and passed on to a hypothetical new generation, that sample will probably contain only a few of the colors present in the original population of beans.

With each successive generation, some alleles may be lost because of this random sampling, which is independent of any selective pressures. Once an allele is lost (that is, it is not passed on to the next generation) it is gone forever, unless it arises again by a mutation (a very unlikely event), or it is reintroduced in the population by an immigrant. The smaller a sample, the greater the likelihood that some alleles will not occur in the genetic material that is passed on to succeeding generations. Therefore, in small populations, genetic drift is likely to lead to a substantial loss of a population's original genetic diversity. This can happen when a new population is established by a small number of colonists.

Genetic drift can also occur when a population experiences a sudden and dramatic reduction in numbers (for example, from exploitation or from natural causes). When this happens, the population is said to have gone through a *genetic bottleneck*. The population that remains after a bottleneck is very likely to have lost some of the alleles that were present in the population before it declined. Even if the population rebounds to its former abundance later, the number of alleles in its gene pool will remain low.

It is hard to document a connection between a past population crash and low genetic diversity, however, because usually genetic data are not available for a population from the period before it declined. But biologist Diana Weber and her colleagues did just that for northern elephant seals (Weber et al., 2000). Genetic diversity is extremely low in this species (Bonnell and Selander, 1974). This suggests that the ancestors of elephant seals went through one or more genetic bottlenecks in the past, which seems likely because heavy exploitation in the nineteenth century caused their population to plummet to an estimated 20 to 100 individuals. In a before-and-after genetic study, Weber's team tested the hypothesis that the genetic diversity of elephant seals was higher before the population declined. They analyzed genetic material from bones and study skins in museums and blood samples from a live seal population in the 1990s. This allowed them to compare estimates of genetic diversity in elephant seals that lived from about 1,000 years ago to genetic diversity in the late twentieth century. As they had predicted, the results showed higher genetic diversity before the population declined.

6.5.2.2 Risks from Low Genetic Diversity

In a population with low genetic diversity, there are not a lot of genetic differences for natural selection to act upon. That reduces the potential for a population to adapt to a changing environment. For instance, if a population is exposed to a novel disease, those individuals with alleles that confer resistance to that disease will have a better chance of surviving and reproducing than individuals that do not have those alleles. In a population with little genetic diversity, the probability that any specific alleles, for instance the alleles for resistance to the disease in question, will be present is low.

In small populations, *inbreeding*, the chance that individuals will mate with close relatives, increases. Closely related parents have one or more ancestors in common, and therefore they have a high probability of sharing some alleles. Often those alleles will result in harmful traits.

Many diseases and other harmful conditions are caused by recessive alleles. That means those traits will only be expressed in an individual if that individual receives a copy of the harmful allele from each parent (unless the trait is sex-linked). In populations that are large and genetically diverse, it is rare for an individual to have two parents with the same harmful mutation. But in inbred populations, an individual's parents, which often are closely related, may carry the same alleles for harmful conditions. If an individual inherits one copy of a harmful allele from each parent, the harmful trait will be expressed. For this reason, the offspring of closely related parents often inherit harmful traits. This happens in small populations in the wild (Box 6.10), as well as zoo populations and some human populations (such as some royal families and some small, isolated communities).

Box 6.10
Is There Evidence That Populations with Low Genetic Diversity Are at Risk of Becoming Extinct?

Evidence suggesting that low genetic diversity may not enhance extinction risk: By the early 1990s, the northern elephant seal population exceeded 120,000 animals and was still growing exponentially, despite its low genetic diversity (Stewart et al., 1994). Although this result surprised researchers, similar situations have been reported in species as varied as birds, insects, and other mammals. The answer to this paradox lies in the connections between genetic variation, evolutionary potential, and environmental change. For instance, if individuals in a population with low genetic diversity do not have alleles for resistance to a particular disease, that will not be a problem as long as that disease is not present in their environment. But they will be vulnerable if, at some future time, the disease emerges within the area occupied by that species.

Evidence suggesting that low genetic diversity enhances extinction risk: A group of Finnish biologists tested the hypothesis that inbreeding affects a population's chances of becoming locally extinct. A network of small, isolated populations of the Glanville fritillary butterfly on the Åland Islands in southwestern Finland provided an excellent setting in which to study this question in a large group of wild populations (Figure 6.2). The team monitored the persistence of hundreds of Glanville fritillary populations. Because their sample size was so large, the investigators were able to study a range of population sizes. (Some populations consisted of only a single group of larvae produced by a single pair of butterflies.) They also measured the heterozygosity of females from 42 of those populations.

During the four years of the study, they observed an average of 200 local extinctions per year. Seven of those extinctions were from the 42 populations that were sampled genetically. This was a large enough sample for them to statistically test the relationship of inbreeding (as measured by heterozygosity) to extinction.

They found that extinction risk was significantly higher in the inbred populations. What's more, they were able to discover the specific population processes that were associated with extinction. In populations that died out, larval survival was lower, eggs hatched at lower rates, and female life spans were shorter than in populations that persisted (Frankham and Ralls, 1998; Saccheri et al., 1998).

There is no single answer to the question: how much genetic diversity is enough to provide evolutionary security. In each case, interactions between environmental changes (including human-caused changes) and a population's genetic diversity influence its ability to adapt and persist (Reed, 2010).

Inbreeding in the wild does not always result in genetic problems. In fact, mating between individuals that are too dissimilar genetically can have disadvantages. This is because the members of local populations usually have adaptations to local conditions. Hence, they possess alleles for traits that are advantageous in their particular environment. If individuals from a locally adapted population mate with individuals from a different population, those favorable alleles might be diluted.

Because of the complex ways that genetic diversity influences population risks, conservation biologists carefully consider potential genetic consequences when they breed captive animals or transplant individuals from small or declining populations. Otherwise, they might cause problems from inbreeding on the one hand or compromise local genetic adaptation on the other. Sections 7.6.1 and 7.6.2 describe some examples of these challenges and ways that they can be addressed.

6.5.3 Risks from Chance Fluctuations

Small populations are vulnerable to chance fluctuations in birth and death rates and in sex ratio. In addition, they are vulnerable to environmental fluctuations and to catastrophes that can drive their survival and reproduction down to dangerously low levels (Box 6.11). They will be especially vulnerable to such chance events if they have ultimate risk factors such as low reproductive rates or if they are in isolated or fragmented habitat.

Box 6.11

Vulnerability to Environmental Change and Skewed Sex Ratio in a Remnant Population of the Tuatara

One population of the *tuatara* (from a word meaning spiny back in the language of the Indigenous Maori), a large, lizard-like reptile of New Zealand, faces such a constellation of risk factors (Figure 6.5). Tuatara conservation is important because this species is the only living representative of a once diverse group of reptiles dating back to the time of the dinosaurs.

Figure 6.5   Tuatara. Credit: ivan-96 / DigitalVision Vectors / Getty Images.

If the tuatara becomes extinct, that whole group will be gone forever. Its loss would therefore mean a more substantial loss of genetic diversity than the loss of a species with many close living relatives.

Although tuataras once lived throughout New Zealand, the species is now extinct except on some offshore islands (Figure 6.2). The smallest of the island populations occurs on North Brother Island, in a 4-ha reserve that supports about 350 tuataras. A combination of ultimate and proximate risk factors results in a high risk of extinction for this population. Already vulnerable because of their low biotic potential due to late onset of sexual maturity (at 15 to 20 years old) and intervals of several years between breeding events, the North Brother Island population faces risks from its small population size and climate change, which may have caused the population to have a skewed sex ratio. As in some other reptiles, including turtles, the sex of tuatara embryos is determined by the temperature of the nest during incubation. In a population that is rare to begin with, a skewed sex ratio will mean that there will be few individuals of the less abundant sex.

Using data from long-term mark-and-recapture studies combined with estimates of changes in air temperature and models of population trends, Grayson et al. (2014) concluded that the proportion of females in the North Brother Island tuatara population has declined. On the basis of their data, they predicted that this population will eventually become entirely composed of males, which of course will result in its extinction.

6.5.4 Risks in Isolated or Fragmented Habitats

Many of the risks that small populations face are exacerbated on islands and in habitat patches. In both these situations, populations are small and isolated. To make matters worse, if a population does go extinct on an island or in an isolated habitat patch, it may be difficult for colonists to get there to establish a new population.

In the 1960s, the biologists E. O. Wilson and Robert MacArthur developed a model of local extinction and colonization rates on islands (MacArthur and Wilson, 1967; MacArthur, 1972). The MacArthur–Wilson model predicts that if we exclude the development of new species (which is rare), the number of species on an island is the number of species that colonized the island by arriving and establishing successful populations minus the number of species that disappeared from it by going extinct.

Wilson and MacArthur suggested that two geographic features influence the number of species on an island: its size and its isolation (the distance from a source of colonists). MacArthur and Wilson's model predicts that extinction will occur more often on small islands than on large islands (because small islands support smaller populations than large islands and have fewer habitats), and islands near a source of colonists will be colonized more often than islands that are far from sources of colonists.

According to the theory of island biogeography based on the MacArthur–Wilson model, the number of species on an oceanic island will eventually reach an equilibrium when colonization and extinction rates balance each other out. The model predicts that when an island is at equilibrium, extinction and colonization will continue to occur, but the number of species will remain constant because the extinction rate and the colonization rate will be equal. At equilibrium: (1) small, isolated islands will have relatively few species because colonization will be rare and extinction will be common, and (2) large islands located near sources of colonists will have relatively high numbers of species because colonization will be common and extinction will be rare. The change in species composition that occurs on an island as some species disappear and others arrive is referred to as *turnover*.

Of course, rates of extinction and colonization are different for different groups of organisms. Bats, birds, some insects, and some plants (for instance those with seeds or fruits such as coconuts, which can tolerate long periods of immersion in salt water) cross oceans and colonize islands relatively often. Large mammals do not. We have already noted that this has important consequences in terms of the vulnerability of island biota to introduced species.

Some evidence relevant to the MacArthur–Wilson model is summarized in Box 6.12.

Not long after the model was proposed, ecologists realized that MacArthur and Wilson's theory about the processes of extinction and colonization might apply to populations in patches of habitat on a mainland as well as to islands. Researchers studied the dynamics of naturally occurring discontinuous habitats – such as caves, lakes, mountaintops, clumps of plants, and so on – and applied the insights gained from those studies to habitat fragments modified by human activities.

Many habitats today are discrete patches surrounded by metaphorical oceans in an inhospitable landscape matrix of lands that have been deforested, cultivated, or converted

Box 6.12

Some Sources of Evidence about Rates of Colonization and Extinction on Islands

A report published in 1917 listed species of birds known to breed at that time on the Channel Islands off the coast of southern California (Figure 6.1). Half a century later, the ecologist Jared Diamond surveyed the same islands and compared his results to the earlier census (Diamond, 1969). MacArthur and Wilson's model predicts that repeated colonizations and extinctions would have occurred on the islands during the period between the two censuses, and that extinction and colonization would have balanced each other out, so that the total number of species on each island would not change. As predicted, Diamond did not find some of the species that had been observed earlier, and he found others that had not been reported previously. Furthermore, for most islands the number of reported species did not change very much. He concluded that extinction and colonization were in equilibrium for the bird fauna of the Channel Islands.

Diamond inferred extinction and colonization from repeated censuses of the birds, but he did not observe these phenomena directly (which would have been very difficult). There are other possible explanations for the discrepancies between successive censuses, however. The observers might have missed birds that were present during the first census. Species that were overlooked initially but found later would be considered colonists, although they had been there all along. Similarly, species that were found originally but missed later would be scored as extinctions although they might still be there. These problems could make it appear that colonization and extinction had occurred more often than they had.

To address this sort of problem, Wilson and the ecologist Daniel Simberloff designed an experiment that allowed them to test the model directly. The experiment allowed them to look for experimental evidence of the sort of local colonization, extinction, and species turnover the model predicts. At the start of their experiment, they censused the terrestrial *arthropods* (invertebrates with jointed, external skeletons and segmented bodies, such as insects, spiders, and centipedes), on six small islands of mangroves in Florida Bay (Figure 6.1). Each island consisted of one to several mangrove trees of a single species. Simberloff and Wilson killed all the islands' invertebrates and other land animals by wrapping the islands in plastic sheeting and spraying them with insecticide. If this procedure was thorough, any species that was found on the islands in censuses conducted during the three years after the treatment must have arrived there as colonists, and any species that disappeared after it reached the island must have gone extinct.

Scores of invertebrate species were detected during the censuses. Although Simberloff and Wilson used conservative standards for concluding that extinctions had occurred, they estimated that there had been 12 local extinctions (about 1.5 extinctions per island per year). Thus, they concluded that there had been turnover in species abundance (Simberloff, 1976).

in other ways. The equilibrium theory of island biogeography predicts that small, isolated habitat patches will have more extinctions and fewer episodes of colonization than large patches that are not isolated. Of course, the number of species that live in a patch also depends upon the quality and the variety of habitats that it contains. In addition, several other factors – such as the potential colonists' capacity for dispersal, their ability to tolerate environmental conditions in the matrix, and their sensitivity to the presence of people and vehicles – influence colonization. Because of differences in those factors, different groups of organisms vary in their tendency to cross unfavorable habitat. Even a two-lane road presents an impenetrable barrier for some organisms, whereas others can easily fly over thousands of kilometers to reach suitable habitat.

If organisms cannot or will not move through the habitat surrounding a patch, then colonization cannot occur, but extinction might continue. Data on the distributions of small mammals at high elevations in the mountains of the southwestern USA and the *Great Basin* (the area of the USA between the Rocky Mountains and the Sierra Nevada, Figure 6.1) suggest that that is what happened there. In that region, the number of species of small mammals in high-elevation habitat patches depends on the size of the mountaintop. Small patches have fewer species of small mammals than large ones, which is consistent with the hypothesis that in the past, the small patches had more extinctions than the large patches (like true islands). Because animals and plants that live in high-elevation habitats are adapted to a cool climate we would expect that under current climate conditions species adapted to that environment might not be able to move very far through the warm, lower-elevation habitat between the mountaintops. For mice, squirrels, and other small mammals, the distance between patches of suitable habitat on different mountaintops might be insurmountable, although those distances might not be a problem for large mammals such as deer and elk or for migratory birds.

Thus, unlike the islands described by MacArthur and Wilson, mountaintop colonization by small mammals is not related to the distance from a source of colonists. When the mammologist James Brown applied the MacArthur–Wilson model to habitat patches, he found that under existing climatic conditions colonization is apparently impossible for small mammals on mountaintops, and therefore extinction and colonization are not in equilibrium in that setting (Brown, 1971; Patterson, 1984).

On the other hand, studies of migratory bird populations on mountaintops in the same region do not show a strong relationship between area and species diversity. The low-elevation habitat is not a barrier to them. In other words, for species that can move readily between mountaintops, extinction and colonization rate are not related to patch size or distance between patches (Johnson, 1975; Brown, 1978).

Concern about an imbalance between colonization and extinction led ecologists to search for landscape features that promote colonization and minimize extinction. This topic is discussed in more detail in Section 8.3, where we shall see how island biogeographic theory has been applied to the design of nature reserves in efforts to maximize the effectiveness of protected areas.

6.6 Implications: Connections between Evolution and Conservation

Ultimately, the conservation of life forms means conserving genetic material. Since genes exert their effects through their responses to specific environments, conservation requires knowledge of the ways in which natural selection favors genetic material that is adaptive in specific environments and the significance of genetic diversity, or the lack of it, for the long-term persistence of populations. Regardless of whether the goal is to maintain abundant species of wild plants and animals, to rescue species that are in trouble, or to minimize the evolution of resistance to pesticides, it is important to know why species are plentiful or rare and what their future prospects look like.

We will now turn our attention to ways in which these concepts are applied to the protection and restoration of populations and species (Chapter 7) and ecosystems (Chapter 8).

7

Strategies

Protecting and Restoring Species

7.1 Overview of Options: Strategies for Preventing Extinctions

In this chapter, we will examine measures to protect and restore animal and plant populations at risk of becoming extinct. To do this, conservationists need to discover and address the causes of specific declines. That may involve efforts to reduce the causes of rarity and to boost small populations directly.

To begin with those endeavors, it is necessary to find out what species are at risk and why.

7.2 Keeping Track of Species in Trouble

The first step in doing something about imminent extinctions is getting species-specific data on the scope of the problem. What species are at risk of becoming extinct? What species are in trouble and where?

Shortly after the end of World War II, delegates from several governments and conservation organizations met in France to discuss strengthening nature conservation through international cooperation. The group set up a new organization that became the International Union for Conservation of Nature (IUCN). In 1964, the IUCN began publishing a database of the conservation status of many species, along with reports detailing population trends and conservation measures. The resulting documents highlight the plight of declining species (Box 7.1) as well as some that are stable or increasing but need to be managed with awareness of potential future threats.

Today, many other conservation organizations, as well as nations, provinces, states, and districts, catalog biodiversity in their jurisdictions. These inventories provide scientific information to guide policies aimed at halting species declines. Frequently these lists are mandated by treaties or laws.

Box 7.1

Sturgeon: Ancient Fish at Risk

Sturgeon are large-bodied, scaleless fishes that inhabit some rivers in Asia, Europe, and North America. They are a *primitive species or group* (meaning that they appear early in the fossil record). Unlike other fishes, sturgeon are partially covered with rows of bony plates. Some species can live up to 100 years and reach several meters in length. They typically don't begin breeding for several decades and don't spawn every year. Most species of sturgeon are *anadromous*, spending most of their life at sea and migrating up rivers to spawn.

Sturgeon have been harvested for centuries (Figure 7.1) for their meat and caviar (eggs), which is prized as a luxury food. (Caviar of one species sold for 8,000 US dollars per kilo in 2009.) Commercial and sport fishing of some species of sturgeon is permitted in some jurisdictions, and they also form the basis of *wildlife-based tourism* in places where it is possible to see them spawning. Sturgeon are culturally important for some Indigenous peoples (Beck, 1995).

The IUCN recognizes 28 species of sturgeon, nearly a third of which are *critically endangered* (facing an extremely high risk of extinction in the wild). Both ultimate and proximate factors contribute to their vulnerability. Because they take a long time to reach sexual maturity and they breed infrequently, they are slow to recover from population declines. During their long lives, they can accumulate pesticides in their tissues. The species that are anadromous must have access to river habitat in which to breed as well as lakes or seas where juveniles mature. These ultimate risk factors of sturgeon biology set the stage for declines in the face of proximate risks. Dams that prevent them from migrating between their spawning and feeding grounds wipe out local populations (although individuals may survive for many years after their final trip upstream).

Figure 7.1   Sturgeon-skinning workshop. Nineteenth century engraving. Credit: clu / Digital Vision Vectors / Getty Images.

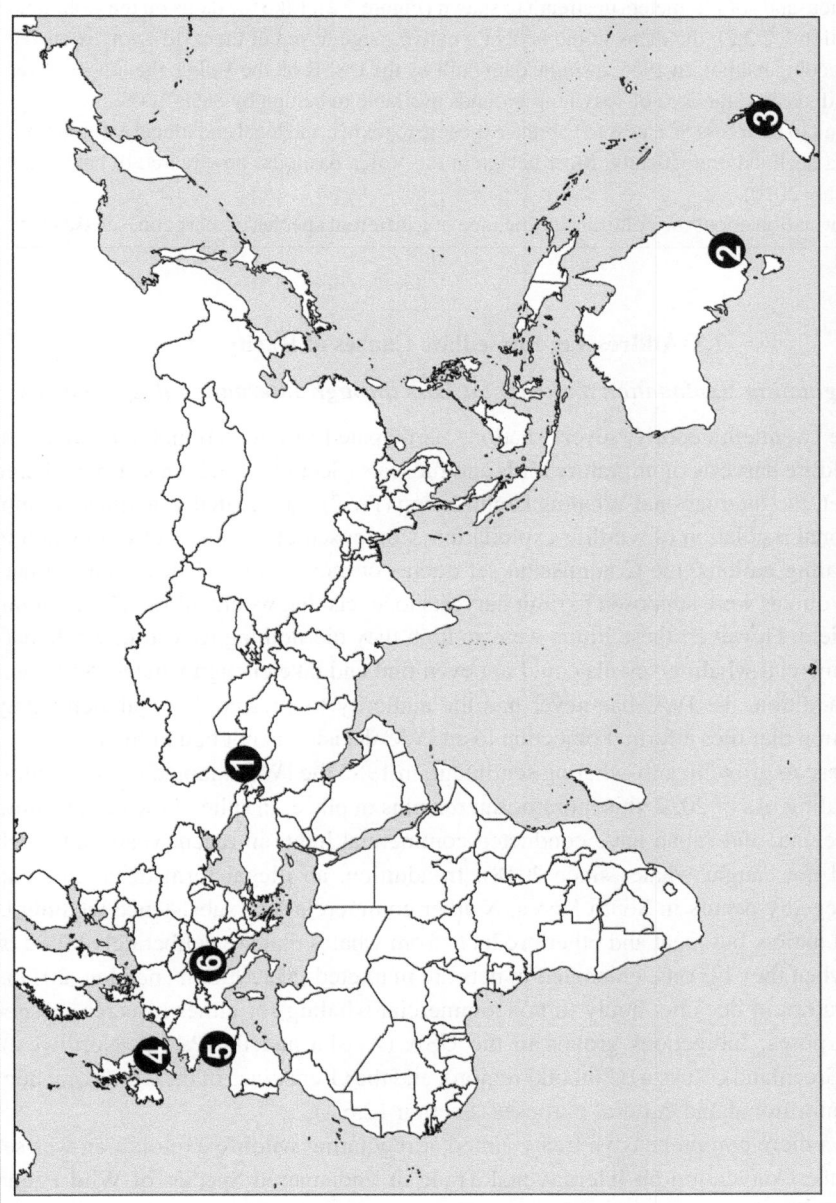

Figure 7.2 Locations of: 1, center of distribution of beluga sturgeon; 2, range of brush-tailed rock-wallaby, Australia; 3, range of saddleback, New Zealand; 4, red kite release sites, Great Britain; 5, remnant population of brown bears in the Pyrenees; 6, source of brown bears in Slovenia (for release in the Pyrenees). Map created by Eva Strand using Esri, DeLorme World Countries Generalized Data & Maps for ArcGIS 2013, with permission.

One species of sturgeon with a particularly poor prognosis is the beluga or European sturgeon of Central Europe. Currently wild native populations of beluga occur in only two rivers, although historically they inhabited many rivers that drained into the Black and Caspian seas, and they migrated thousands of kilometers upstream to spawn (Figure 7.2). Like the dams on the Columbia River (Section 5.2.2.2), the dams in the beluga's native range have cut them off from extensive areas of breeding habitat. In 1955, a single dam built by the USSR on the Volga, the longest river in Europe, decreased the area of spawning grounds available to beluga by 88%–100%.

As a result of their loss of access to habitat, exposure to toxins, and legal and illegal exploitation, wild beluga declined dramatically. Most beluga in the Volga basin are now reared in hatcheries (Gesner et al., 2010).

For information about the cultural significance of a different species of sturgeon, see Box 9.5.

7.3 Addressing Immediate Causes of Rarity

7.3.1 Regulating Exploitation and Habitat Loss through International Agreements

Early in the twentieth century, several nations participated in international agreements to regulate wildlife harvests of migratory birds and fur seals (Sections 1.3.2.4 and 1.5.3). Three decades later, the International Whaling Commission (IWC) was created as another attempt at international regulation of wildlife exploitation. Composed of members of both whaling and nonwhaling nations, the Commission set quotas on how many whales could be harvested. The quotas were supposed to limit harvests to levels that would allow for maximum sustained yield. However, these limits were so high they did nothing to restrict whale harvests. Commercial whaling vessels could not even find and take enough whales to fill their quotas. In addition, the IWC has never had the authority to enforce its regulations. Any member nation that files a formal objection to an IWC decision is not bound by it.

In response to growing anti-whaling sentiment, in 1982 the IWC adopted a ban on commercial whaling. As of 2022, this moratorium remains in place. In spite of the moratorium, Norway, Iceland, and Japan have conducted commercial hunts in recent years (although Iceland had not caught whales since 2018). In addition, an illegal form of commercial whaling allegedly occurs in South Korea. Neither commercial nor subsistence whaling is legal in that nation, but meat and other products from whales that are deliberately killed or left to die when they become entangled in nets are marketed (MacMillan and Han, 2011).

The moratorium does not apply to noncommercial whaling for either research or subsistence purposes. Indigenous groups in the USA (Alaska and the Pacific Northwest), Denmark (Greenland), Russia (Chukotki region), and the Grenadines in the Caribbean hunt whales for nutritional and cultural purposes (Section 12.6.1).

In 1973, a more comprehensive treaty aimed at regulating wildlife exploitation was set in motion: the Convention on International Trade in Endangered Species of Wild Fauna and Flora (CITES) (pronounced SEE tees). Representatives of 80 nations agreed on the text of this treaty, which was aimed at preventing species of wild plants and animals from becoming extinct or endangered because of international trade. As of 2021, 184 parties (183 nations and the European Union) had ratified this agreement.

CITES regulates trade in wild plant and animal species on the basis of their vulnerability to commercial exploitation. Enforcement of the CITES regulations is politically complex, often requiring international cooperation. It is also expensive, dangerous, and logistically complicated. In most cases not enough personnel are funded to be able to make a dent in the problem. Because poaching for the international market offers high profits, fines and jail sentences are seldom sufficient to deter offenders, whereas the incentives for corruption or non-compliance are high (Box 7.2)

Box 7.2
Illegal Cactus Trade

The cactus family is a large group of plants with high ecological and cultural importance. Cacti range in size from a few cm to trees that reach heights of 20 m. They occur in many environments, from desert to moist tropical forest and from sea level to mountain tops. The saguaro, a familiar icon of desert landscapes (Figure 7.3) is considered a *keystone species* in some desert ecosystems (Section 7.7).

Cacti typically have adaptations that allow them to survive periods of low water and high temperature. Hundreds of species of mammals, birds, reptiles, and insects interact with cacti. Their fleshy stems allow them to store water and nutrients, which provide critical resources for many animals and people. For centuries, people have utilized cacti for food and drinks, dyes, medicines, psychoactive drugs, fodder, and even building materials. Today cacti are also valued for landscaping, as house plants, and for scientific research.

Figure 7.3   Many birds use saguaro cacti, including this crested caracara. Credit: Walter Niederbauer / 500Px Plus / Getty Images.

Nearly one third of the world's approximately 1,500 cactus species are at risk of extinction. Many of these occur at low population densities and have low reproductive rates (Santos-Díaz et al., 2011). The main threats cacti face are habitat destruction and illegal harvesting.

Over 500 cactus species occur in Mexico, and about three fourths of those are endemic (Taylor, 1997). These endemics are, of course, at high risk of extinction because of their restricted distributions. In 1991, Mexico signed on to CITES. Trade in some cactus species is legal but regulated in Mexico. This trade requires collection permits and export permits from the Mexican government. Compliance is low, and enforcement is complicated because of confusion about which species can be traded legally.

Although cacti are not usually considered charismatic species, they have passionate devotees (*see Q7.1*). An article in the American magazine *The Atlantic* about illegal cactus trade in Big Bend National Park near the border between the USA and Mexico (Figure 7.4) highlighted this situation:

Some cacti thieves will pull a plant from the wild because, like with tiger bones or rhino horns, someone will pay a lot of money for it. But cacti are also besieged by those who profess to love the spiny plants – the private hobbyists and horticulturalists who collect and show cacti like paintings or purebred dogs….

It's tough to drum up support for cacti conservation. Rhinos and tigers and elephants have sad, relatable eyes to guilt people – and their money – into fighting smugglers, but few people shed a tear at the impending doom of *mammillaria herrerae*, [the "golf ball" cactus] which in the past 20 years has had 95 percent of its population poached from the wild.

(Phippen, 2016)

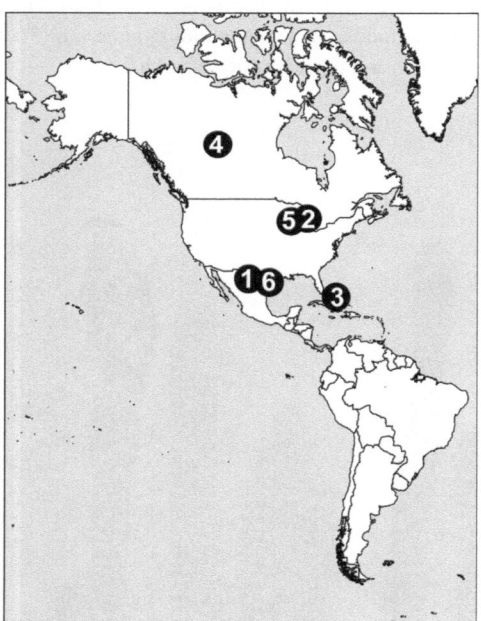

Figure 7.4   Locations of: 1, Big Bend National Park, Texas; 2, Pitcher's thistle distribution and Kirtland's warbler breeding range; 3, Kirtland's warbler winter range; 4, whooping crane breeding range #1; 5, whooping crane breeding range, #2; 6, whooping crane winter range. Map created by Eva Strand using Esri, DeLorme World Countries Generalized Data & Maps for ArcGIS 2013, with permission.

At around the time CITES was established, a different international treaty addressed other aspects of conservation. The Ramsar Convention on Wetlands of International Importance is a treaty aimed at slowing the destruction of wetlands, especially those of importance to migratory waterfowl. (*Ramsar* is not an acronym; it is the name of the city in Iran where this convention was negotiated.) Member nations are required to designate at least one wetland of international significance that will be managed to maintain its ecological characteristics. Over 80 nations have signed this treaty.

By the 1980s, three threads were prominent in discussions about conservation: (1) the diversity of the Earth's life forms is a resource of great value to present and future generations, (2) conservation of that resource depends on its sustainable and fair use, and (3) benefits derived from the use of genetic resources should be shared equitably. Work on these issues by the United Nations culminated in the Convention on Biological Diversity (CBD), a treaty that was opened for signature in 1992. By 2021, all member states of the United Nations except the USA had become parties to the CBD.

7.3.2 Protecting Species and Habitats through Legislation

By the middle of the twentieth century, many countries had adopted legislation prohibiting the harvest of species in trouble. In the United States, the Endangered Species Act (ESA) of 1973 was the first comprehensive federal legislation that explicitly sought to address the needs of species facing extinction. Section 7 of the ESA requires federal agencies to consult with the Secretary of the Interior to ensure that their actions will not jeopardize threatened or endangered species (or subspecies) or their habitats. This provision led to a controversy over a proposed dam on the Little Tennessee River. The discovery of a population of a fish known as the snail darters in water near the dam site triggered Section 7. Because it was likely that construction of the dam would jeopardize the only known population of this fish and cause its extinction, work on the dam was halted. The media painted a picture of an obscure, trivial fish standing in the way of economic development. This contributed to a public perception of the incident as a case of endangered species legislation granting political victories to environmental extremists on behalf of an insignificant species.

Congress eventually avoided the issue by passing legislation that exempted the dam from the ESA, potentially giving the go ahead for wiping out a species in its native habitat. However, populations of the snail darter were transplanted elsewhere, and other populations were eventually discovered, so the dam did not wipe out the snail darter after all. But this controversy set the stage for future confrontations that would be viewed as battles between species protection and economic gains.

Under the ESA, subspecies and "distinct population segments" qualify for protection. For instance, the grizzly bear or grizzly, a subspecies of the brown bear, is abundant in Alaska. In the contiguous USA, however, grizzlies occur only in small, isolated remnant populations at the southern edge of their geographic range. Some of those grizzly populations are listed as threatened under the ESA, although the species as a whole is not endangered.

The agreements and laws discussed so far reflect a trend toward shifting priorities. The Migratory Bird Treaty, the Fur Seal Treaty, and the International Whaling Commission came about in response to intensive exploitation of wildlife for commercial gain, but it was not until the last half of the twentieth century that nations began to regulate the protection of inconspicuous and little-known species. The ethical underpinning of this development was the idea that all species have value.

However, since we cannot help all species, this commitment to the value of all forms of life did not resolve the question of which species are most deserving of our assistance, a point that we shall return to in Section 7.7.

7.3.3 Manipulating Interactions among Antagonistic Species

In addition to addressing human-caused mortality, it is sometimes possible to boost wild populations by manipulating natural causes of mortality, such as competitors, parasites, herbivores, or predators. Boxes 7.3 and 7.4 illustrate cases where natural enemies of rare species were reduced either by being excluded or removed.

Box 7.3

Protecting Young Pitcher's Thistle Plants from Insect Herbivory

Scientists at the University of Nebraska tested the hypothesis that protecting juvenile Pitcher's thistle plants from herbivorous insects could increase population growth in this threatened species. (Despite its slightly confusing name, Pitcher's thistle is not a carnivorous pitcher plant. It is a true thistle, a member of the sunflower family.) Pitcher's thistle occurs only in sandy habitats along the shorelines of the western Great Lakes in the USA (Figure 7.4), where it colonizes open habitats. It is at risk of extinction because of its narrow geographic range and specific habitat requirements.

For several years, a juvenile Pitcher's thistle grows in the form of a cluster of basal leaves. While in this stage, the juvenile plants are exposed to herbivorous insects. At the end of the juvenile phase, each plant sends up a stem that flowers and sets seed once.

The researchers protected young Pitcher's thistle plants from insect herbivory in experimental plots by covering the plots with screen cages that excluded insects. They also controlled insects by spraying the treatment plots with a mixture of water and insecticide. Plants in control plots were sprayed with water only and covered with cages that were open on one side to allow insects to enter. The investigators subsequently recorded insect damage to Pitcher's plants in the experimental and the control plots and estimated the number of viable seeds per flowering head. The experiments were conducted at two dune sites along the shore of Lake Michigan. Insect damage was much higher at one of the sites than the other.

At the site where there was a lot of herbivory, the juvenile mortality of plants that were protected with cages and insecticide was significantly lower than mortality in the control plants. In addition, at that site the estimated production of viable seeds was 338 seeds per plant in plots that had been treated with the exclusion cages and insecticide compared to only 31 seeds per plant in the control plots. Thus, at the site where herbivory was high, protection from herbivores boosted the reproductive success of Pitcher's thistle. The researchers concluded that in some situations excluding natural enemies was a useful tool for assisting rare plant populations (Bevill et al., 1999) (*see Q7.2*).

Box 7.4

Evaluating Evidence: Did Cowbird Control Enhance Kirtland's Warbler Survival?

Kirtland's warbler nests in Michigan and winters in the Bahama Islands off the coast of Florida (Figure 7.4). For breeding, this migrant requires stands of jack pine trees that are 1.5 to 6.0 m tall and 7 to 20 years old. Since jack pines can reproduce only after fire, these warblers can nest only on sites that were burned within a very narrow time frame. In part because of their highly specific habitat requirements, Kirtland's warbler populations became small and localized. The species was listed as endangered in 1967 (Walkinshaw, 1983).

Even in areas where suitable jack pine stands exist, they are transient because they depend on disturbance. As the trees age, they outgrow their usefulness to Kirtland's warblers. Historically, in a landscape where fires were frequent there was a dynamic landscape mosaic, with jack pine stands of different ages. Over a large area, the trees would eventually grow too large to provide breeding habitat for the warblers, but very young stands would also grow and eventually reach the appropriate size. But development and fire suppression created a situation where managers would have to mimic that disturbance regime if the habitat requirements of the warblers were to be met. To address this need, federal and state agency personnel began setting fire to jack pines so that stands of the right age would be available continuously. (Managers subsequently switched from burning to clear-cutting and replanting.)

The breeding warblers face other problems too. One of these is brood parasitism from the brown-headed cowbird, a species of open habitats. Formerly found only in the eastern states, this bird of open country expanded its range westward as forests in the Midwest were cleared. By the middle of the twentieth century, more than 50% of Kirtland's warbler nests contained cowbird eggs.

In 1972, the US Fish and Wildlife Service began removing cowbirds from six Michigan counties within the breeding range of the Kirtland's warbler. Between 1972 and 1981, over 33,000 cowbirds were trapped out of the area. The results of this program, in terms of the rate of brood parasitism and fledging success were striking. Cowbird parasitism declined markedly, and there was a concomitant increase in the fledging success, from fewer than one fledgling per nest between 1931 and 1972 to an average of 2.8 young per nest during the period when cowbirds were controlled.

Yet despite these encouraging results, the Kirtland's warbler population did not increase much. In 1971, the last year before cowbird removal was initiated, 201 singing males were counted during the breeding census. During the decade when cowbirds were removed, an average of 207 singing males were counted in the same area, just slightly higher than the pre-removal number. Thus, if we assume that the number of breeding males is a reliable indication of size of the warbler breeding populations, it appears that cowbird removal did not make much difference (Kelly and DeCapita, 1982). However, we do not know whether the population would have declined if cowbirds had not been removed. From a scientific standpoint, it would have been useful to have a control area in which cowbirds remained. But when trying to rescue such a rare species, that would not have been ethical. Perhaps other factors that were not addressed during this period (for example, mortality during migration or on the warblers' wintering grounds) limited the Kirtland's warbler's breeding population.

This example illustrates the importance of addressing all factors that contribute to a population's decline, which is especially challenging when dealing with migrant species that have widely separated seasonal ranges. It is also an example of the difference between the specific objectives of a management program and its overall goal. If managers had looked only at their objectives, they would have seen that a lot of cowbirds were killed, and they might have judged the operation a success. But that action was aimed at achieving a larger goal, boosting the warbler population, which was not achieved.

Probably as a result of habitat management, the Kirtland's warbler population recovered enough for it to be removed from the endangered species list in 2019. Because of its specific habitat requirement, however, it will continue to depend on active conservation measures.

Brood parasitism is an adaptation that works for cowbirds because the species they parasitize rear the young that hatch from cowbird eggs. This is surprising because to us the eggs and chicks of cowbirds do not look at all like warbler eggs and chicks, at least to our eyes (Figure 5.2). We will see below that scientists have devised ways of taking advantage of this trait to devise strategies for boosting the reproduction of some rare species.

7.4 Managing Intrinsic Limits on Population Growth

Sometimes it is possible to intervene in ways that bypass limitations on a population's productivity through programs that boost birth rate or survival. Such measures are stopgaps intended to temporarily increase the size of a rare population until it is large enough to persist without assistance.

7.4.1 Improving the Odds for Little Things

The survival of young individuals in the wild is low in most populations. Conservation biologist Malcolm Hunter put it succinctly when he wrote that "one of the fundamental laws of nature is that little things tend to die quickly" (Hunter, 1996b:315). Young plants and animals get eaten by predators; they lack adaptations for obtaining the resources they need, and they are not good at defending themselves or dealing with environmental fluctuations. They dry out, starve, or die from getting too hot, too cold, too wet, or too dry.

7.4.1.1 Headstarting

Anything that increases juvenile survival might increase population growth. Furthermore, since some species provide no care at all for their young and those that do are generally capable of producing far more young than they can care for, it stands to reason that we should be able to increase reproductive output by taking seeds, eggs, or young from the wild, rearing them, and subsequently returning them to their natural habitat. This approach is termed *headstarting*, by analogy with early childhood education that strives to boost children's chances of positive outcomes later in life. The long-standing horticultural practice of growing plants from seeds to a stage where they can be transplanted outdoors is a variation on this theme that is used in assisting reproduction in some rare plants.

In several parts of Latin America and South and Southeast Asia, the eggs of marine turtles are removed from their nests and taken to nearby hatcheries where they are protected from predators and poachers until they hatch and the young are released. Where turtle eggs that are taken from the wild for headstarting would die from other causes, such as predation or poaching, if they were left in the nest (in other words, if nestling mortality is compensatory), then headstarting them might increase hatchling survival and boost turtle reproductive output. This, of course, assumes that the young turtles also live long enough to reproduce.

7.4.1.2 Saving Nestlings from Sibling Aggression

In some species of wild birds, it is possible to identify eggs that are unlikely to result in fledged young. This may result from species-specific characteristics of nestling behavior or of parental care. Because of an unusual feature in the development of raptors, it is often possible to predict which nestlings will not survive. Hawks, owls, falcons, and eagles usually begin incubating after the first egg in a clutch is laid. As a result, the egg that is laid first also hatches first, and the juvenile from the first egg starts growing before any of its siblings hatch. Sometimes the oldest, largest nestling subsequently attacks its younger siblings. Because the younger, smaller nestlings get less food and are also the object of aggression they usually die. This behavior is known as *fratricide* (brother killing), or *cainism*, after the biblical character Cain who killed his brother Abel.

Because birds of prey have predatory instincts that lead them to attack smaller animals, it is not hard to see how cainism in birds of prey might come about. But why wouldn't it be eliminated by natural selection? Could there be any evolutionary advantage to a behavior in which a nestling kills its own kin? The answer to this puzzle involves (1) the size differential among nestlings and (2) potential limitations on survival due to variable food supply. In the absence of fratricide, all the chicks might starve in lean years. But if the first chick to hatch ("Cain") is larger and stronger than its siblings and attacks them (thereby reducing competition for food), the chance that at least that one chick will obtain enough food to survive improves. In other words, where food supply is a limiting factor in some years, fratricide could be adaptive in lean years because it increases the probability that the oldest nestling will survive and have a chance to pass on its genes.

In some eagle species, the younger chick (Abel) can be rescued, raised in captivity or in the nest of another species, and returned to the nest later, when Cain is no longer aggressive. This approach has been used with eagles in parts of Europe (Cade and Temple, 1995).

As with all interventions, managers should weigh the risks and benefits of this strategy carefully. It should be used only where there is clear evidence that food is limiting reproductive success. Chicks should be taken from the nest only if there is no possibility that they will survive in the wild, and care should be taken to minimize the chance that the nest will be abandoned as a result of disturbance from removing and later replacing Abel.

7.4.2 Replacement Clutching

Many birds produce a second clutch if the first one disappears. This adaptation increases the probability that reproduction in the wild will occur even after a clutch is lost. When dealing with species that produce more than one clutch, biologists can remove the first clutch, and sometimes even a second one, and rear it artificially, hoping that the parents will produce and rear another. This technique, known as *replacement clutching* or *multiple clutching*, has been used to increase the reproductive output of the California condor (Box 7.5).

Box 7.5

Replacement Clutching to Increase California Condor Numbers

Condors are large vultures of the western hemisphere. There are two living species, the California condor of the USA and the slightly larger Andean condor of South America. Because of their massive size (they have wingspans of several meters), condors are considered charismatic megafauna, although because of their bare heads and plain plumage they hardly qualify as beauties of the bird world by most standards (Figure 7.5).

Historically the geographical range of the California condor extended in western North America from British Columbia to northern Mexico, but by the early twentieth century, persecution, habitat loss, and poisoning had eliminated the birds everywhere except in a few parts of California.

Condors soar on warm thermal updrafts and scavenge on carrion. Like many other large secondary consumers, they are at risk because they require large expanses of remote habitat

Figure 7.5   California condor. Credit: Weili Li / Moment / Getty Images.

and because they are high on the food chain. California condors are poisoned by lead ammunition and other toxins that they ingest while feeding on the remains of game killed by hunters or on ground squirrels that are shot as nuisance animals. Regulating the use of lead ammunition is controversial because hunters favor it because it is inexpensive. Consequently, federal and state regulations and proposals for restricting the use of lead shot are often challenged or overturned (Druzin, 2020).

By 1981, only 22 condors remained in the wild. A few years later, US Fish and Wildlife Service personnel captured all remaining wild California condors and added them to a captive breeding program with the goal of ultimately releasing captive-bred birds and re-establishing wild breeding populations.

Condors lay a single egg per clutch, and if the chick survives, the young bird is dependent on its parents for so long that they cannot reproduce the following year. Because of their low reproductive rate, condor populations increase slowly at best. But this limitation can be circumvented artificially by removing the first egg, incubating it, and raising the nestling in captivity. Between 1983 and 1986, biologists artificially incubated 16 California condor eggs. Ten pairs replaced the stolen egg by laying a second one, and after those eggs were removed three pairs produced a third egg.

As part of a cooperative effort between multiple federal and state government agencies in the USA, zoos, the Yurok Tribe, the federal government of Mexico, and several non-governmental organizations, eggs obtained through replacement clutching gave rise to a captive condor population that provided birds for reintroduction to the wild. After being released, captive-bred birds from those efforts established three breeding subpopulations, one each in California, Arizona, and Mexico. (The locations of these are kept confidential to prevent egg theft.)

Despite this progress, lead poisoning remains a serious threat to wild condors. Unless this problem is addressed, condor populations probably will not be able to sustain themselves (BirdLife International, 2020a).

7.4.3 Cross-Fostering with Surrogate Parent Species

Cross-fostering uses the same strategy as the cowbird: placing the eggs or young of one species in nests where another species will devote resources to nurturing the imposter as if it were their own (Box 7.6).

Even though most fishes, amphibians, reptiles, and invertebrates lay eggs, they are not candidates for cross-fostering because adults in those groups usually do not care for their young. Cross-fostering is not appropriate for most mammals either, because their offspring spend a long period within the female's uterus. However, in *marsupials* (kangaroos, opossums, the koala, and their relatives) the mammalian reproductive system differs in a way that allows for cross-fostering.

Marsupials give birth when their young are at a very early stage of development. The front limbs of a newborn marsupial are developed enough to allow it to climb along the mother's fur to her teats, where it attaches and begins nursing. In most marsupials the teats are in a pouch that protects the offspring while it nurses and grows.

This arrangement allows for a unique method of cross fostering because young of a donor species can be placed in the pouch of a surrogate marsupial species where they might continue to develop until they finish nursing (Box 7.7).

Box 7.6

Cross-Fostering Whooping Cranes with Surrogate Sandhill Crane Parents

Cranes are large, long-legged wading birds that occur on every continent except Antarctica. Two species are native to North America: the endangered whooping crane and the more common sandhill crane. Historically, whooping cranes bred in Canada and the USA and wintered along the Gulf of Mexico, but numbers dwindled as a result of hunting, habitat loss, human disturbance, and the collection of eggs and birds for museum specimens. By the middle of the twentieth century, only 18 whooping cranes remained, all in a single flock that bred in Canada's Northwest Territories and wintered along the Gulf Coast of Texas.

In an effort to increase the numbers of whooping cranes and at the same time to establish a second population, researchers at the Fish and Wildlife Service cross-fostered whooping crane eggs in sandhill crane nests. Although whooping cranes typically produce clutches of two eggs, they rarely raise two chicks. Researchers reasoned that the second egg of a whooping crane clutch could be removed and placed in a sandhill crane nest without decreasing reproduction in the endangered whooping cranes. This conservation strategy assumed several things: (1) Disturbance from removing whooping crane eggs would not lower reproductive success by causing nest desertion or predation on the remaining egg; (2) surrogate sandhill crane parents would incubate and provide appropriate parental care to whooping crane eggs and chicks; and (3) when they were old enough to breed, whooping cranes would form pair bonds and mate with other whooping cranes, eventually establishing a self-sustaining population.

Things went according to the plan at first. The sandhill cranes proved to be good surrogate parents. They incubated the whooping crane eggs and reared the young that hatched. But rates of reproduction and survival were disappointing. Although nearly 300 whooping crane eggs were placed in foster nests between 1975 and 1988, by 1991 there were only 13 whooping cranes in the new flock. And, of critical importance to the success of this venture, the fostered whooping cranes failed to breed. Difficulty in finding mates in the small population and the inability to form pair bonds with members of their own species contributed to this problem (Ellis et al., 1992). Because of those difficulties, the cross-fostering program was discontinued.

Nevertheless, whooping cranes eventually increased in population and distribution (Figure 7.4) because of successful reintroductions of captive-bred birds (Section 7.5.1).

Box 7.7

Cross-Fostering between Marsupials: Rock Wallabies

The brush-tailed rock-wallaby is a small marsupial named for its long, bushy-ended tail (Figure 7.6). Although it was widespread in Australia when Europeans arrived, this species was subsequently killed for the fur trade and targeted as a pest. Its geographic range has contracted so that the brush-tailed rock-wallaby now occurs only in southeastern Australia (Figure 7.2), where it inhabits rocky cliffs, ledges, and caves (Woinarsky and Burbridge, 2016). It lives in colonies of up to about 30 individuals in patches of suitable habitat separated by expanses of ground that lack shelter from introduced foxes and other predators. Because the brush-tailed rock-wallaby cannot safely disperse between these isolated habitat fragments, assisted reproduction combined with reintroductions might improve the prospects of this species.

Scientists in Australia reasoned that reproduction could be assisted by cross-fostering brush-tailed rock-wallabies in the pouches of related species. Accordingly, they transferred 14 young

Figure 7.6   Brush-tailed rock-wallaby. Credit: Tier Und Naturfotografie J und C Sohns / Photographer's Choice RF / Getty Images.

brush-tailed rock-wallabies weighing 1 to 106 g into pouches of surrogate mothers from other wallaby species. When the young were placed in pouches of mothers whose own young were greater than or equal in size to the cross-fostered newcomers, the transferred young grew and were weaned at the same age as the surrogate mothers' own young. However, two brush-tailed wallaby young that were transferred into a pouch that contained smaller young both died. Since the amount and composition of the milk that the mother produces change as the young animal develops, the researchers speculated that those surrogate mothers might not have produced enough milk of the appropriate composition to support growth of the foster brush-tails (Taggart et al., 2005).

7.5 Enhancing or Restoring Population Size and Geographic Range: Reintroduction

In order to rescue species in trouble, breeding programs have to be paired with successful reintroduction to the wild. This means augmenting populations where they have declined within their original range (*reintroduction*) or establishing populations in areas of previously unoccupied suitable habitat (*translocation*).

In either case, it is helpful to know how the composition of groups of released individuals could affect their chances of becoming established. For example, the sex ratio of a founder group is an important consideration. The social behavior of individuals in a founder group is another factor that might influence the success of translocations or reintroductions. Box 7.8 describes research designed to study the latter question with a species of small, ground-nesting birds in New Zealand.

Box 7.8

Evaluating Evidence: Testing the Hypothesis That Familiarity with Neighbors Influences Reproductive Success in a Translocated Songbird

The saddlebacks of New Zealand are two species of forest birds named for an orange rectangle (saddle) that extends across their back (Figure 7.7). Like New Zealand's other native birds, saddlebacks evolved without mammalian predators, and therefore they lack adaptations to avoid many predators. Once widespread in New Zealand, saddlebacks are now confined to islands where predators are absent (Figure 7.2). They are especially vulnerable because they nest near the ground and their young hop around noisily. (If these birds had coevolved with predators, natural selection would most likely have eliminated this behavior long ago!)

Saddleback conservation involves establishing additional populations in patches of suitable habitat. To get information on the most favorable composition of founder populations of the North Island saddleback, researchers at two universities in New Zealand studied the role of social behavior within groups of birds translocated to an island in Lake Rotorua on North Island.

Data from studies of other bird species suggested that individuals interacting with familiar neighbors might have higher reproductive success than those that interacted with unfamiliar neighbors, perhaps because birds in the latter group would spend more energy on aggressive interactions. For example, during a study of male red-winged blackbirds in the USA, breeding males with familiar neighbors fledged significantly more offspring than males without familiar neighbors (Beletsky and Orians, 1989).

Accordingly, the researchers in New Zealand designed an experiment in which they created two founder groups of 18 saddlebacks each and released them on the island. One of the groups was made of up birds from a single donor patch, while the other consisted of individuals taken

Figure 7.7   Saddleback. Note the orange patch, or saddle, across the back. Credit: Alvin Setiawan / 500Px Plus / Getty Images.

from several patches. To test the hypothesis "that founder groups will do better if they are made up of individuals that are familiar with one another," they released the founder groups in areas of suitable habitat at different locations on the island, and monitored their behavior and reproduction (Armstrong and Craig, 1995:134).

At first, familiar birds formed pair bonds more rapidly, but by the beginning of the breeding season, rates of pairing in the two groups were similar, as were subsequent reproductive outputs. Thus, the results of this study did not support the hypothesis that familiarity between birds within founder groups influenced the potential reproductive success of translocated populations. However, the researchers cautioned against assuming that their results could be extrapolated to other species because familiarity might be more important in species with stronger bonds:

Based on the evidence for saddlebacks, there is no reason to believe that using familiar individuals will improve the outcome of translocations. However, it is important to recognise the limitations of the available data …. It would ideally have involved several translocations of both single- and mixed-patch groups, but this was impossible given that only one island was available and that was small.

(Armstrong and Craig, 195:139)

7.5.1 Transitioning from Captivity to the Wild

Moving organisms from one place to another is challenging. Although we hear about cases where introduced species thrive and become pests, deliberate or accidental introductions often fail. To increase the odds of success, the captive environment should be as close as possible to the environment that released organisms will face. This is relatively easy with plants if they are kept in a nursery where temperature, humidity, water, soil type, nutrients, and light/dark cycles can be controlled. Immediately before their release, conditions can gradually be adjusted to approach the conditions they will experience when they are moved outdoors, so that young plants can adjust to their changing environment.

It is less easy to expose animals reared in captivity to the environment they will encounter in the wild, especially if they are around humans instead of parents from whom they might otherwise learn how to obtain food, select habitats, migrate, and choose mates. For predatory birds and mammals, hunting skills learned from parents are often critical. Consequently, it may be necessary to have a transition period in which captives are gradually weaned from dependence on people. One way to address this challenge with birds of prey is a technique called *hacking* (Zimmerman, 1975). In this method, developed by falconers, a young, captive-reared raptor is placed in a shed and fed regularly. After the bird becomes accustomed to the feeding schedule, a window or door is opened so that the fledgling is free to leave. Scheduled feedings continue for as long as the bird returns for food. In this way, the animal obtains nourishment it can depend on until it learns to hunt on its own. The US Fish and Wildlife Service used a similar technique when they reintroduced endangered red wolves in North Carolina (Box 6.4).

Birds generally form a sense of their species identity on the basis of the parent(s) they see early in their development. If they are not reared by their own species, they may form inappropriate attachments because of this *imprinting*. This is what happened with early efforts to breed whooping cranes in captivity. To avoid this, researchers later developed elaborate strategies for preventing visual contact between captive-bred whooping cranes and people, so that young cranes would select adult whooping cranes (not people) as mates. Handlers used puppets and crane costumes to disguise themselves as large, white birds, and when it came time to migrate, the pilots of ultralight aircraft that led young cranes along their migratory route dressed in white.

7.6 Managing Genetic Diversity in Reintroduced Populations

The low genetic diversity of captive populations presents a different kind of challenge when rare species are released in the wild. Breeding programs often involve populations that have gone through a genetic bottleneck (Section 6.5.2). Genetic diversity in these populations is restricted even further because they are descended from small numbers of individuals that were brought into captivity. Even if their numbers increase dramatically because of successful breeding, populations derived from a small group of captive organisms might have a reduced chance of long-term survival in the wild due to restricted evolutionary potential.

The consequences of moving organisms, and therefore genes, are far-reaching. *Genetic rescue* (the restoration of genetic diversity) and the prevention of inbreeding are important. However, it is also important to use individuals that are adapted to the environment where they will be released (Boxes 7.9 and 7.10). Conservationists have to balance these considerations when they release or move rare organisms.

Box 7.9
Reintroducing the Red Kite to Parts of Its Former Range in the British Isles

The red kite is a hawk of European fields, pastures, and woodlands. It feeds primarily on small-to-medium-sized mammals and carrion, including dead sheep. This species was once common in the British Isles and throughout much of continental Europe. But, like other birds of prey, kites were poisoned, trapped, and shot. Bounties were paid for their carcasses, and collectors took their eggs. As a result, kite numbers dwindled, and their distribution shrank.

Kites were legally protected in Britain as early as 1880, but they did not recover (Davis and Newton, 1981). By 1917, the red kite was extinct in England and Scotland, although a small population persisted in Wales. Genetic variation was low in the Welsh population, perhaps because of a bottleneck in the past. The occasional presence of young birds with reduced pigmentation in their feathers, a trait caused by a homozygous recessive allele, suggested that the Welsh population was inbred.

In 1989, the British government and private organizations jointly initiated a program to reintroduce red kites to parts of their former range in the British Isles. The Welsh population was too small for it to provide nestlings for reintroduction. However, a few nestlings were raised in captivity from eggs taken from Welsh nests that were considered vulnerable to egg collectors. To ensure that the parents continued to incubate the remaining eggs, the donor eggs were replaced by dummy eggs. (Investigators' hypothesis that those eggs were likely to be taken by collectors was supported when some of the dummies were in fact stolen.) Thus, removing the eggs to captivity proved to be an unusual case of compensatory mortality. If the researchers hadn't taken the eggs, collectors would have, so the removals by researchers did not increase egg loss over what it otherwise would have been.

Major genetic differences between the kite populations of Great Britain and continental Europe were thought to be unlikely because the species had once been part of a large, interconnected population. For this reason, the European continent was considered a suitable source of young birds for reintroductions in England. Initially, nestlings were taken from kite populations in Sweden and Spain because those populations were relatively close to the British Isles and had high enough productivity to be able to tolerate some removals. Young were collected from widely scattered donor nests, to maximize the genetic variability among the transplants.

To minimize negative consequences of captive rearing, contact with people was limited. When the young were 10 to 12 weeks old, they were released, but food was still provided at the release locations.

Survival was high in the new populations. After the reintroductions, the red kite population in the UK increased rapidly, although mortality from poisoning (much of it from banned chemicals) continued to be a threat. Examinations of kites that were found dead revealed that secondary poisoning from consuming poisoned rodents (Box 5.2) and direct poisoning from ingesting baits left out for predators were common (Molenaar et al., 2017).

Between 1992 and 1995, 162 young were fledged, and the newly established populations in England and Scotland appeared to be as productive as the donor populations (Evans et al., 1994, 1997). As of 2020, re-established populations of the red kite persisted throughout most of England and Scotland (BirdLife International, 2020b).

7.6.1 Minimizing Inbreeding by Managing Kinship in Reintroduced California Condors

A more precise assessment of the risk of inbreeding is possible where detailed pedigrees and genetic data provide information on the degree of relatedness between individual birds in the captive population. Captive California condor flocks are derived from a small number of founders that are the ancestors of the captives. Each year, the chicks produced by the captive condors are either placed in one of the wild populations or kept in the captive population. To minimize the likelihood of inbreeding, geneticists use data from pedigrees to estimate the degree of kinship between each chick that hatched in captivity and each member of the wild populations. That information then influences decisions about each chick's placement. For example, female 49 had the lowest kinship with members of the California population and the highest kinship with members of the Mexico population, so she was placed in the California population. Although chicks are ideally released into the population with which they have the lowest kinship, other considerations – such as the sex of the chick, the size of the recipient population, and the need to maintain genetically valuable chicks as future breeders in the captive population – also influence these decisions. Practical considerations such as the ability to obtain permits for export to Mexico also play a role (Frankham et al., 2017).

Although avoiding inbreeding is an important consideration when planning introductions, that approach can also cause problems. If a genetically differentiated population that is adapted to local conditions receives individuals from a population that evolved in in a different part of the species' range, then locally adaptive genes in the recipient population might be diluted if managers' only concern is avoiding inbreeding. This *genetic swamping* could cause the loss of adaptations to local conditions. The next example illustrates how this problem can be addressed if geographically specific genetic data are available.

7.6.2 Genetic Management to Preserve Local Adaptations

Box 7.10

**Using Genetic Information to Guide the Selection of Source Regions
for Reintroducing Brown Bears in Southwestern France**

The brown bear, the most widely distributed of the world's bears, occurs across Eurasia and northern North America (where it is known as the grizzly). It is abundant in Russia, Canada, and Alaska, but populations in the southern part of its range are small and fragmented. By the 1990s, two small, isolated populations remained in the Pyrenees, a mountain range along the border between France and Spain (Figure 7.2), where they inhabited steep slopes, alpine meadows, and valleys.

When French scientists considered reinforcing the population of brown bears remaining in southwestern France, they faced a decision about what part of the vast brown bear range should provide source animals. Fortunately, data on bear DNA from 17 countries in Europe allowed scientists to identify regions with the closest relationship to the French bears.

The geographically closest relatives of the Pyrenees bears occurred in isolated groups in France, Spain, and Italy, but these bears, too, were becoming rare. For this reason, the French Ministry of the Environment contemplated introducing bears from larger but more distant populations in Russia and Romania. However, when geneticists Pierre Taberlet and Jean Bouvet from the Université Joseph Fourier in France compared bear DNA from the French Pyrenees and 14 other European countries, they concluded that the bears from Russia and Romania were not as closely related to the Pyrenees bears as were those from closer localities. On the basis of this information, the officials searching for a source of bears to transplant turned their attention from the eastern European bears to those that were closer to the Pyrenees (Taberlet and Bouvet, 1994).

The researchers suggested that bears from either Slovenia or Sweden would be appropriate sources of donor bears. Because the estimated population of 350 bears in Slovenia at that time was managed as a game species, it seemed likely that it could withstand the removal of a few bears for movement to the Pyrenees. An area of suitable habitat where bears had previously lived was selected to receive the newcomers. Accordingly, managers moved two females that the researchers named Melba and Ziva and a male dubbed Pyros from Slovenia to the Pyrenees in 1996 and 1997. About a decade later, another five Slovenian bears (four females and one male) were released in the Pyrenees. These formed the nucleus of a population that numbered about 20 bears by 2012.

Although the human population density of the region is low, the reintroduction is not without critics because the bears kill an estimated 200 to 300 sheep each year. The French government compensates shepherds for their losses, but only if tooth and claw damage to the dead animals can be demonstrated (Bland, 2012).

Judging from the rapid increase in the bear population, the reintroductions seem to have been successful. Further genetic analysis, however, revealed a cause for concern. Pyros was the ancestor of most of the bears that were born after he arrived. Although that meant that Pyros clearly had high evolutionary fitness, this situation was worrisome because of the potential for future inbreeding in this small, isolated population (Palazón et al., 2012).

7.7 Setting Priorities: Which Species Should We Try to Save?

Saving dwindling species from extinction takes a lot of work and funding. Clearly, these resources are not available for all species. This means we have to make decisions about which species we will help.

Science cannot tell us what species we should prioritize. However, science can shed light on some of the potential consequences of prioritizing some forms of life over others. Sometimes, the protection of a single species might protect many other organisms as well. If such focal species can be identified, and the assumption that one protected species can serve as a stand-in for many others is valid, then protecting that species could serve as a useful shortcut, ensuring that maximum benefits result from helping a single species. There are several ways in which this might work.

Indicator species are species that reflect the fate of other attributes of an ecosystem (Spellerberg, 2005). Like the proverbial miner's canary, an indicator species might provide information about air or water quality or the locations of specific environmental zones. *Umbrella species* are those that require large expanses of habitat, usually because they themselves are large and have vast home ranges. In protecting the habitat of such species, managers hope to protect habitat for many smaller, less well-known species too. This is likely to be the case if the range of the umbrella species encompasses many habitats. A species or subspecies that serves as a symbol for conservation, usually because of its charismatic appeal, is termed a *flagship species*. Large animals often serve as flagship species. The giant panda, northern spotted owl, koala, and Florida panther are examples. Species that play a pivotal role in ecosystem functioning, out of proportion to their abundance or biomass, are said to be *keystone species*. For example, many species depend on beavers, kelp (Section A.2.1.1), or saguaro cactus in their respective ecosystems. If a keystone species disappears, other species might go too.

A single species or subspecies can sometimes play more than one of these roles. The northern spotted owl is considered an indicator of old-growth forest (Box 9.4). It is also an umbrella species because it requires large areas of forest, and it is definitely a flagship for conserving ancient forests. Similarly, the giant panda is an umbrella species and a flagship species.

Care should be exercised in designing management strategies around indicator, flagship, umbrella, or keystone species, however. A species should not be used as an indicator of other species unless there is clear evidence that populations of the indicator and the species it serves as an indicator for rise and fall in tandem. Species that are monitored for political or social reasons – for instance, because they are designated as threatened or because they have charisma – should not automatically be assumed to indicate trends in populations of other species (Landres et al., 1988).

If the species in these categories really do what we hope they do, then protecting them should be a good use of resources. It is important to remember, however, that often hypotheses about the relationships between species in these categories and other species are not tested (Simberloff, 1998; Margules and Pressey, 2000).

Ideally, the species we prioritize should have high importance for conservation and for science. Endemic species or subspecies are usually priorities for conservation because if they are lost from their local range, they will also be globally extinct. Species such as the tuatara (Box 6.11) that are the only living members of a taxonomic group also have high biological importance (*see Q7.3*). However, species that are targeted for protection and restoration also reflect a host of considerations that are not necessarily related to these criteria. Politics, economics, and culture all influence priorities. For example, conservation tends to focus on large vertebrates, especially birds and mammals. Groups of organisms that are poorly known or lacking in appeal or just very tiny get neglected.

7.8 Evaluating Efforts to Protect and Restore Species

The examples in this chapter illustrate ways of managing species with the goal of preventing extinctions, through combinations of monitoring, legal protection, and management to promote recovery. Laws and treaties have halted declines in many species and contributed to some recoveries. Yet by the late twentieth century it became clear that many species were still at risk and that number was increasing. Protection sometimes came too late to halt declines and was often focused too narrowly on regulating exploitation without addressing habitat loss, invasive species, or other factors contributing to declines.

Helping species that are in trouble often involves complex interventions. Cross-fostering, reintroduction, headstarting, and replacement clutching entail handling wild individuals and often maintaining them in captivity as well. For this and other reasons, these intensive forms of intervention are controversial.

Although offsite conservation in zoos, captive breeding facilities, or botanical gardens seems like a logical way to increase the size and range of dwindling species, it involves risks at every stage of the process, from the initial removal of plants and animals from the wild, through maintaining them in captivity, to eventually returning them to their native habitats. Taking individuals from the wild is problematic because populations in need of assistance are small to begin with. Furthermore, activities associated with removal may disturb the donor population, disrupt breeding, and introduce diseases, parasites, or invasive species. Taking individuals from the wild can even facilitate habitat destruction if removal from the wild (without successful establishment at a different site) removes an obstacle to commercial development. Maintaining wild species in captivity also entails a risk of mortality in captivity and semi-domestication.

Another practical problem is the challenge of finding habitat where a population will eventually be able to sustain itself. Efforts to save species in trouble are likely to fail unless they involve a comprehensive program to address all factors that limit a species' recovery. No amount of intervention will succeed if the environment contains toxins or doesn't provide sufficient food or includes non-native species that outcompete natives.

Breeding projects may divert attention and resources from the important issue of what causes species to decline in the first place. The money and effort spent on breeding are not available for other projects such as habitat conservation (Chapter 8). Finally, offsite management has unforeseen impacts on ecosystems and cultures.

To summarize, critics allege that offsite conservation is expensive, time consuming, and ineffective and diverts resources from more important issues and conservation priorities.

On the other hand, proponents of offsite conservation point out that it has important advantages. For some species, offsite intervention is the only hope. It can allow species to get through a crisis, and it may eventually assist in maintaining or re-establishing sustainable wild populations that do not depend on extreme conservation measures.

Offsite conservation arouses public sympathy for declining species, especially large, appealing animals like pandas, whooping cranes, and condors. Such projects have public relations and educational value. They can raise awareness of the plight of rare species and the need for habitat conservation.

Although captive populations typically have genetic problems due to past bottlenecks and small numbers of founders, progress in conservation genetics is addressing these problems. Information obtained using advanced techniques of genetic analysis combined with information about the history, geography, ecology, and behavior of small populations allows for precise genetic management geared toward maximizing genetic diversity, minimizing risks from inbreeding, and enhancing the chances that introduced individuals will be adapted to the sites where they are reintroduced.

The most effective species conservation involves multi-pronged approaches that combine regulation, enforcement, and education. In addition to protecting species, protection of ecosystems is essential. The next chapter looks at ways of doing this.

8

Strategies

Protecting and Restoring Ecosystems

This chapter considers lands and waters set aside specifically to protect and restore their natural features. In so doing, it brings up some fundamental questions. What areas should we protect and why? Who and what are we protecting for; who and what are we protecting from? What conditions do we want to maintain? What is natural? What, if anything, do we want to return to?

Science can help answer some related technical questions about how to accomplish our goals. What reserve designs are likely to preserve the most biodiversity? How might we restore or mimic missing interactions? What is the evidence in support of different answers to these questions? But answers to questions about why we should pursue those goals depend on values. Science can provide information about what outcomes are probable given specific actions, but science cannot answer questions about what should be done.

8.1 What Is a Reserve?

A *reserve* or a *preserve* (I use the terms interchangeably) is something (such as cash, food, energy, rights, or land) that is set aside or held for the future. Reserves are set aside to protect part of the natural world from something (usually human impacts that are viewed negatively) or to maintain something. Regulations on use are more stringent within reserves than outside of them. Thus, setting aside a preserve for conservation involves a formal or informal requirement of restraint in accordance with rules: Some aspects of resource use within the boundaries of a preserve are limited under some circumstances. Just as the designation of a preserve denies certain rights to some groups under some conditions in some places, it also involves the retention of the rights of certain groups to use those places in certain ways or at specific times (Mascia and Claus, 2009). This means that designing and managing preserves are political, as well as scientific, endeavors. For a reserve to be effective, some entity must have the power to compel compliance with regulations. That power may come from formal rules, from community norms, or both.

Typically, naturalness is a criterion for preserve status, and site selection is often guided by a search for pristine conditions. Defining "natural" and "pristine" is far from straightforward, however. What reference conditions should guide us in deciding what is pristine? Are all human impacts unnatural (Section 9.5.)?

Reserves are set aside for myriad purposes, with varied degrees of protection – ranging from regulated use of economically valuable resources to preservation of areas with minimal human impact – and allow a spectrum of uses ranging from resource exploitation to recreation to research. Even in the strictest reserves, some low-impact uses – such as research, worship, or recreation – are usually permitted.

Reserves range in size from a single tree to millions of hectares. They can be designated by kings or communities, churches or secular governments, organizations, or individuals.

8.2 Historical Background

We don't know when the first reserves were designated, but we do know that they have been around for millennia. In large, stratified societies, rulers decreed restrictions on use within certain areas. Reserves, often in the form of sacred groves or other places, were and in some cases still are also designated within small societies, in some cases by local rulers, in others by community bodies (Box 1.1).

In the USA, the initial impetus for establishing protected areas on a large scale came from two strands of the conservation movement: a utilitarian push to protect forests from excessive exploitation and a preservationist push for protection of nature from exploitation and development (Section 1.5.1). Those forces led to the designation of national forests on the one hand and national parks on the other.

In the late nineteenth and early twentieth centuries, several nations began designating large areas as reserves for the protection of nature (Box 8.1).

Box 8.1

The Beginning of National Reserve Networks in North America and Russia

USA and Canada: Toward the end of the nineteenth century, explorers in the Rocky Mountains of the USA and Canada were impressed by scenic vistas and geothermal phenomena that they encountered. Hot springs and geysers, where water heated by molten rock was driven up to the Earth's surface by underground pressure, were of particular interest. While some saw commercial potential in the privatization of these impressive, high-elevation landscapes, others argued against private ownership. In both countries, worries about private development led to the designation of government reserves to protect these landscapes for the public. In the USA, Yellowstone National Park in the territories of Wyoming and Montana was set aside in 1872 as a "public park" under the control of the Secretary of the Interior.

During the following decade, a similar situation developed in Canada, although on a much smaller scale. Several individuals filed claims to the hot springs at Banff in the Canadian Rockies. To prevent private exploitation, an area of slightly more than 26 km^2 was vested in the Crown in 1885 (Figure 8.1), effectively preventing private ownership. Eighteen months later, Canada's Parliament enlarged the park to 670 km^2 (Lothian, 1987; Allin, 1990). Both Canada and the USA subsequently expanded protection to establish national park systems encompassing hundreds of thousands of square kilometers.

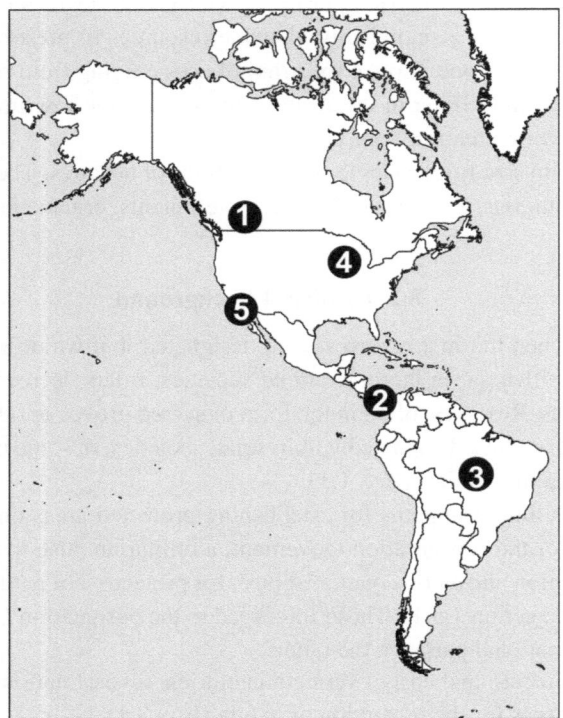

Figure 8.1   Locations of:1, Banff National Park, Alberta, Canada; 2, Barro Colorado Island, Panama; 3, Cerrado region of Brazil; 4, Madison, Wisconsin; 5, Sweetwater National Wildlife Refuge, California. Map created by Eva Strand using Esri, DeLorme World Countries Generalized Data & Maps for ArcGIS 2013, with permission.

Russia (Union of Soviet Socialist Republics, USSR, from 1922–1991): In the 1890s and early twentieth century, several scientists in tsarist Russia advocated setting aside large areas of ecosystems that were thought to have been minimally impacted by human use, arguing that these *zapovedniks* should be standards of nature which could provide baseline information for understanding the impacts of people. From the outset, the fate of the vast belt of steppe stretching across Eurasia (Section A.1.5) was of particular concern to scientists. Because its rich, dark soil (known as *chernozem* or black-earth) was deep and fertile, much of the steppe had been plowed. In 1894, the soil scientist Vasily Dokuchaev wrote about the critical importance of protecting relatively undisturbed examples of this ecosystem:

Unfortunately our virgin black-earth steppes, with their unique charm, limitless expanses and unusual organisms … are disappearing with astonishing rapidity …. It is all the more regrettable and distressing that our steppes … have never been systematically studied or regularly monitored …. This is not only of scientific interest but also of enormous and universally recognized practical importance.

(Quoted in Shtilmark, 2003:11)

A few decades later, the *Tsentralno-Chernozemny* (Central Chernozem) zapovednik (now a biosphere reserve) was established in western Russia near the border with Ukraine (Figure 8.2), which at that time was part of the Soviet Union.

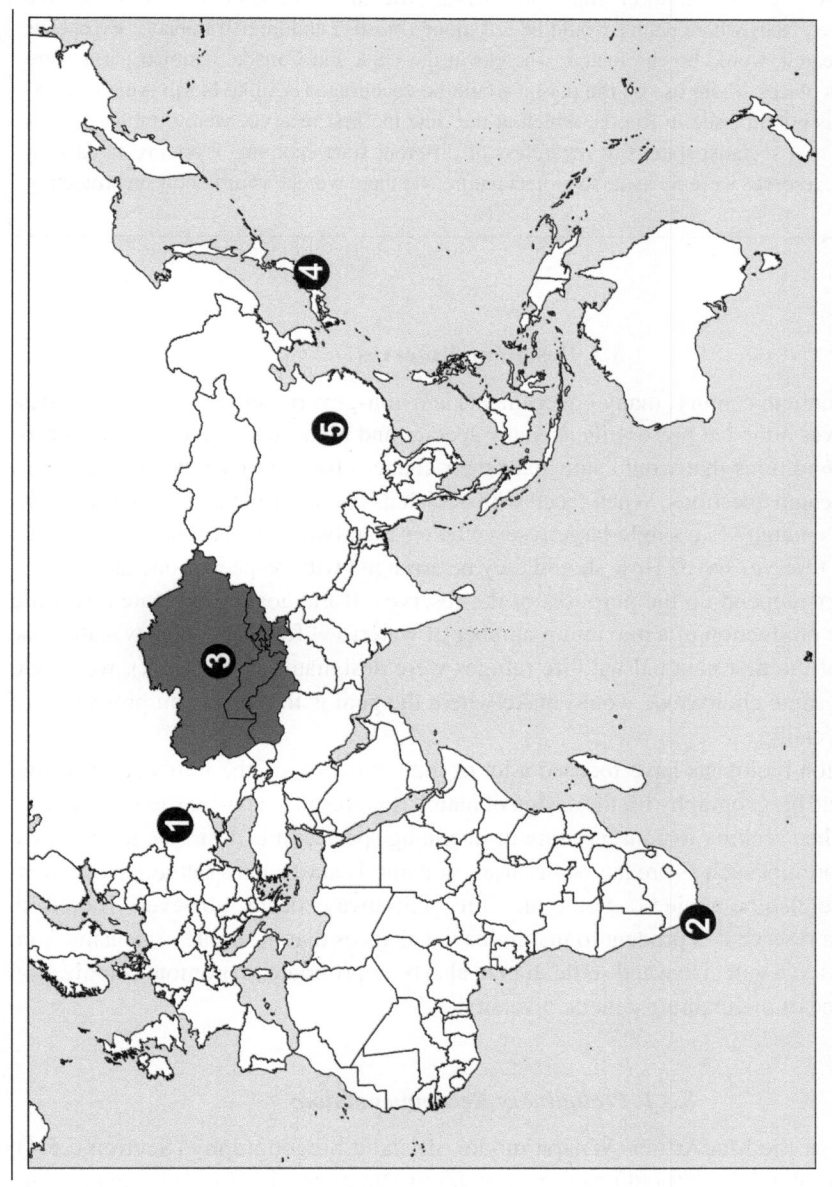

Figure 8.2 Locations of: 1, Central Chernozem (Black Earth) Biosphere Reserve, Russia; 2, Cape Floristic Region, South Africa; 3, Mountains of Central Asia; 4, Japan; 5, Daba Mountains, China. Map created by Eva Strand using Esri, DeLorme World Countries Generalized Data & Maps for ArcGIS 2013, with permission.

In spite of the stated goal of minimizing human impacts in the zapovedniks, management in some of these reserves included poisoning wolves, introducing valuable fur-bearing species, and managing game through hunting (Shtilmark, 2003).

The zapovedniks and the North American reserves differed in their goals. Zapovedniks were intended to be places where nature would be left alone (mostly) and human impacts, except for scientific research, would be eliminated, whereas in the USA and Canada, national parks were envisioned as places where use by the public would be encouraged. Unlike North America, privatization was not an issue in Russia, which at the time the first reserves were established was transitioning to a socialist state. But regardless of different state economic systems, in each of these nations, reserves were set aside to protect nature, yet there were tensions between protection and use.

8.3 Designing Reserves

In the late twentieth century, many governments and non-governmental organizations designated reserves aimed at preserving as many species and high-quality ecosystems as possible under conditions that would safeguard their chances for long-term persistence. This raises some design questions. When faced with decisions about setting land or waters aside, how much is enough? Is a single large reserve better than two or more small ones? What shape should reserves have? How should they be arranged with respect to one another?

The answers depend on the purposes of the reserves. If the goal is to create a wildlife refuge for the production of a maximum number of waterfowl for hunters (as was the case when some of the first national wildlife refuges were designated in the USA), we would not make the same choices we would make where the goal is the long-term protection of biological diversity.

Conservation biologists have focused a lot of their attention on the implications of the theory of island biogeography for the design of nature reserves in island-like habitat patches. In general, when seeking to conserve rare or declining species, it is desirable to minimize sources of mortality such as predators, diseases, or natural catastrophes (although overpopulation should also be avoided). Some mortality will always occur, however. Therefore, in designing a reserve it is prudent to maximize the chances that desirable organisms from outside the reserve will arrive and settle. It may also be appropriate to promote colonization for the purpose of maintaining genetic diversity.

8.3.1 Preliminary Recommendations

On the basis of the MacArthur–Wilson model of island biogeography (Section 6.5.4), the American physiologist and ornithologist Jared Diamond (1975) made several recommendations for the design of nature reserves with the goal of maximizing opportunities for colonization and reducing the chances that desirable species will go extinct (Figure 8.3).

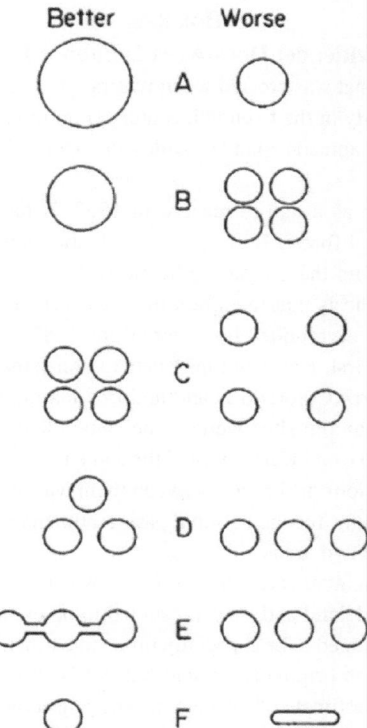

Figure 8.3   Diamond's recommendations for the design of nature reserves. Diamond suggested that in each of the paired designs below, the configuration on the left would be likely to have a lower extinction rate than the configuration on the right. From Diamond (1975), with permission from Elsevier.

- Reserves should be as large as possible (Figure 8.3A).

     This recommendation is derived from data showing that large areas support more species than small areas and on inferences about the historical ecology of islands that were at one time located near a large land mass.
- A single large reserve will conserve more species than several small reserves with the same total area (Figure 8.3B).

     This recommendation is based on two assumptions. First, that it is always best to conserve as many species as possible in a reserve. Second, that large areas generally contain more species than small areas (if they have the same habitats). It follows from the first recommendation that reserves should be as large as possible. The choice between a large preserve or several small ones is often referred to by the acronym SLOSS for Single Large or Several Small reserves.
- Separate reserves should be designed in ways that maximize movements between them.

Box 8.2
Evaluating Evidence: Does Area Influence Extinction Rate?

Barro Colorado is an island that was created when waters rose around a hill as part of the formation of the Panama Canal early in the twentieth century (Figure 8.1). Because it was surrounded
by water only recently, this manmade island provides an opportunity to test ideas about the effect
of area on extinction.

The island was set aside as a nature reserve in 1923. It has become one of the world's
most studied areas of tropical forest. Hence, it is a valuable natural laboratory (although it is
a sample of one; and therefore there is no replication). In the 1970s, the American ecologist
James Karr tested the hypothesis that reduction in area caused extinctions of birds on Barrow
Colorado Island after it was surrounded by water (Karr, 1982). He studied the effects of area
on extinction in two ways. First, Karr used mist nets (a non-lethal method of trapping birds or
bats) to sample birds on Barro Colorado Island and in a national park a few kilometers away
on the mainland. The two sample sites were at the same elevation and in similar vegetation
(low-elevation tropical forest), and Karr sampled the same microhabitat (forest undergrowth) at
both locations. The only obvious difference between them was area: 16 km^2 for Barro Colorado
Island, compared to 22,000 km^2 for the national park on the mainland. Second, Karr compared
lists of bird species from the two areas.

Fifty species of birds of mature forest that were known from the nearby mainland were not
recorded on Barro Colorado Island and were presumed to be extinct there. Bird species that required forest undergrowth seemed to be especially prone to extinction. Karr speculated that many
other species might have been originally present but not been recorded in the forest on Barro
Colorado, so he believed his estimate of extinctions was conservative.

Karr considered several alternative hypotheses about what might have caused this difference
in the bird faunas of the two sites. He assumed that before the island was created any birds on the
mainland would have been able to reach Barro Colorado because they were very close and there
were no barriers between the sites. So, he did not think access was the cause of the difference. He
also considered it unlikely that there would have been substantial differences in the kinds of forest
birds at the two sites before isolation occurred.

However, although habitats at the sample site on the mainland and at Barrow Colorado were
similar, there was evidence that over time the habitat mosaic became less diverse on the island.
As the forest on Barro Colorado aged, the ground cover became sparser, which might have influenced the susceptibility of ground-dwelling birds to predation, and therefore they might have
been vulnerable to becoming extinct.

In light of these findings, Karr concluded that island size was a major factor that caused extinctions of forest birds on Barro Colorado Island but that predation and changes in habitat also
contributed to the high rate of extinction among birds of mature forests on Barro Colorado Island
(*see Q8.1*).

The next three recommendations are based on the principle that maximum
connectivity between reserves is best and therefore reserves should be designed to
promote movements between them. This should be accomplished by arranging reserves
so that they are close together (Figure 8.3C), clustered rather than arranged in a linear
fashion (Figure 8.3D), or connected by corridors (Figure 8.3E).

• Reserves should have a low ratio of edge to interior habitat (Figure 8.3F).

   This recommendation assumes that the goal is to provide habitat for organisms that are sensitive to threats such as predators or invasive species, which are greatest at habitat edges. For species that thrive at the junctions between different habitats, the opposite principle will apply, but because most edge-dependent species are tolerant of disturbance, they generally do not depend on reserves.

To summarize, Diamond's recommendations for the design of nature reserves based on the MacArthur–Wilson model of island biogeography were that reserves should be large, close together, connected to each other, and compact.

8.3.2 Details

8.3.2.1 Size

Large reserves have advantages over smaller ones in certain situations. Because they have relatively little edge in relation to total area, large reserves tend to contain habitats that are buffered from the negative influences of predators or weeds in the surrounding landscape. Very large reserves also may meet the area requirements of large carnivores and other wide-ranging species. Finally, large reserves provide more types of habitats, thus perhaps allowing long-term adjustments to changes in climate or other features of the environment.

   Nevertheless, some biologists have pointed out that there are situations in which a single large reserve might not be the best conservation option. If a species has specific habitat requirements that are found in only a few small, widely spaced patches, that species will likely derive more benefit from protection of those small, isolated patches than from a reserve that contains vast expanses of unsuitable habitat. Furthermore, when a disease or a catastrophe such as a fire or a flood occurs, it might be better to have several populations in separate reserves, to avoid putting all our metaphorical eggs in one basket. (Of course, if we are dealing with fire-dependent species such as jack pine, we will not want to minimize the chance of fire.) In a large preserve, an entire population or group of subpopulations could be wiped out from such an event.

8.3.2.2 Shape

The amount of a patch's edge also depends on its shape (Figure 3.5). If two patches of the same shape differ in size, the smaller one will have a higher ratio of perimeter (edge) to area (interior) than the larger one. Long, thin patches or those with irregular borders have high proportions of edge.

8.3.2.3 Connectivity

Diamond's third, fourth, and fifth principles relate to the distribution of reserves across a landscape. Because many organisms are unable to move through the habitats surrounding reserves, conservation biologists are interested in *corridors*, habitats that bridge the gaps between reserves (Harris, 1984).

The term corridor has been used to mean several things, including linear strips of habitat that physically connect reserves, chains of discrete patches along a migration route, and underpasses below or bridges over highways. The ability to travel between preserves depends not only on the type of habitat in the landscape matrix but also on the amount and type of human disturbance.

Not all species need corridors. Species that cannot move through the habitat between preserves may benefit from corridors, but corridors also have potential disadvantages. They might facilitate the spread of unwanted organisms such as introduced species or diseases (Hess, 1994), and disturbances such as fire. Corridors may allow genetic swamping of small populations (Section 7.6.1). Because of these mixed potential outcomes from corridors, their value is controversial (Noss, 1987; Simberloff and Cox, 1987).

To summarize, the arguments for large, connected preserves have intuitive appeal, but there are many factors to consider including interspecific interactions, dispersal abilities, habitat requirements, disturbances, and ultimate and proximate vulnerabilities (Simberloff and Abele, 1976a, b; Williams, 1984). Such considerations may outweigh the advantages of reserves that are large and connected.

8.3.3 Landscape Considerations

8.3.3.1 Managing Connections between Subpopulations

Many species are organized into *metapopulations*, fairly independent local subpopulations that are connected by occasional movements of individuals between them. Movements between subpopulations in a metapopulation can play a significant role in the regional dynamics of a species for two reasons. First, if some populations dwindle to very low numbers, from which they are unlikely to recover on their own, they may be rescued by immigrants from a nearby population (Brown and Kodric-Brown, 1977). Second, if some of the subpopulations that make up a metapopulation disappear entirely, their habitats may be recolonized by individuals dispersing from other parts of the metapopulation. The only difference between these two scenarios is that in the case of a rescue the dispersers arrive before all members of the original population are gone, whereas with recolonization the colonists do not get there until local extinction has occurred. Of course, sometimes it is difficult for the field biologist to distinguish between these two situations.

Because habitat quality varies across a landscape, reproduction may exceed mortality in some areas but not in others. Productive habitats can serve as *sources* of individuals that disperse into less productive habitats, which are termed *sinks* (Pulliam, 1988). Unfortunately, in highly developed landscapes it has become harder for organisms to move between subpopulations within a metapopulation, so habitats are more likely to function as sinks than as sources.

8.3.3.2 Movements of Organisms across Landscapes

The management of species that move between ecosystems goes back to the Migratory Bird Treaty (Section 1.5.3). But the issue asserted itself with new urgency as concerns about the effects of habitat fragmentation and about the needs of wide-ranging animals intensified.

The protection of habitat for migratory or nomadic species poses special challenges for preserve design and management. Appropriate habitat must be available at essential points throughout the annual cycle. Many birds that migrate long distances must have *stopover* sites (sometimes termed *staging areas*) where they can bulk up on high-energy foods to fuel the rest of their journey.

8.3.3.3 Patch Dynamics

Large reserves that include vegetation in many stages of succession are likely to have high species richness because different organisms are adapted to different stages. Sites with vegetation at early successional stages provide reservoirs of organisms that can colonize open patches after disturbances occur. For example, riparian zones along rivers, streams, and creeks experience frequent flooding, scouring, and sediment deposition. These processes remove or damage existing vegetation and create seedbeds for new vegetation to become established. Consequently, colonizing species tend to inhabit streams and riverbanks. When a riparian zone occurs within a matrix of late-successional habitat which is not disturbed as frequently, the riparian ecosystem provides places where pioneer species are maintained. In mature forests, riparian zones harbor sources of colonists that move into forest openings such as gaps that are created by fallen trees (Agee, 1988).

Similarly, when a disturbance occurs in a reserve, we would expect immigrants from surrounding habitats to colonize the disturbed patches. The American tropical ecologist Daniel Janzen (1983), tested this hypothesis by recording vegetation that became established in an opening created when a tree was blown down in Costa Rica's Santa Rosa National Park. Although the forest was considered "pristine," the plants that became established were those of adjacent anthropogenic habitats such as roads, pastures, fencerows, and fields. Thus, the consequences of a natural disturbance in the forest reserve depended on the organisms in the landscape around the reserve.

8.3.4 Long-Term Considerations: Reserve Design for Changing Environments

Because the Earth is dynamic, the distributions of organisms shift as environmental conditions change. This has happened many times in the past in response to changes in geology and climate. Continents drifted to different latitudes, the tilt of the Earth's axis shifted, mountain ranges arose or were eroded, shifts within the Earth's crust caused land masses to rise from beneath the seas or became submerged, glaciers formed and melted, and sea level rose and fell. Many of these changes were gradual, acting over thousands or millions of years. Others, such as volcanic eruptions or collisions from asteroids, were sudden.

In the past, when climate and geology changed gradually, natural selection might favor adaptations to the altered environment. If the climate became colder, a population of rabbits might eventually develop a thicker coat or other adaptations to cold conditions. If the population's gene pool lacked the necessary mutations, they might survive by moving to lower elevations or farther south. If the population could neither adapt nor shift its range, it would eventually go extinct. One strategy for dealing with this situation is to protect areas with pronounced environmental gradients, so that organisms will be able to shift their distributions in response to long-term environmental changes without having to move through the intervening matrix (Hunter et al., 1988).

The area where two habitats come together is termed an *ecotone*. Ecotones may consist of areas with different physical environments or differences in the history of adjoining patches. The transition zone at timberline, where forest borders tundra, is an ecotone, as is the boundary between a patch of forest that burned 20 years ago and one that burned 200 years ago.

Ecotones may be abrupt or gradual. Steppe vegetation often occurs where the climate is too dry to support the growth of trees but wet enough for perennial grasses to dominate. In the Midwest of the USA, steppe spans an east-west climatic gradient. The western steppe is relatively dry, and moisture increases to the east. At its eastern margin, the steppe typically borders temperate deciduous forest. In this transition zone, conditions are marginal for trees, but trees can survive in microhabitats that are a little wetter. Thus, trees extend hundreds of kilometers westward into the midwestern steppe along water courses, forming forested projections known as *gallery forests*. In contrast, the ecotone between steppe and forest in the southern Andes of Argentina is less than 50 km wide (Veblen and Lorenz, 1988). Ecotones between forest and steppe are not static; they continually shift in response to changing conditions. Droughts and fires inhibit tree regeneration and favor steppe expansion. On the other hand, under moist, fire-free conditions steppe contracts and forest expands.

Organisms that are adapted to a wide range of environmental conditions or have good dispersal abilities are the least affected when the environment changes. Poor dispersers with narrow tolerance ranges, like the small mammals on mountaintops discussed in Section 6.5.4, face higher risks than those that can move long distances. One implication of this for conservation is that preserves in ecotones are likely to be valuable for organisms with low mobility as the Earth's climate changes.

8.3.5 *Practical Constraints*

In practice, theoretical considerations do not tell the whole story when preserves are created. Scientists do not just sketch a reserve on a blank slate and direct policy makers to carve it out of the landscape. Reserves are superimposed on existing land uses, political and economic interests, ownership patterns, land-use potential, and traditions. So, the design of nature reserves is never as abstract as the discussion above might suggest. It must take into consideration political, economic, and social constraints, and often theoretical debates such as SLOSS become moot in the face of practical realities.

8.4 Setting Priorities: Which Ecosystems Should We Save?

It is not possible to protect everything, so it is necessary to prioritize areas for protection. In preservationist conservation, the goal is often to protect areas that are similar to reference areas which represent relatively "pristine" conditions. In the Western Hemisphere, many conservation biologists advocate using pre-Columbian conditions as a *benchmark*, or standard, for naturalness. This is based on the common premise that because precontact Native American populations often existed at low densities, had limited technology, and were spiritually connected to the natural world, their impact on the nonhuman world was relatively minor. Although there were and undoubtedly are differences in the environmental impacts of Europeans and Indigenous peoples, this distinction between the environmental effects of native groups and Euro-Americans is problematic because it implies that some people (Europeans and their descendants) are not part of the natural world, but others (Indigenous peoples, such as pre-Columbian Native Americans and First Nations groups) were. To avoid the problems that stem from regarding some people as more natural than others, some conservation biologists argue that all human influence is unnatural and therefore conditions with no human influence should be the standard for naturalness (Hunter, 1996a). This too is problematic (Section 9.5).

In Section 7.7, we considered some ways of prioritizing species conservation using individual species that are indicators, umbrellas, flagships, or keystones as surrogates for groups of species. If we are correct in assuming that individual species in those categories can act as stand-ins for entire groups of species, then protecting one species should indirectly protect many others. The approach to conserving ecosystems uses a similar strategy. Ecosystem conservation strives to conserve species diversity by protecting key ecosystems. There are several ways to prioritize ecosystems for protection.

As with species, when we are conserving ecosystems, we need to know what kinds of ecosystems we have, where they are, and how they are doing. Some conservation NGOs have prepared lists that provide a framework for keeping track of ecosystems (Olson et al., 2001; Spalding et al., 2007). Major types of ecosystems are described in the appendix.

8.4.1 Coarse and Fine Filters

Sometimes preserves are set aside for certain species, but since it is impossible to target every species for conservation efforts, an alternative approach is to try to protect communities. If we have a good classification system that identifies all communities, if we are good at identifying examples of those communities that are in good shape, and if our efforts to protect them are effective, then most of the species and interactions those communities contain should be protected. This has been termed a *coarse-filter* approach. The filter is the classification system for identifying communities, usually on the basis of vegetation (Nature Conservancy, 1982). One advantage of this approach is that it may conserve poorly known species that would otherwise be overlooked. Many species, particularly microorganisms that perform important ecological functions, are not well known. They cannot possibly be protected with a species-by-species approach (Noss and Cooperrider, 1994).

A coarse filter approach sometimes uses *biodiversity hotspots* (regions with high concentrations of biological diversity). Many of the species in biodiversity hotspots are threatened or endangered. Some examples of such regions are described below (Box 8.3).

Box 8.3
Examples of Biodiversity Hotspots

Cape Floristic Region: The Cape Floristic Region located at the southwestern tip of Africa (Figure 8.2) has a *Mediterranean climate* characterized by cool, wet winters and dry summers. The area is topographically diverse, extending from coastal lowlands to mountains. This diversity supports about 9,000 species of plants as well as many endemic animals. This hotspot includes a type of chaparral (Section A.1.4) known as *fynbos*. About two thirds of the plants in the Cape Floristic Region are endemic. Most of the region is prone to fire because of the dry summers that are characteristic of Mediterranean climates. Plants with edible underground storage organs are common and have been an important source of energy for hunter-gatherers in the region.

Cerrado: The *Cerrado* is a vast tropical savanna of plateaus covered by grasses, forbs, and scattered trees in central Brazil, with small areas in neighboring Paraguay and Bolivia (Figure 8.1). Frequent fires in this environment have selected for plants with adaptations to frequent burning. There are over 12,000 native plant species in the Cerrado, making it the most species-rich savanna in the world. Many plants of the Cerrado are not only endemic to the region but occur only at a single site! Many ethnic and cultural groups and their descendants have inhabited the Cerrado, including several Indigenous groups, explorers, runaway slaves, and settlers.

Mountains of Central Asia: This biodiversity hotspot includes mountains and high plateaus in parts of seven countries: five former Soviet republics and parts of China and Afghanistan (Figure 8.2). The cold, high-elevation environments of this hotspot harbor the wild ancestors of many fruits (such as apricots, apples, grapes, cherries, and pomegranates), grains, spices, and vegetables as well as 16 endemic species of tulips.

Japan: Japan is an archipelago of nearly 7,000 islands (Figure 8.2). It encompasses varied topography ranging from coastal plains to treeless mountaintops. Japan supports high species diversity, including about 5,600 species of plants, nearly a third of which are endemic. Although much of the climate is moderate, the mountains of Honshu and Yakushima encompass one of the snowiest and one of the wettest places on earth, respectively. The Japanese macaque or snow monkey lives farther north than any other wild monkey.

Hotspots often occur on islands (Japan) or in mountainous regions (Cape Floristic Region; Mountains of Central Asia). Mountains and islands both provide isolated locations within which local adaptations evolve, eventually leading to the formation of new species. Islands, by definition, are surrounded by barriers to terrestrial organisms that cannot cross water. Mountains provide high- and moderate-elevation habitat islands that are separated by low-elevation areas which may prevent movements between peaks (Section 6.5.4). In addition, in mountain environments small differences in elevation, slope, and exposure result in pronounced differences in temperature, solar radiation, moisture, and soil type over short distances. This variation favors the development of complex habitat mosaics that often support high concentrations of diverse plants, animals, and cultures found nowhere else. When interchange among the local human populations of mountain environments or of islands is low, unique languages and cultures may develop. In other words, the features that promote the development of species diversity also promote the development of cultural diversity, a point that we shall return to in Section 12.4.1.

Of course, some species might not be adequately protected by a coarse filter. Therefore, a comprehensive strategy for preservationist conservation should complement a coarse filter with measures focusing on the needs of specific organisms that would be missed. This strategy, which strives to aid species that fall through the cracks of coarse filter management, is termed a *fine-filter* approach.

Standards for evaluating fine-filter conservation are straightforward. Success can be judged on the basis of whether the target species attain levels that are likely to persist into the future. The IUCN has objective criteria for ranking species and subspecies as vulnerable, threatened, or endangered. These standards provide a roadmap for how to evaluate the status of a species.

In comparison, evaluating coarse-filter conservation is more complicated because monitoring the status of ecosystems is far more complex than monitoring the status of species. There are many ways to measure the state of an ecosystem, but there is no agreement on which is best.

8.4.2 Gap Analysis

The state of Hawaii in the USA comprises the most isolated archipelago in the world. These volcanic islands are biologically and physically diverse, with elevations ranging from sea level to over 4,000 m. The resulting intricate mosaic of habitats was the setting for the evolution of many endemic species. However, as one researcher noted, "the stimulating evolutionary insights provided by Hawaiian plants and animals are tempered by the bleak prospects for their continued survival" (Scott et al., 1986:1).

In historical times, a larger proportion of endemic bird species has become extinct in Hawaii than in any other comparable region on Earth, and many extant birds are currently endangered (Atkinson and LaPointe, 2009). Multiple factors contributed to the demise of the Hawaiian avifauna, especially habitat modification and introduced species, including pigs and other ungulates, rats, mongoose, cats, and mosquitoes that transmit avian pox and malaria to birds.

When wildlife biologist J. Michael Scott and his coworkers surveyed endemic forest birds on five of Hawaii's largest islands (Scott et al., 1986), they discovered some troubling things. First, huge amounts of money and effort were spent on programs to save a small number of highly endangered species. Scott likened this situation to trying to stop a massive hemorrhage but waiting until it was too late to do much good. Second, protected lands often failed to include the habitats of species at risk. When the team mapped areas that they deemed important for the long-term survival of native forest bird communities and associated rare species and then compared those maps to maps of protected areas, they found that on much of the island of Hawaii most of the best forest bird habitat was not protected. The situation was even worse on some of the other islands. On the basis of these findings, the researchers delineated areas they considered high priorities for additional protection.

Out of these insights, the Gap Analysis Program was born. In this approach, the distributions of selected groups of organisms are mapped and compared to the locations of protected areas. When the maps are superimposed, gaps in protection are revealed. Priority

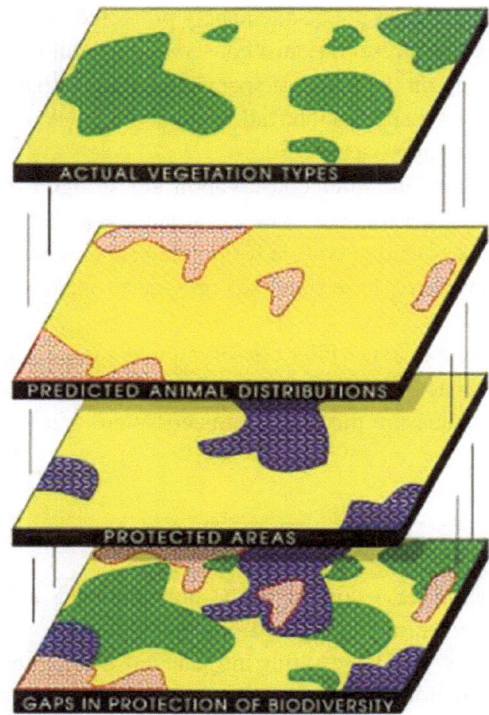

Figure 8.4   Gap analysis. Maps of a hypothetical study area showing how the distributions of vegetation types of interest, animal groups of interest, and land ownership can be used to identify gaps needing protection. Top map: actual distribution of selected vegetation types (green); second map from top: predicted distribution of selected groups of animals (pink); third map from top: distribution of protected areas (blue); bottom map: top three maps superimposed. Areas that are green or pink but not surrounded by blue are identified as priorities for protection. From Scott et al. (1993), with permission from Elsevier.

areas for biodiversity protection – places where there are many species, or concentrations of endemic species, or rare or endangered organisms – can be identified (Figure 8.4). Next, the gaps are compared to maps showing land ownership, so that potential areas for protection can be identified. Thus, gap analysis provides a tool for prioritizing and focusing habitat protection efforts, with the goal of efficiently using scarce conservation resources (Scott et. al., 1993) (*see Q8.2*). Gap analysis is widely applied in varied political and ecological contexts (Box 8.4).

Although it is a valuable tool, gap analysis can produce misleading results because it provides a snapshot of habitat protection at only one point in time. It can underestimate the level of protection faced by some communities while overestimating the security of others because it does not incorporate site history. Focusing on a community's current level of representation obscures past events. Communities that are sensitive to human activities have probably already disappeared or nearly disappeared from

Box 8.4
Using Gap Analysis to Identify Protection Priorities in China's Upper Yangtze River Basin

China has one of the world's highest concentrations of biodiversity, as well as a vast network of more than 2,000 nature reserves. However, the degree of protection varies. This creates a need for research designed to identify places where protection and biodiversity do not coincide.

In forests of the Upper Yangtze River Basin of eastern China (Figure 8.2), 110 scientists used gap analysis to map priority areas for the conservation of six indicator groups: birds, mammals, amphibians plus reptiles, insects, fungi, and flowering plants (Wu et al., 2006). They ranked potential conservation areas according to biological importance and habitat integrity. This procedure yielded 16 conservation areas that were ranked high on both those criteria. Within each of the 16 areas, the diversity of species in the indicator groups was compared with the percentage of the area that was in nature reserves.

Maps of these priority conservation areas for the region revealed important conservation areas with low protection. For example, only 8% of the Daba Mountains (*Daba Shan*), one of the highest ranked priority areas, was protected. This region of mixed evergreen and deciduous temperate forest marks a transition between temperate zones to the north and subtropical conditions to the south, an unusual configuration that is responsible for the persistence of several rare species.

The *Daba Shan* harbors many ancient endemic plant species including the dawn redwood, a deciduous conifer. This species arose about 100 million years ago and was at one time distributed across much of the northern hemisphere. Today the nearest relatives of the dawn redwood are the giant sequoia and the coast redwood of California. The dawn redwood was thought to be extinct until it was discovered in the *Daba Shan* in the 1940s.

Most dawn redwood populations died out during the Pleistocene, but some survived in the Daba Mountains where mountains blocked the glaciers, and warm slopes provided refuges from the ice age climate for the dawn redwood as well as other primitive plant species.

intensively used lands. Since they now remain only on protected lands, it will appear that they are well protected, even though they are much reduced in comparison to their former distribution. On the other hand, communities that are well adapted to human activities will be abundant and widespread outside of protected areas and poorly represented within them. They give the impression of being poorly protected because they are found mostly outside of preserves. Yet they may not need protection because their species are well adapted to anthropogenic disturbance. In parts of the Pacific Northwest of the USA, early and mid-successional forests are widespread, but they are not in need of special protection because they regenerate quickly after disturbances, and therefore they are not likely to become rare in the foreseeable future (Cassidy et al., 2001). To address this type of problem, care should be taken to interpret the results of gap analysis in the light of ecological and historical contexts.

Gap analysis and similar geographic approaches to conservation planning provide useful information if they are developed from good data and sound assumptions. In many situations, however, the data needed for gap analysis are not available. On the plus side, gap analysis programs can spur efforts to collect relevant information (Scott et al., 1993; Dudley and Parish, 2006).

Comparisons of gap analyses for different groups of organisms can shed light on what groups are not well represented in preserves. Cacti (Box 7.2) are the fifth most threatened major taxonomic group in the world, but scientists who conducted a global gap analysis of the diverse and highly threatened cactus family found that 261 out of 1,438 species (18%) of cacti were not present in any protected area. When the researchers considered only members of this group that the IUCN considered most vulnerable, they found that the percentage of threatened cacti that were not protected was even higher. Nearly one third of cacti in that category were not protected in a preserve. This was greater than the global percentage of threatened amphibians, birds, or mammals in that situation (Goettsch et al., 2019).

The most critical assumption underpinning gap analysis is that the indicator species or communities that are used are good surrogates for aspects of biological diversity which are not measured directly. But the species for which distribution data is available are not necessarily the ones that are the best indicators. Butterflies are often included in gap analysis because their geographic ranges have been mapped, their ecology is often well studied, and the presence of a particular butterfly species may indicate that its host plant is close. That does not necessarily mean, however, that other forms of biodiversity are also nearby (Margules and Pressey, 2000). One way of addressing this problem to some extent is to include vegetation mapping in gap analysis. The distribution of plant communities integrates information about many environmental variables such as soil texture, depth, and drainage; geology; climate; and site history – information that is useful for understanding the distribution of biodiversity.

Gap analysis is "rooted more in the sciences (both biological and social science) than in chance or politics" (Dudley and Parish, 2006:1). But although it provides a rigorous methodology for examining the distribution of some kinds of biodiversity in relation to protection, its application still involves subjective choices. Gap analysis is a technique for identifying priorities, but it must itself prioritize indicators of biodiversity. We cannot map the distributions of all organisms, so we must make choices about suitable indicator species or groups of species. Science can provide information that is helpful when addressing questions about what groups of species should be the focus of gap analysis. But groups of people with different priorities may nevertheless have different answers to those questions.

8.5 Managing Reserves

Once a preserve has been established, what next? Should managers simply build a fence and watch what happens? Is it appropriate to try to modify preserve environments and, if so, how much and what kinds of management are called for?

The answers to these questions depend on the goals of the protected area being managed. The IUCN organizes protected areas into a framework of categories that differ in what is being protected and what kinds of uses are allowed (Table 8.1).

Table 8.1   *Types of protected areas recognized by the IUCN* (Dudley and Parish, 2006). Level of protection decreases and allowed uses increase from top to bottom of table

Category	Principal goal
Ia Strict nature reserve	Science or wilderness protection
Ib Wilderness area	Wilderness protection
II National park	Ecosystem protection and recreation
III Natural monument or feature	Conservation of specific natural features
IV Habitat/species management area	Conservation through management intervention
V Protected landscape/seascape	Landscape/seascape conservation or recreation
VI Protected area with sustainable resource use	Sustainable use of natural resources

If left alone, the environment within a reserve changes. Succession and disturbances will alter it unless managers deliberately modify those processes. So will long-term changes in geology and climate, albeit on a longer time scale.

Preserves are never truly natural systems, for several reasons. First, some elements of the original landscape (species, structures, interactions, and processes) are invariably missing. Furthermore, reserve boundaries rarely encompass all the ecological elements and landscape features necessary to preserve an intact and fully functioning ecosystem. Even a large national park such as Yellowstone does not include all the requirements of its wildlife. The park is part of a much larger ecosystem. Few large carnivores remain. The elk and bison that summer in the park do not have enough winter forage to meet their needs within the park's boundaries. Should managers intervene to supply or compensate for these missing elements?

Second, before they were set aside, almost all reserves were affected by people, either directly by the presence of Indigenous inhabitants within the boundaries of the future reserve or indirectly through flows of matter and energy from outside those boundaries. Third, most reserves have been invaded by alien plants and animals that have altered ecosystem structure and function. Fourth, preserve biota are impacted by human-caused atmospheric changes such as ozone thinning, climate change, and radioactive fallout. Fifth, all visitors – researchers, park rangers, hunters, bird watchers, and others – have impacts. Even forms of use such as research and non-consumptive recreation, which do not involve extracting resources, have impacts.

In any case, very few reserves are managed solely for their natural features. Usually, people make demands on them as well, ranging from recreation to resource extraction. This means that managing reserves involves managing people: controlling foot traffic so it doesn't destroy sensitive vegetation, cause erosion, or disturb nesting animals; managing vehicular traffic to prevent congestion and air pollution; disposing of human waste and garbage; enforcing regulations about where people can go and whether they can collect anything; and providing interpretive information to visitors. In addition, the public often resists attempts to let nature take its course in natural areas. People usually want fires to be put out, and they don't like animals to starve. The public wants trails to be maintained. They may want injured or diseased animals to be rescued, and the public is likely to oppose animal control. Visitors seeking excitement may want to get close to dangerous animals or to push their physical limits. Such experiences contribute to the value of their experience but create headaches for park personnel and risks for less adventuresome visitors. (Adventure-seekers may also impose financial burdens on reserve management when they get in situations where they need to be rescued, especially if they are in inaccessible terrain.)

People who have traditionally harvested resources within areas that became preserves may depend on those harvests. This situation is especially acute in parts of the developing world, where in many cases people have been removed from their homelands to create wildlife preserves, a problem we shall return to in the next chapter.

8.6 Restoring Ecosystems

Although protection through the creation of reserves is a central feature of preservationist conservation, many ecosystems have already been degraded to such an extent that there is not much good-quality habitat left to protect. This fact has given rise to the discipline of *restoration ecology*, in which scientific knowledge is applied with the goal of recreating ecosystems as they existed at some time in the past. Ecological restoration is especially useful in situations where existing preserves are inadequate but are surrounded by degraded habitat that is capable of being restored. In such cases, restoration of neighboring lands can sometimes increase the effectiveness of reserves.

Much knowledge of what works in restoration results from trial and error. In the process, restoration ecology provides valuable knowledge about the structure and functioning of the restored ecosystems (Box 8.5).

Some nations, such as the USA and Canada, have government mandates that require restoration to address anticipated environmental impacts from projects such as dams, airports, and highways. If a project is expected to cause a loss or degradation of habitat, *mitigation* (compensation for or alleviation of negative effects) may be required. Mitigation can take several forms, including restoration of the affected environment, restoration of an area of comparable habitat somewhere else, or creation of similar habitat to compensate for loss of habitat at the affected site.

The mitigation process is often used where wetlands are lost or altered. Substantial losses of wetland ecosystems have occurred because of drainage, diversion, and fill. The assumption is that the value of a restored wetland will offset the loss. If that assumption is correct, mitigation should not cause a loss of wetland functions and values.

The idea behind this approach has intuitive appeal. But it is problematic for two reasons. First, we don't have unambiguous standards for measuring change in the quality or health of a wetland (or any other ecosystem). There is no clear standard that will tell us if a wetland has been improved. Although regulating agencies and scientists have developed a variety of rating systems that are useful in a lot of contexts, they involve subjective decisions.

Box 8.5
Long-Term Restoration of Native Grassland and Forest in Wisconsin
In 1934, Aldo Leopold and John Curtis began research on restoring native ecosystems at the University of Wisconsin–Madison Arboretum (Figure 8.1), which lies at the ecotone of North America's midwestern steppe and temperate deciduous forest (Sections A.1.3 and A.1.5). Their work led to the restoration of several hundred hectares of land that had been altered by logging, agriculture, grazing, and development.

Several insights about species assemblages and disturbance dynamics emerged from this project. Restoring plant populations was relatively straightforward, but the assumption that the appropriate animals would find the restored habitats and colonize them was not borne out. As a result, important interactions were missing. In the deciduous forest, the distributions of two native forbs, wild ginger and bloodroot, were affected by the absence of an ant species that normally would have dispersed their seeds. Because the ants were missing, seeds were not moved far from the parent plants, and unusually dense patches of young plants developed (Jordan, 1988).

A different insight emerged from work on returning abandoned fields to prairie. Efforts to plant native species into the sod of non-native Kentucky bluegrass and Canada bluegrass were stymied by weedy species. A series of experiments with controlled burns subsequently revealed that fire reduced the density of the bluegrasses and favored the spread of some of the native prairie species. This demonstrated the importance of disturbance as a tool for prairie restoration and stimulated further research on the role of fire (Curtis and Partch, 1948).

The University of Wisconsin Arboretum now contains native communities in which most of the pre-restoration plant species are present and non-native plants are largely absent. However, the spatial scale of the restored ecosystems imposes some constraints. Although fire can be used, the prairie remnants at the arboretum are not large enough to sustain herds of wild bison.

Second, mitigation does not always achieve its goals. It is difficult or impossible to re-establish a functioning ecosystem with all its original components. So, it is hard to tell if the promises implied in mitigation have been fulfilled. This is true even where mitigation has clear objectives with explicit measures of success, as in the next example (Box 8.6).

Although restoration that is mandated as part of mitigation can be a valuable research tool, mitigation may increase the likelihood of habitat loss. If a developer is allowed to degrade a habitat in return for a promise to restore or create similar habitat elsewhere, and if mitigation fails to do so, as is often the case, the result is a net loss of habitat.

Efforts to restore missing elements can operate at a variety of scales. An example of a continent-wide approach to ecological restoration is described in Section 8.7.

As a result of habitat loss and exploitation many large animals, including many carnivores as well as herbivores such as bison, have disappeared from much of their range. Consequently, interactions with these large animals are now missing from much of the landscape. The next section covers an approach to restoration that proposes to address this situation.

Box 8.6

Ecosystem Restoration for Endangered Species in a California Salt Marsh

In 1984, as a result of construction that damaged a marsh which provided habitat for two endangered birds (the light-footed clapper rail and the California least tern) and an endangered plant (the salt marsh bird's beak), the California Department of Transportation was required to provide habitat for the endangered organisms by restoring a salt marsh within the Sweetwater National Wildlife Refuge in San Diego Bay (Figure 8.1).

Permits for this project stipulated specific performance standards for the endangered species populations. After several years, these were met for one of the birds (the tern) and the endangered plant. But although the area came to resemble an unmodified wetland in terms of water level and plant composition, it failed to attract nesting clapper rails. Closer analysis revealed that even several years after restoration was begun, cordgrass – an important component of the vegetation at the marsh – had not reached its full height. This meant that it did not provide adequate nesting cover for breeding clapper rails. Further research suggested that the cause was inadequate nitrogen, which rapidly leached from the site's sandy soils. That problem was addressed by applying fertilizer, but although the cordgrass grew taller, within a few years it was attacked by insects. Additional research identified a missing predatory beetle that normally would keep the herbivorous insects in check.

After nine years, researchers had learned a great deal about the functioning of the marsh, but the site still failed to attract one of the endangered species for which it was designed. Each time a missing link in the functioning ecosystem was identified and replaced, another missing piece connected to it became obvious (Zedler, 1988; Haltiner et al., 1997).

These results were disappointing and expensive for the California Department of Transportation, but from a scientific standpoint, the project was very valuable. Each stage was treated as an experiment designed to test hypotheses about ecosystem functioning. The results suggested additional questions and testable hypotheses (Zedler, 2000).

8.7 Rewilding: A Landscape Approach to Restoring Large Animal Populations

Some conservationists articulate a vision of landscape-scale restoration focused on *rewilding*, the reintroduction of large animals in their former ranges (Foreman, 2004). This approach is based in part on the assumption that large carnivores and herbivores are usually umbrella species (and therefore expanding their geographic ranges should also protect the habitat requirements of many other species) and keystone species (so protecting them should conserve crucial ecological processes). Although rewilding can involve large herbivores such as bison as well as carnivores, much attention is focused on carnivores because they are at the top of food chains and might have consequences that cascade throughout ecosystems.

Advocates of rewilding also point to intangible reasons for bringing back large, wild animals. In an essay on *Rewilding and Biodiversity*, conservation biologists Michael Soulé and Reed Noss (1998:24) wrote: "Wilderness is hardly 'wild' where top carnivores, such as cougars, jaguars, wolves, wolverines [a large member of the weasel family], grizzlies, or black bears, have been extirpated. Without these components, nature seems incomplete, truncated, overly tame. Human opportunities to attain humility are reduced."

Pleistocene rewilding expands upon this vision by suggesting that conservation should strive to recreate species assemblages similar to those that occurred in Eurasia and North America at the end of the last ice age, before widespread megafaunal extinctions occurred (Section 6.4.1.1). This would mean introducing either extant species that are descended from extinct Pleistocene species (and might still be close genetically) or modern-day species that are ecologically (but not necessarily genetically) similar to extinct ones. For example, living giant tortoises have been introduced to an island in the Indian Ocean as surrogates for extinct giant tortoises. Proponents of Pleistocene rewilding suggest that Bactrian camels in the genus *Camelus* might serve as a proxy for extinct camels from the genus *Camelops* (Donlan et al., 2005).

This proposed restoration of conditions from the Pleistocene is controversial on both theoretical and practical grounds. Advocates suggest that this kind of rewilding would restore evolutionary potential that was lost when important interspecific interactions ceased because of the megafaunal extinctions. Critics argue that ecosystems have changed in the thousands of years since the end of the last ice age, and that rather than restoring pre-existing ecosystem functions, introductions of non-native species could have catastrophic consequences (Rubenstein et al., 2006).

Carnivores can have far-reaching ecological effects on multiple trophic levels. Several changes occurred after wolves were reintroduced to Yellowstone National Park. Willows and quaking aspens along streams increased, as did populations of beaver. Changes in water tables and river channels also occurred. In scientific literature and popular media, the wolf introduction is credited with boosting populations of animals ranging from beetles to bears, and with transforming "not just the ecosystem ... of the Yellowstone National Park, but also its physical geography" (Sustainable Human, 2014) (*see Q8.3*).

However, some carnivore biologists (Mech, 2012; Allen et al., 2017) suggest that this idea has been accepted uncritically. In a journal article with the title "Can we save large carnivores without losing large carnivore science?" they point out that many studies and popular interpretations assume that where reintroductions of carnivores to their historic ranges were followed by ecological changes, those changes must have been caused by the reintroductions. But drought, winter weather, fire, climate change, hunting pressure, changes in predation by bears and cougars, and other factors also contributed to the changes that followed the arrival of wolves in Yellowstone.

Cheetahs, lions, African elephants, and some other animals that are candidates for reintroduction are currently rare or endangered. Rewilding advocates suggest that conservation of those species might get a boost from reintroduction efforts, while critics contend that it would be unwise to divert attention from ongoing efforts to conserve them in their current ranges.

Another practical question is how the public would respond to the introduction of non-native carnivores. Proponents of Pleistocene rewilding contend that "humans have emotional relationships with large vertebrates that reflect our own Pleistocene heritage ... [therefore] Pleistocene re-wilding would probably increase the appeal and economic value of both private and public reserves" (Donlan et al., 2005:913). Others suggest that there would be substantial opposition to rewilding, at least where it involves carnivores, because of concern about the potential for attacks on people and domestic animals. Reintroductions of bison, especially where they are central to Indigenous cultures, are less controversial.

The rewilding controversy highlights a central ethical issue in preservationist conservation: What is our responsibility to the nonhuman world? According to Soulé and Noss (1998), people have a responsibility to address "ecological insults" caused by past overhunting, extirpation, and land abuse. In this view, not accepting this responsibility amounts to arrogance and moral failure. We will return to this important issue in the next chapter.

8.8 Evaluating Ecosystem Conservation

According to a report put out by the Secretariat of the Convention on Biodiversity Diversity, between 1970 and 2010, the worldwide extent and biodiversity coverage of protected areas increased. But when the group analyzed 31 indicators of trends in biodiversity (at the levels of genes, populations, species, and ecosystems), they found that most indicators of the state of biodiversity declined during that period (Butchart et al., 2010). Determining the reasons for this is important but difficult because of the variety of ecological, socioeconomic, political, and cultural settings in which preserves occur.

The purpose of site evaluation is to infer how a course of action changed that site's fate compared to what would have happened in the absence of the project. This involves speculation because we cannot know what might have happened and because there are many confounding factors. Comparisons across time (before-and-after studies) or space (comparisons to nearby control areas) might help with this challenge, however. If data on baseline conditions before reserve designations or site restorations are available, those data can be compared to data collected after the change. Often such information is not available, however. It would be useful if this kind of information were required before preserves were established, but that is rarely the case.

Once a preserve has been designated or restoration is underway, conditions within and outside a project can be compared if a similar site can be found. This is challenging because of biases in the locations of preserves. Scientists at Duke University in North Carolina used statistical methods to test the hypothesis that the locations of protected areas in 147 countries were biased toward remote places (which they termed "rock and ice"). They found that in comparison to unprotected areas, most national protected area networks were likely to be located at higher elevations, on steeper slopes, and at greater distances from roads and cities (Joppa and Pfaff, 2009). This finding suggests that the placement of reserves tends to be biased toward protecting a small set of landscapes and the biodiversity they contain. Furthermore, protected areas in mountainous terrain might not have been exploited even if they had not been protected, because, as noted above, they tend to be inaccessible.

Analyses of data from remote sensing of vegetation have shown that tropical forest cover is declining within protected areas. This could be interpreted to mean that protected areas are not making a difference, but comparisons with areas outside reserves reveal that on average deforestation is higher outside the reserves than within them (Geldmann et al., 2019).

Assessment is even more complex when we consider social, cultural, ethical, and aesthetic impacts. These often-overlooked factors are important because the fate of protected and restored areas is affected by how they are viewed by stakeholders (Higgs, 1997; Wohl et al., 2015).

We have seen that preserve design, protected area management, and ecological restoration usually involve efforts to move toward minimizing the effects of people. In the next four chapters we will look at challenges to that approach and the development of a *stewardship approach* geared toward conserving complex and resilient ecosystems which integrate conservation of biological and cultural diversity.

We will begin with developments that set the stage for these new developments.

Part III

Promoting Biocultural Diversity and Resilience:
A Stewardship Approach to Conservation

9

Historical Context

New Opportunities and Challenges

Utilitarian and preservationist approaches to addressing environmental problems have been woven through the conservation movements of the Western world for nearly two centuries. By using those approaches, scientists and managers made crucial contributions to conservation, but by the late twentieth century changing social, economic, and political conditions along with new scientific insights and trends in ethics and philosophy presented challenges not fully addressed by those strategies. Excessive exploitation, habitat conversion, invasive species, and toxins in the environment led to escalating threats to biodiversity. Dramatic growth of the human population and improved standards of living added to the stresses on ecosystems.

By the late twentieth century, players with a variety of perspectives and agendas had joined conversations about what conservation should entail. Indigenous rights activists, advocates for animal rights and rights of nature, ecofeminists, scholars in the social sciences and humanities, legal experts, and representatives of non-governmental organizations, national governments, and international development agencies offered diverse perspectives and competing agendas.

Although there are important differences in the interests these new players bring to the table, there are common threads. Many advocate a transformative change in the way people relate to the natural world. Some promote *sustainable development* (long-term use that meets present needs without compromising the needs of future generations). That focus goes along with attention to the social, political, and cultural consequences of conservation, particularly for the survival of threatened cultures and marginalized groups. At the same time, the relevance of local and traditional ecological knowledge attracted the interest of scholars in several disciplines, and studies of the effects of natural disturbances led to reassessments of classical ecological theories. The current scale of environmental changes infuses a sense of urgency into these discussions.

Many stakeholders now seek an approach that strives to integrate conservation of complex ecosystems and cultural diversity. This approach developed in response to six kinds of issues, which we consider in this chapter. The first issue is a practical one. Protected areas and species recovery programs have not stemmed the ongoing loss of biodiversity (a practical issue). Second, prevailing theories of ecological dynamics led to conservation strategies focused on controlling a few processes, with unanticipated consequences (a scientific issue).

Third, excluding people from the decision-making process and from using resources they have traditionally used creates resentment, erodes support for conservation, and contributes to inequitable power dynamics that undermine conservation (socioeconomic and political issues). Fourth, policies that aim to protect biodiversity often failed to deal in a fair way with the needs of people who depend directly on natural resources, and those policies sometimes undermine human rights and cultural diversity (ethical issues). Fifth, the idea that people are not a part of nature has been challenged (a philosophical issue). And finally, the rights of Indigenous peoples to self-determination and control of their resources are increasingly recognized in international declarations and court cases (legal issues).

These problems are, of course, interrelated. Excluding people from their resource base creates socioeconomic and political problems, but it is also problematic on ethical grounds. If we re-evaluate our philosophical position regarding our place in nature, that has ethical implications. If our scientific understanding does not reflect conditions in the natural world, that has philosophical implications as well as practical consequences for the effectiveness of strategies based on that understanding. Conforming to international declarations has ethical and legal implications. But, for the sake of simplicity, these six types of problems are considered separately below.

9.1 Practical Considerations: What Is Feasible?

Although efforts to protect and restore species and ecosystems have been effective in some settings, extinctions and ecosystem degradation continue. Furthermore, it is simply not possible to assist more than a small fraction of the Earth's biodiversity using the strategies described in Chapters 7 and 8. Even with efficient and effective methods of setting priorities, some species and ecosystems are left out or cannot be effectively protected.

9.2 Scientific Considerations: Science for a Complex and Rapidly Changing World

9.2.1 Consequences of Halfway Technologies: Interfering with Whatever Little Turtles Do

In an influential paper on *Sea Turtle Conservation and Halfway Technology* in the journal *Conservation Biology*, the ecologist Nat Frazer argued that many interventions designed to aid declining species are *halfway technologies*. Halfway technologies in medicine are things we do to compensate for a disease or to postpone death when we don't understand the disease process. A heart transplant might prolong the life of the patient, but it does not cure the underlying disease. Frazer argued that many conservation measures likewise fail to address the causes of species' declines:

When we define the impending extinction of a sea turtle species solely in terms of there being too few turtles, we are tempted to think of solutions solely in terms of increasing the numbers of turtles…. Programs such as headstarting, captive breeding, and hatcheries may serve only to release more turtles into a degraded environment in which their parents have already demonstrated that they cannot flourish.

(Frazer, 1992:179)

In addition to suggesting that such technological fixes treat symptoms but not underlying causes, Frazer objected to these manipulations for a more fundamental reason: They prevent turtles or other species that receive special care from serving ecological roles in their environment. He suggested that the death of the very individuals whose mortality we strive to prevent may fulfill ecological functions that we do not understand or even suspect. Although they are not contributing to the next generation of sea turtles, young turtles that die in the wild might be providing food for parasites or predators; they might limit the extent of beds of seagrass by grazing on them; or they might decay and provide detritus to microorganisms that form the basis of marine food chains. By "interfering with whatever little turtles do," headstarting could disrupt ecological relationships in unknown ways:

Do we assume that the millions of little turtles that used to come off our beaches play no important role simply because we rarely see them playing it? We like to see big turtles nesting, and we like to see eggs being laid and little turtles hatching and entering the ocean, and we like to see large juveniles feeding in our seagrass beds and reefs and coastal waters. But we almost never see little turtles doing whatever little turtles do in their natural environment ....

The halfway technology of headstarting ... prevents the turtles, while they are being held in captivity, from performing whatever ecological function they normally serve in the natural environment.

(Frazer, 1992:181)

Frazer's point is akin to the point Leopold made in his essay *Thinking Like a Mountain*: Our appreciation of the roles that organisms play in ecosystems is often superficial (Section 4.3.3). Leopold was concerned with the implications of utilitarian management, but Frazer points out that even preservationist management often involves manipulating ecosystems to favor a narrow subset of organisms or processes of interest to people, with ecological consequences that are poorly understood (*see Q9.1*).

9.2.2 The Anthropocene: New Conservation and Novel Ecosystems?

With both utilitarian and preservationist conservation we assume that we can control nature by tinkering with a few processes. But the natural world is not always stable and predictable. Ecosystems can suddenly flip into unexpected, and often irreversible, states. Ecosystems that tip readily into a new state rather than returning to their prior condition are said to have low *ecological resilience*. Anthropogenic change is increasing the likelihood of such transitions.

People have altered cycles of matter and energy, facilitated extinctions, disrupted interspecific interactions, introduced toxic synthetic chemicals, changed the chemistry of the atmosphere, changed disturbance regimes, and caused widespread introductions of invasive species at unprecedented scales. The pace of anthropogenic change makes it even harder to understand the functioning of the contemporary biosphere and the consequences of our interventions. Because anthropogenic changes have such pervasive effects, some scientists now use the term *Anthropocene* to denote the current geological epoch.

The implications of these changes for conservation are hotly debated. Some scientists suggest that today's conservation strategies are inadequate for tackling the realities of the Anthropocene. They suggest that we need a *new conservation* that pays more attention to finding ways to simultaneously maximize the preservation of biodiversity and the improvement of human well-being. This, they contend, entails broadening the concerns of conservation to include poverty alleviation and other sociopolitical and economic factors. It also means recognizing the conservation value of habitats that are outside protected areas, what the journalist Emma Marris calls "a global, half-wild rambunctious garden" (Marris, 2011:2; Kareiva and Marvier, 2012). Proponents of this approach argue that sometimes we have to manage for *novel ecosystems* which are so fundamentally altered that they cannot be restored to historical conditions. They conclude that in some cases even the goal of eliminating exotic species should be abandoned (Kareiva et al., 2012) (*see Q9.2*).

Critics of the new conservation maintain that it compromises conservation objectives, surrenders to demands for unlimited economic growth and development, and treats nature as "a warehouse for human use" (Miller et al., 2014:509). They point to the varied and widespread harmful impacts of many non-native species and suggest that "the contention that novel ecosystems are inevitable and perhaps desirable encourages any tendency to delay prevention and redress of various harmful environmental impacts" (Simberloff, 2015:50).

Some scientists and development experts suggest that rather than arguing about whether to prioritize parks or people, we should do both. In fact, many conservation projects are touted as *win-win strategies* that will conserve biological diversity and improve human well-being. This is a laudable goal and one that appeals to funding agencies. But it is complex. Win-win scenarios are not always possible. Compromises are necessary, tradeoffs between the interests of groups of people with different priorities are inevitable, and evaluating tradeoffs requires rigorous ways to assess the complex impacts of different options.

The new-conservation controversy is about values as well as methods. Proponents of that approach argue that conservation needs a more human-friendly vision. Its critics contend that the new approach refuses to acknowledge that people "act unjustly by displacing other species or degrading their habitats" (Noss et al., 2013:242). This issue – what are the ethical obligations of people to nonhuman life forms? – is at the heart of many controversies that we will encounter in this and the following chapters.

9.3 Socioeconomic and Political Considerations: Contrasting Interests and Perspectives

9.3.1 Causes of Environmental Damage

Recall that concern about human population growth has a long history in Western culture (Section 5.3.1). Lloyd and Malthus speculated about potential connections between population and resource depletion as far back as the eighteenth and nineteenth centuries. But in the mid-twentieth century, ecologists became particularly concerned about the effects of human population growth on tropical forests. The idea was that in poor countries, especially those in the tropics, where a great deal of biodiversity occurs, population growth caused more and more land to be cultivated, and as the amount of land suitable for agriculture dwindled, forests were cleared and steep, highly erodible hillsides were farmed (Croat, 1972).

Ecologists interpreted this correlation in time as evidence that population growth was a major driver of deforestation and other negative environmental impacts. However, political ecologists, who look at the effects of economic structures and power relations on the environment, argue that this interpretation is an oversimplification. They suggest that the relationship between human population growth and habitat degradation depends on land ownership and other socioeconomic and political factors that influence access to resources (Boxes 9.1 and 9.2) (Bryant, 1997).

Box 9.1
Causes of Deforestation in El Salvador

The political ecologist William Durham at Stanford University has analyzed the causes of deforestation in El Salvador (Figure 9.1), one of Central America's most densely populated nations. The growth curve for El Salvador's human population after 1800 approximates exponential growth (Figure 9.2). In the period between 1770 and 1892, the population of El Salvador increased by 1.38% per year, and estimated forest cover declined. These parallel trends, if viewed without considering other things that might have affected deforestation, might seem to support the idea that growth of the human populations is the principal cause of environmental degradation.

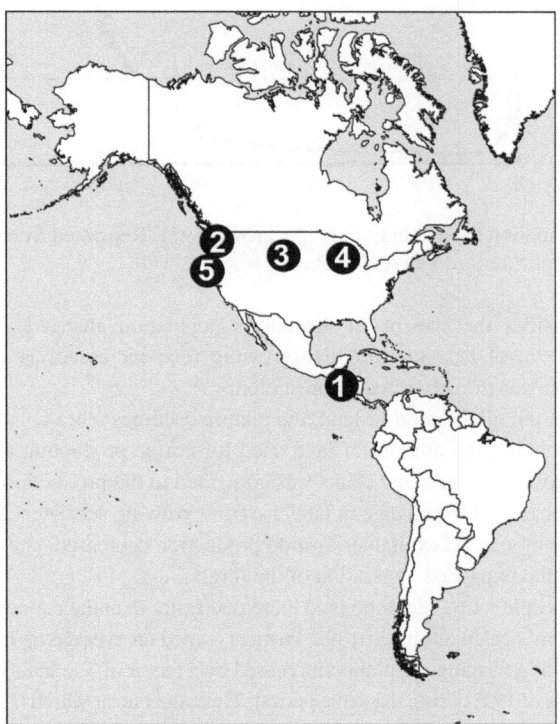

Figure 9.1 Locations of: 1, El Salvador; 2, Pacific Northwest, USA, and southwestern coast of Canada; 3, Devils Tower, Wyoming; 4, Menominee Indian Reservation, Wisconsin; 5, south end of Puget Sound, Washington State. Map created by Eva Strand using Esri, DeLorme World Countries Generalized Data & Maps for ArcGIS 2013, with permission.

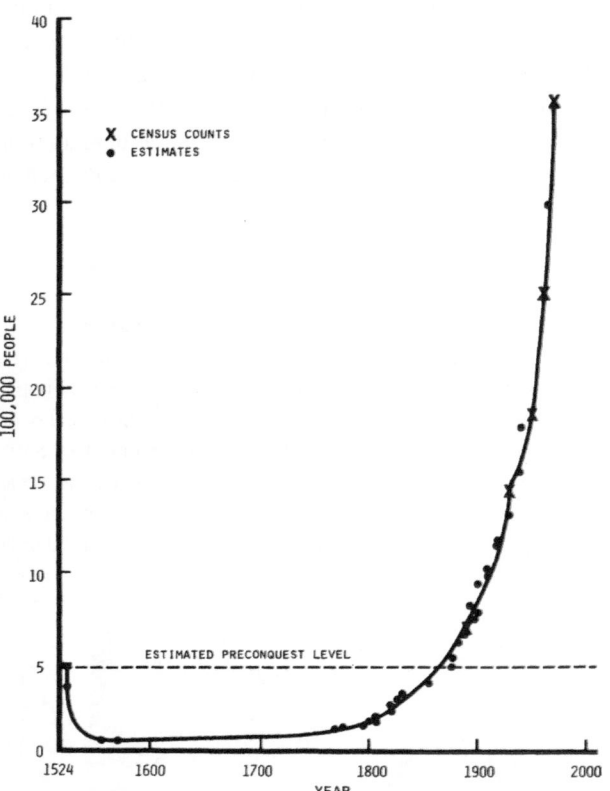

Figure 9.2   Population growth in El Salvador, 1524–1971. Reprinted from Durham (1979), with permission of Stanford University Press.

Other things besides the size of El Salvador's population changed as forest cover was declining. The amount of land converted to growing food increased, as we would expect if increased population was the driver for deforestation.

But if we look at the distribution of land, the picture becomes more complex. In the last half of the nineteenth century, the amount of land used for coffee production expanded. Land that had been used to grow corn, rice, and beans was converted to the production of coffee and other crops for export. Because of the change in land use from growing subsistence crops to plantation agriculture, the amount of land available for food production decreased, and small-scale farmers moved further into and expanded cultivation of the forests.

The number of people with little or no land increased faster than the country's total population. According to Durham's calculations, in 1892, farmers owned an average of 7.4 ha. Between 1892 and 1971, El Salvador's farming population increased by a factor of 3.8, but average land holdings decreased by a factor of 19.5 during the same period. Thus, the rate at which land holdings declined was more than five times greater than the rate at which the population increased. By 1971, half of the agriculturally active population was either landless or owned less than 1.0 ha of land.

In other words, although the growth of El Salvador's population contributed to deforestation, changes in land use and land ownership also played a role.

The work of Kathleen Homewood, a British anthropologist, and her coworkers provides another analysis of the contribution of socioeconomic and political factors to environmental problems (Box 9.2) (*see Q9.3 and Q9.4*).

Box 9.2

Causes of Declines in Wildlife Habitat and Abundance in East Africa

In a paper on *Long-term Changes in Serengeti-Mara Wildebeest and Land Cover: Pastoralism, Population, or Policies?*, Homewood et al. (2001) reported on a long-term study of interacting factors that potentially affect ecological change in East Africa's *Serengeti-Mara Ecosystem*, the area within which millions of wildebeest make their annual migrations (Figure 9.3) and pastoralists traditionally graze their livestock. During the wet season in this region, grasses on the plains provide adequate forage for wildebeest and other large grazing animals to survive and reproduce, but during the dry part of the year they do not. At the start of the dry season, these animals migrate away from the plains and toward the northwest, where the tall grasses of open woodlands sustain the population until the rains return.

Most of the rural population of the Serengeti consists of traditional livestock herders; many also engage in some farming and some work in wildlife-based tourism. Rapid growth of the human population and the spread of cultivation are widely held to have caused habitat loss and declines in the wildebeest population within this region. Homewood and others framed this interpretation as a hypothesis, which they tested in studies of the ecologies of wildebeest and human communities in the savanna that straddles the border between the nations of Kenya and Tanzania. The political geography of the region allowed this group to conduct a natural experiment because Kenya and Tanzania differed in their economic systems. At the time of this study, Tanzania had a centrally planned economy, whereas much of Kenya's land was privately owned. Livestock herding, wildlife-based tourism in nature reserves, farming, and hunting occur on both sides of the border, but mechanized commercial farming was permitted only in Kenya.

Homewood and others synthesized information from a variety of sources on multiple variables that potentially affect wildlife populations and land use in the area (Table 9.1). Their data showed that over a 20-year period, wildebeest had declined in Kenya but not in Tanzania. In Kenya, the amount of land used for commercial wheat farming increased during the study period, while the wildebeest population declined. Areas that were converted to commercial farming were fenced to exclude wildebeest. Since the wildebeest had previously used the converted farming area during their breeding season, the fencing reduced their breeding habitat. This did not happen in Tanzania, where there was no commercial farming.

Because the expansion of commercial farming occurred during a period when wildebeest numbers declined, the researchers reasoned that the factors that caused the increase in commercial farming might have contributed to the wildebeest decline. In Kenya, favorable market conditions and government policy decisions that encouraged private land tenure were associated with individuals' decisions to convert to commercial agriculture (which in turn was associated with negative impacts on wildebeest numbers). But neither growth in the herder population nor trends in cattle numbers were associated with the wildebeest decline in Kenya (Homewood et al., 2001). Thus, the researchers' answer to the question they posed in the title of their paper was that policy, not pastoralism or population, was the principal cause of declines in wildlife habitat and wildebeest numbers.

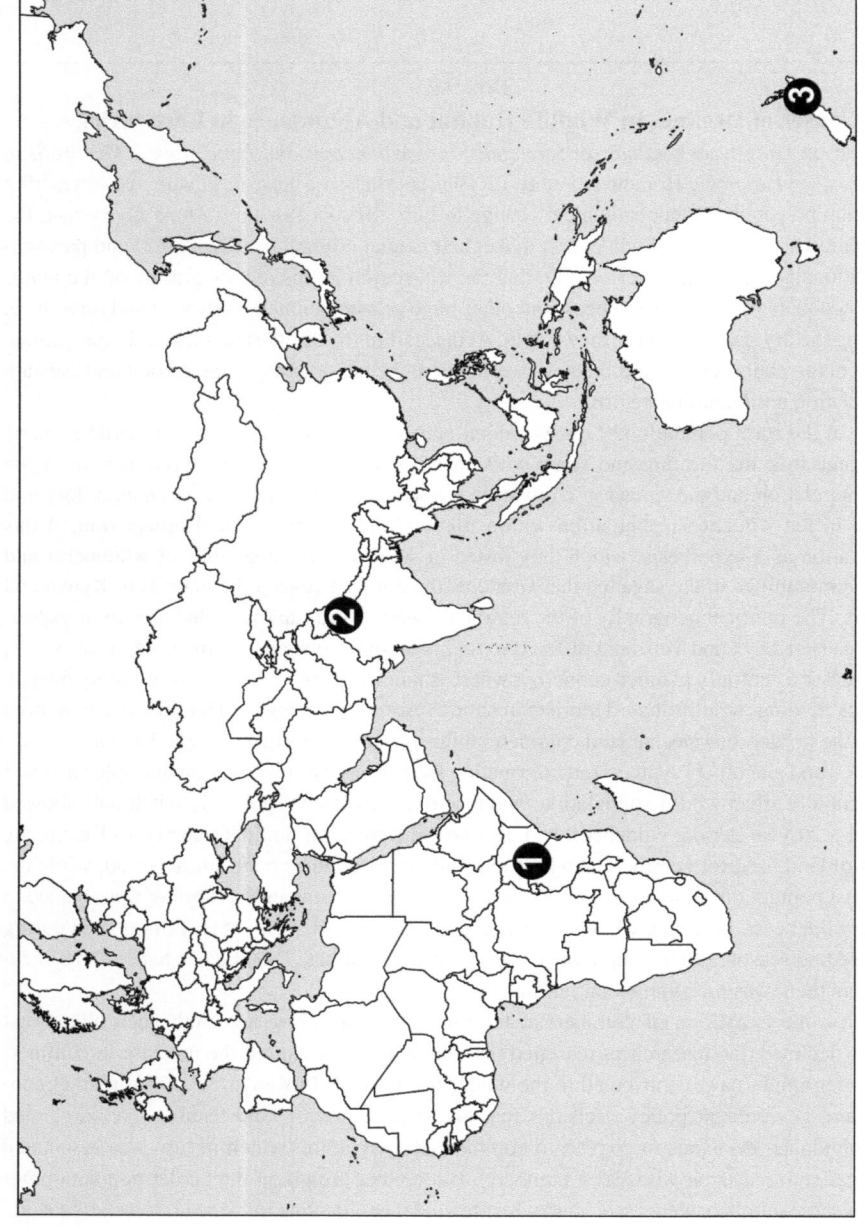

Figure 9.3 Locations of: 1, Serengeti-Mara Ecosystem; 2, Uttarakhand, India; 3, Whanganui River, New Zealand. Map created by Eva Strand using Esri, DeLorme World Countries Generalized Data & Maps for ArcGIS 2013, with permission.

Table 9.1 *Sources of information on explanatory and response variables assessed in a study of the causes of ecological change in the Serengeti-Mara Ecosystem (Homewood et al., 2001).*

Variables studied	Assessed from
Explanatory Variables	
Ecological variable	
Climate trends	Rainfall in wildebeest dispersal areas over time, obtained from monthly weather data
Plant productivity (biomass)	Vegetation surveys
Socioeconomic variables	
Human population growth	National census data
Livestock population growth	Livestock population records
Human well-being	Household surveys, archives (reproductive history, mortality, fertility, income, settlement size, migration, education, land-use choices, economic returns)
Economic policies	Archived data on policies that affect benefits of different land uses: agricultural subsidies, permitted land uses, land tenure, market access
Response variables	
Wildlife population trends	Trends in numbers of wildebeest and other large grazing mammals, calculated from aerial surveys
Land cover	Change in area of large farms and small farms from Landsat time-series data

The emphasis that political ecologists placed on connections between poverty and environmental degradation generated pushback. Some ecologists suggested that political ecologists' focus on socioeconomic factors neglected ecological considerations (Walker, 2005).

9.3.2 Resource Conflicts

The next two examples, one from India (Box 9.3) and one from the USA (Box 9.4), highlight situations in which conflicts about livelihoods and forest resources have occurred between groups with different interests.

Preservationist conservation was problematic in both these situations. The Chipko protestors were interested in restoring their rights to the forest. They were not interested in strict protection. The Indian ecologist Ramachandra Guha contends that a strategy of environmental protection in places that are off limits to people is inappropriate for places

Box 9.3
Resistance to Commercial Forestry in India's Unquiet Woods

In Uttarakhand, a state in northern India where tropical forest dominates steep terrain along the foothills of the Himalayas, the traditional economy consists of grain cultivation, cattle raising, and the collection of forest products (Figure 9.3). Women in this society contribute a substantial portion of the agricultural labor and also collect wood and fodder from the forest. In the early twentieth century, protests against Forest Department regulations were a recurrent feature in this region. Traditional forest management clashed with the scientific forestry that had been imposed by British colonial administrators. Under the latter, local supplies of forest products dwindled so much that people in some villages had to walk long distances to obtain fuelwood.

The situation worsened in 1970 after severe flooding during the monsoon caused the worst floods in living memory. Floodwaters and landslides destroyed bridges, roads, buses, farms, and houses, killing hundreds of people and animals. Villagers saw a link between large-scale clear-cutting and floods, landslides, and erosion. In this context, a grassroots, non-violent social movement for the protection of forests and villagers' rights to manage them emerged from India's "unquiet woods" (Guha, 1989b).

In 1964, a local cooperative organization working for village self-rule formed with the goal of developing employment opportunities in forest-based industries, in accordance with the philosophy of Mahatma Gandhi. Subsequently, the group asked the Forest Department for an allotment of ash trees to provide wood for the manufacture of farm implements in their local workshop. The agency refused the request but granted a much larger allotment to a sports factory. In response, when contractors arrived, villagers hugged the designated trees. Their resistance succeeded in preventing the harvest.

Tree hugging caught on in other villages, and the *Chipko* ("to hug") movement was launched. The following year, when men were lured away from the village of Reni by a government ruse, women blocked access to nearly 2,500 trees that were slated for auction (Bhatt, 1990). The movement spread, gained wide international support, and succeeded in substantially slowing commercial forestry in the region.

The Indian ecofeminist Bina Agarwal has pointed out that because the women differed from the men in their responsibilities, interests, needs, and knowledge, they also had a different perspective on the environmental issues connected to Chipko (Agarwal, 1992). Control of critical resources was especially important to the women because they provided most of the labor for traditional forestry. When the male councilors of one village unanimously agreed to convert some of their oak forest to a potato-seed farm because it would bring in cash, women had a different understanding of the potential gains and losses. They opposed the project because it would destroy an area that provided them fodder and fuel and would therefore require them to travel an additional 5 km to collect what they needed (Agarwal, 1992). Although their labor would increase, the women were unlikely to benefit from the new cash income, which would go to the men. In addition to their opposition because of these concerns, they objected to the fact that they had not been consulted. "When we undertake all the work related to forests," they asked the district authorities "why was our opinion not taken?" (Bhatt, 1990:14).

Box 9.4
Conflict over Old-Growth Forest in the Pacific Northwest

The northern spotted owl occurs in temperate coniferous forests, especially old-growth forests in the Pacific Northwest of the USA and the southwestern coast of Canada (Figure 9.1). These 350–750-year-old forests are dominated by massive Douglas-fir trees (Franklin et al., 1981). The spotted owl is considered an indicator species (Section 7.7) for this structurally complex and productive ecosystem. Scientists have estimated that a viable population of northern spotted owls requires large expanses of late-successional habitat.

The Pacific Northwest in the USA includes extensive forests on public lands, where federal government policies regarding resource use and management are controversial. Most of the habitat that would meet the owls' needs is on those public lands in areas that also provide jobs in rural communities where both the forest products industry and the area of old-growth forest had declined markedly during the late twentieth century.

In 1990, the northern spotted owl was listed as threatened under the US Endangered Species Act. This meant that federal agencies which managed land within the owl's range were required to manage for its conservation. In response to litigation and counter-litigation from both environmental groups and the timber industry regarding the federal government's responsibility for protecting owls, scientists developed a comprehensive management plan for Pacific Northwest forests. The goals of the plan, which was adopted in 1994, were to ensure the viability of the spotted owl while allowing some timber harvest on federal lands. It called for extensive reserves in riparian zones and late-successional forest but permitted regulated timber sales in the matrix between reserves. Both sides of the debate criticized the plan.

The debate over the fate of the Northwest's ancient forests was acrimonious, characterized by lawsuits, injunctions, and appeals, as well as civil disobedience, violence, and bitterness. Although the media portrayed the conflict as a choice between protection of a single species versus jobs, many historical factors – including automation of the timber industry, escalating foreign competition, and shifts in the age structure of the region's forests from mostly old, large trees to younger trees because of prior overharvesting – contributed to job losses.

People on each side of the dispute valued the forest but in different ways. Rural residents tended to see the forest as a source of jobs and a lifestyle. They identified with the Wise Use movement of the American West, which held that making a living from working in the woods required increased access to public lands, an expansion of private property rights to land, and reduced government regulation. Environmentalists saw the forest as a source of aesthetic and spiritual inspiration as well as ecosystem services, including endangered species protection, and identified with environmental NGOs. Their concerns focused on a broad critique of the nation's forest policy because it allowed timber harvests that they considered economically and ecologically unsustainable and which depended on government subsidies (Proctor, 1996; Prudham, 1998).

---

like India, which are densely settled by peasants and tribals. For these groups, defense of their traditional livelihoods against commercial interests is more important than protection of supposedly pristine places. According to Guha,

The ecological battles presently being fought in India have as their epicenter the conflict over nature between the subsistence and largely rural sector and the vastly more powerful commercial industrial sector .... These include opposition to large dams by displaced persons, the conflict between small artisan fishing and large-scale trawler fishing for export, the countrywide movements against commercial forest operations, and opposition to industrial pollution among downstream agricultural and fishing communities.

(Guha, 1989a:80–81)

In the PNW, loggers and environmentalists both saw the conflict as a choice between protection versus use (although the scientific team that developed the management plans sought to craft recommendations that integrated the two). In an essay titled *Are You An Environmentalist Or Do You Work For A Living*? (a saying from a bumper sticker sold in the logging town of Forks, Washington), the environmental historian Richard White pointed out that these opposing views of nature reflect different views about work (White, 1996) which contributed to the impasse between the two groups. Loggers felt that their appreciation of the forest was grounded in family traditions of working in the woods. White suggests that environmentalists tended to condemn types of work such as logging and large-scale farming that directly harvest resources (although small-scale farming is an exception). This attitude toward work often goes along with a preference for physically taxing leisure activities such as backpacking, skiing, or kayaking, and the idea that natural places should be used be used only for recreation. This approach, White suggested, leaves environmentalists in the position of "patrolling the borders" of the natural world (White, 1996:185). It contributes to a perception shared by many poor people, people of color, and Indigenous people that mainstream environmentalists are elitist and do not care about the concerns of marginalized groups, preferring to focus on environmental issues in places where people do not live. White's critique echoed concerns about the ethical implications of conservation movements.

9.4 Ethical Considerations

9.4.1 Environmental Justice

In 1987, the Commission for Racial Justice of the United Church of Christ issued a report on *Toxic Wastes and Race in the United States*, which found that Blacks and minority ethnic groups in the USA were disproportionately exposed to hazardous wastes in their neighborhoods (Chivas, 1987). A subsequent report in 2008 concluded that although there had been some improvement, the problem had persisted (Bullard et al., 2008). Reaction to this situation crystallized in the *environmental justice movement*, a new form of environmental activism.

The environmental justice movement has also spawned movements to address inequities in other contexts. In poor countries, environmental protections are generally weaker, working conditions are more dangerous, and environmental hazards are greater. In some regions, such as the western Amazon, opposition to health hazards from contamination is coupled with resistance to the extraction of resources such as oil (Orta-Martínez and Finer, 2010). In other contexts, the recognition that extreme weather events, rising sea level, and similar manifestations of climate change affect people in poor communities and in island nations disproportionately has led to a movement for *climate justice* that involves advocacy for mitigation, adaptation, and compensation.

As the environmental justice movement focused attention on environmental problems affecting marginalized groups, another issue related to the social impacts of conservation came to the fore: the displacement caused by the designation of protected areas (Agrawal and Redford, 2009).

9.4.2 Conservation Refugees

It's no secret that millions of native peoples around the world have been pushed off their land to make room for big oil, big metal, big timber, and big agriculture. But few people realize that the same thing has happened for a much nobler cause: land and wildlife conservation. Today the list of culture-wrecking institutions put forth by tribal leaders on almost every continent includes not only Shell, Texaco, Freeport, and Bechtel, but also more surprising names like Conservation International (CI), The Nature Conservancy (TNC), the World Wildlife Fund (WWF), and WCS (The Wildlife Conservation Society). Even the more culturally sensitive World Conservation Union (IUCN) might get a mention.

(Dowie, 2005)

Except in extremely remote areas, most nature preserves were formerly inhabited by, or at least provided resources for, local people. By the late-twentieth century, Western conservation had resulted in the eviction of millions of people from their homelands as part of widespread efforts to protect nature. Local and marginalized people, most of them Indigenous, were excluded from their territories and prevented from using natural resources that were vital for their survival (Spence, 1999; Chatty and Colchester, 2002; Dowie, 2005, 2009). The magnitude of the suffering caused by this *fortress conservation* (Brockington, 2002) has been compared to wars. Not surprisingly, the welfare of "conservation refugees" deteriorates dramatically after they are displaced. Livelihoods are lost, intergenerational transfer of traditions is disrupted, and the frequency of wildlife attacks on crops and people increases.

Conservation policies that cause displacement are based on the assumption that human activity degrades ecosystems, an attitude expressed in the legal definition of wilderness in the USA as a place "where the earth and its community of life are untrammeled by man" (Wilderness Act of 1964). In this view, eliminating people from natural areas (except for approved uses) is good for nature. Clearly, this is problematic for people living in those areas.

To the extent that preserves halt ecosystem degradation, they theoretically benefit everyone on the planet through ecosystem services such as safeguarding watersheds, moderating climate change, and reducing extinction rates. Since all parts of the Earth are interconnected, and decisions about conservation have effects that are far removed in space and time from their source, people living far from preserves may feel that they have a stake in decisions about protected areas. However, people who experience immediate deprivation when a preserve is set up rarely feel that potential long-term benefits are appropriate compensation. Although local people sometimes receive income from jobs as park guards or guides, and local communities get money from tourist expenditures, these are not necessarily enough to make up for lost homelands or livelihoods (Box 12.2).

To better understand the roots of the preservationist emphasis on protected areas that exclude people, it is useful to examine Western culture's assumptions about the relationship of people to nature and how those ideas influence decisions about how to interact with and manage the natural world today.

9.4.3 Should Trees Have Standing?

In 1972, the American legal scholar Christopher Stone wrote an article entitled *Should Trees Have Standing?* in which he proposed that nonhuman entities have rights and, therefore, they should have legal standing in court cases (Stone, 1972). He pointed out that over the course of human history legal concepts of rights have expanded. Whereas slaves, women, children, and some ethnic groups were once denied legal rights, international law now recognizes the right of all people to self-determination. Stone suggested that rights should be expanded further, to include "forests, oceans, rivers and other so-called 'natural objects' in the environment – indeed, to the natural environment as a whole" (Stone, 1972:456) (*see Q9.5*).

Stone argued that the US Supreme Court should recognize rights in a way that could "contribute to a change in popular consciousness" and benefit the "future of the planet as we know it" (Stone, 1972:500,501). At the time Stone was writing, the Sierra Club – a non-governmental organization founded by John Muir – had tried to sue to prevent Disney from developing a ski resort in a California mountain valley. The Court rejected the organization's lawsuit because the group could not prove that it would suffer damage from the project. Justice William Douglas dissented, however. In his opinion, he suggested that such problems could be avoided if the environment had legal standing to sue for its own preservation.

Initially, Douglas's suggestion was mocked. A 72-line poem in a legal journal (Naff, 1972) envisioned bizarre consequences:

> If Justice Douglas has his way –
> O Come not that dreadful day –
> We'll be sued by lakes and hills
> Seeking a redress of ills
> How can I rest beneath a tree,
> If it may soon be suing me? ...

Nonetheless, Stone's suggestion generated considerable interest, particularly among advocates of a biocentric approach to nature that is based on the inherent value of natural entities regardless of their economic worth (Section 5.3.5). Granting legal rights to nature is a strategy aimed at counteracting powerful interests that weigh short-term material and financial gain over other values. This approach has been used in several settings. Many municipalities in the USA now grant rights to entities such as wetlands, streams, and rivers. The Yurok Tribe has granted personhood under tribal law to the Klamath River in northern California. Rivers in New Zealand and India now have legal rights. In addition to these cases where rights are granted to specific natural entities, nature has also been granted general rights in some contexts. In 2008, Ecuador adopted a new constitution which states that "Nature, or *Pacha Mama*, where life is reproduced and occurs, has the right to integral respect for its existence and for the maintenance and regeneration of its life cycles, structure, functions and evolutionary processes" (Political Database of the Americas, 2011). Two years later, delegates to a World Peoples Conference on Climate Change and the Rights of Mother Earth adopted a Universal Declaration on the Rights of Mother Earth which states that "Mother Earth and all beings of which she is composed have ... the right to life and to exist," "the right to continue their vital cycles and processes free from human disruption," and a series of specific rights such as rights to water, clean air, and integral health (Global Alliance for the Rights of Nature, No date).

These documents do not specify who will serve as the legal guardians of nature. It is often assumed that Indigenous peoples will play that role. However, in two legal cases in Ecuador that were settled in favor of nature, that was not the case. One case was brought by the state, which sued to stop illegal mining. In the other case, an American couple with riverfront property successfully opposed a road project that would have deposited excavated material and rock in the river.

Granting legal rights to nature is seen as way that people can act as nature's guardians in legal proceedings. Generally, cases that involve granting legal rights to nature involve conflicts over proposed development, pollution, or resource extraction. However, the best way to represent nature's interest is not always obvious (Tanasescu, 2017). Two hypothetical scenarios described below highlight potential situations where conflicting interests could claim to represent nature.

Setting fire to trees is a destructive act. Is burning therefore a violation of the rights of trees? Or a forest's rights? Fire suppression harms trees that need fire in order to reproduce, as well as game and plants that provide acorns, roots, and berries which have cultural importance for some Native American groups. Does fire suppression violate their rights? Fire suppression also allows trees to expand into and replace steppe and savanna (Sections A.1.5 and A.1.7). What are the implications of a rights-of-nature approach for those ecosystems?

Devils Tower is a rock formation in northeastern Wyoming (Figure 9.1). This national monument has spiritual value for two groups: Native American tribes of the region and some recreational climbers. Because climbing can take place throughout the year, tribal leaders requested a voluntary ban on climbing during the month of June, a period of important ceremonies at this sacred place. While many climbers comply, not everyone does. If Devils Tower had been granted legal rights, would the outcome of this controversy have been different? Who is entitled to speak for nature in this case?

New Zealand is attempting to avoid such problems by specifying ahead of time who can represent nature. In 2017, a committee consisting of representatives of the Indigenous community and the Crown was created to represent the Whanganui River (Figure 9.3). It will be interesting to see how further legal developments address the ethical and practical challenge of appointing legal guardians of nature.

9.5 Philosophical Considerations: People and Nature

9.5.1 What Is Natural?

Two views of the relationship of people to the natural world are prevalent in Western culture. Both see people as separate from nature, but they differ in how much control humans should exert. The preservationist perspective is that nature will take care of itself and will be better off if people leave it alone. The utilitarian view is that people should dominate nature by manipulating it for their own use. In these perspectives, people either degrade nature or they improve upon it, but in either case, people are not considered part of nature. Of course, these are extremes; no person or institution can completely control nature; nor can anyone live in a world free of human impacts. We shall see in this chapter and the three chapters which follow that contemporary developments in ecology and anthropology, as well as currents in social movements such as environmental justice, Indigenous rights, and the rights of nature, have influenced conversations about our place in the natural world.

European settlers in the eastern United States found forests that seemed dark, foreboding, and full of fierce beasts such as wolves and bears. Westward expansion across the North American continent fostered a set of attitudes toward the natural and cultural environment that involved conquest. This was expressed as *Manifest Destiny*, the idea that it was the responsibility of Euro-Americans to tame the frontier. Settlers sought to control nature, as well as the people living *in* it, creating landscapes and societies similar to those of Europe. This mindset assumed that Euro-American settlers encountered an uninhabited, "virgin" continent.

A similar attitude toward controlling nature prevailed when Europeans colonized Mexico, the Caribbean, South America, Africa, Oceania, and parts of Asia. Nature was to be made orderly. In 1905, a Commissioner of British East Africa wrote that "large parts of South America and Africa" should be made to submit to discipline: "marshes must be drained, forests skillfully thinned, rivers be taught to run in ordered course and not to afflict the land with drought or flood at their caprice" (quoted in Collett, 1987:139). Likewise, the colonists regarded Africans as inferior people who should be converted to civilized Christian farmers wearing European clothing.

The Jesuit missionaries in New France (which later became eastern Canada and New England) were an exception. Although they worked to convert the Indians to Catholicism, the Jesuits were relatively sympathetic to native people, regarding them as innately good individuals living in egalitarian communities. In the seventeenth and eighteenth centuries, Jesuit ideas influenced French intellectuals, who saw a *noble savage* in the missionaries' accounts. To Europeans who were dissatisfied with the status quo, Indigenous people of North America represented a pure state of nature. The idea that civilization led to alienation from nature gained traction when it was popularized by poets, painters, and philosophers of the Romantic Movement in Europe and America.

From the end of the Middle Ages until the middle of the nineteenth century, European thinkers had generally held that to be human "was to be something special and splendid." In contrast to this, the Romantics did not consider human civilization splendid. They saw the natural world as something good and harmonious from which people were excluded. Anthropologist Matt Cartmill has suggested that the nonhuman world came to be seen as "sacred and sane, whereas people and their works ... [were] inherently cruel, disordered, and sick" (Cartmill, 1983:71).

A corollary to this is the idea that "wilderness stands as the last remaining place where civilization, that all too human disease, has not fully infected the earth" (Cronon, 1996:69). As the concept of wilderness became an important part of the Romantics' response to their disillusion with modern life, the cultural symbolism associated with wilderness shifted. Instead of seeing wild places as hostile, romantics came to see wild places as sublime refuges from civilization and places to renew contact with God. To American romantic writers such as Henry David Thoreau and John Muir, wilderness was a sacred temple where one could overcome separation from nature.

By the middle of the twentieth century, alarm about environmental degradation, world wars, genocide, technological change, and atomic bombs enhanced the prevalent despair about human nature. This was accompanied by a sense that wilderness provided the last unspoiled haven from the ills of the modern world (Cronon, 1996). In industrialized countries, the idea of wilderness as a pristine refuge inspired the designation of large areas of habitat that should be kept off limits to machinery and development.

The idea that people are outside of nature and that anything people do in the natural world is bad is reflected in some definitions of biodiversity. The term biological diversity originally referred to species richness (Section 5.2.3), but it has been expanded to encompass life at all levels of organization, from genes to ecosystems, as well as processes that sustain life. In order to track whether biodiversity is harmed, it is necessary to have criteria for what is natural. One way to approach this is to define natural as the absence of all human impacts (Hunter, 1996a). With this definition, any impact or change caused by people is a loss of biodiversity. In a paper published in the journal *Conservation Biology* in 1993, tropical ecologist Kent Redford and anthropologist Allyn MacLean Stearman pointed out that this definition of biodiversity differs from Indigenous conceptions:

To conservationists, a forested landscape [in the tropics] in which all biodiversity is conserved should include healthy populations of jaguars, woolly monkeys, white-lipped peccaries, [etc.] .... It should be an area in which gene frequencies, species, landscapes or process [sic]—be they ecological or evolutionary—are allowed to follow a course largely unaffected by human activity .... It is clear that if the *full range* of genetic, species, and ecosystem diversity is to be maintained *in its natural abundance* on a given piece of land, then virtually any significant activity by humans must not be allowed .... even low levels of indigenous activity alter biodiversity as defined above. ...

[But] In our experience, ... In the indigenous view, preserving biodiversity means preventing large-scale destruction, ... but conserving biodiversity may not preclude ... forms of extractive activities that leave large areas of the forest or other natural habitat altered but still standing.

(Redford and Stearman, 1993:252,253 emphasis in original)

9.5.2 Who Is Natural?

In the dualistic mindset of Western culture, if we are the opposite of natural and nature is good, then people are bad. John Muir was ambivalent about whether some people were more natural than others. Sometimes he expressed admiration of and sympathy for Indians (Worster, 2008). But at other times he judged them to be as "dirty" as the rest of humanity, entitled to inhabit wilderness only if they accepted no material goods (not even dirty rags) from the modern world and used only those materials that he judged to be natural. In his first summer in the Sierra, Muir encountered

an old Indian woman with a basket on her back … [who wore] calico rags, far from clean. In every way she seemed sadly unlike Nature's well-dressed animals, though living like them on the bounty of the wilderness. Strange that mankind alone is dirty. Had she been clad in fur, or cloth woven of grass or shreddy bark like the juniper or libocedrus [cedar or possibly incense cedar] mats, she might then have seemed a rightful part of the wilderness, like a good wolf, or bear.

(Muir, 1911:78–79)

On the other hand, the idea that people are morally tainted sometimes goes along with a tendency to romanticize Indigenous peoples as a source of inspiration for contemporary society. They are believed to exist *in* nature without affecting it (although in fact they may have substantial impacts that create ecosystems very different from what would exist in their absence). In this context, a variant of the noble savage concept has resurfaced, this time as what is sometimes termed the "ecologically noble savage," "ecologically noble Indian," or "ecological guardian" (Redford, 1991; White, 1999). In this view, rather than contaminating nature as Muir suggested in the above passage, Indigenous peoples are seen as living in harmony with their environment because of a spiritual connection to a natural world that has been badly treated by modernity, and because they have acquired ecological knowledge that allowed them to use resources over long periods of time without apparent environmental degradation (Gadgil et al., 1993; Berkes et al., 2000; Anderson, 2005).

For millennia, Indigenous people depended on resources in their local environment supplemented by what they obtained through trade. In the process, they amassed a wealth of information about species' requirements and interactions and the uses of plants and animals. Nomadic pastoralists in Africa, the Middle East, Central Asia, and Artic Eurasia observed sophisticated grazing regimes; hunter-gatherers in North and South America and Australia developed complex cycles of deliberate burning; and farmers in parts of East and South Asia worked communal lands. This does not mean that native peoples always managed their environments sustainably. Nevertheless, it is clear that for millennia many groups of Indigenous peoples have been active managers of their environments and utilized resources in ways that persisted for a long time.

Some scientists and environmental historians argue that native people have a benign influence on their environment only under certain conditions – when they live in sparsely populated regions with large areas of available habitat and lack access to modern technology or commercial markets. This view often goes along with the idea that in order to be true guardians, Indigenous people should remain in a pure, traditional state, refraining from using any modern conveniences. In the words of a Yanomamo Indian objecting strongly to this judgment:

ecologists, missionaries, and the government continue to see us as forest-dwelling animals, without the right to decide anything, without the right to potable water, electricity, or television, without the right to exploit the mineral wealth of Yanomamo lands; mineral wealth that any civilized person would unhesitatingly exploit if it were in their backyard …. Indians no longer want to live as if they were captive in a large zoo to be photographed by tourists.

(Quoted in Redford and Stearman, 1993:252, their translation)

Others argue that early Indigenous peoples brought about mass extinctions of charismatic megafauna (Section 6.4.1.1).

Regardless of their effects on past environments, the portrayal of contemporary Indigenous peoples as custodians of ecological wisdom is problematic. It can be a patronizing stereotype that fails to credit Indigenous people with the ability to deliberately and effectively manage their environment and the right to choose how they want to blend tradition and modernity (White, 1999). Indigenous peoples live in many settings, each with unique historical experiences, practices, and values. They interact with their environment in diverse and complex ways that are not easily captured by simplistic generalizations (Richardson, 2009).

Many Indigenous people and non-Indigenous advocates of Indigenous rights agree that Indigenous communities are more likely to continue environmentally sustainable practices if they are able to maintain their traditional culture. Evolving legal developments since the late twentieth century have promoted that goal in some (but certainly not all) contexts.

9.6 Legal Considerations: Indigenous Rights

The movement for Indigenous self-determination that developed in the 1970s added an Indigenous-rights perspective to Western conservation. In the USA and Canada, the Indian Rights movement was influenced by the Civil Rights movement of the mid-twentieth century. But it also grew out of national and international legal developments.

Indigenous rights affect conservation in several ways. Rights to self-determination, land tenure, water, maintenance of culture, management of natural resources, use of sacred lands, control of traditional knowledge, sharing of benefits from development, and hunting, fishing, and gathering in accustomed places have implications for conservation.

9.6.1 Treaties

In Canada and the USA, federal policy toward Indigenous peoples is based on the concept of *sovereignty* – the coexistence of separate, self-governing states. Many, although not all, Native American and First Nations peoples signed treaties with the conqueror state that spelled out the obligations of both parties. The Indigenous groups that signed treaties ceded large amounts of their ancestral territory but reserved rights to hunt, fish, and gather in their traditional lands and waters.

Treaties between Indigenous peoples and national governments are legally binding contracts between sovereign nations. They supersede state or provincial laws and regulations. However, treaty provisions have often been and sometimes continue to be violated.

Treaty rights to harvest wild plants and animals sometimes conflict with conservation regulations made by state or provincial fish and wildlife agencies and national parks. The treaty rights are rights that the Indians *reserved*. In other words, these are rights that the Indians retained; they are not rights that were granted to them (Blumm and Brunberg, 2006). Nevertheless, Indians hunting or fishing within their ancestral territories have been convicted of violating regulations, for instance by hunting without purchasing a state permit, or out of season, or using prohibited methods such as hunting deer with spotlights or fishing with banned gear. In some cases where Indians brought legal actions challenging those decisions, the resulting cases went through successive appeals all the way to the Supreme Court of the USA or Canada. Some examples of how this played out in the USA are covered in Box 9.5.

Box 9.5
Interpreting Treaty Rights

To be held as Indian lands are held: In 1854, the Menominee Indians of northern Wisconsin and Michigan (Figure 9.1) signed a treaty with the US government that established a reservation on a remnant of their ancestral territory. The treaty stated that these lands were "to be held as Indian lands are held," which meant that Indians had rights to continue to hunt and fish on those lands.

But in 1953, Congress terminated the tribal status of dozens of US tribes, including the Menominee. Eventually, the policy was reversed, and the reservation was reinstated, in large part because of the advocacy of Ada Deer, a Menominee woman whose activism contributed to the birth of a movement advocating Native American self-determination (Wilkinson, 2005).

While the termination was supposed to be in effect, however, the state of Wisconsin took over the regulation of tribal hunting and fishing. The state argued that when their tribe was terminated and the reservation was dissolved, the Menominee lost their hunting and fishing rights on what had been reservation land. The Menominee disagreed, maintaining that they retained their hunting and fishing rights on that land. Three Menominee men who were charged with violating state game laws when they hunted deer on former reservation land challenged their convictions in court. Following several decisions and reversals in higher courts, in 1968 the US Supreme Court ruled that the treaty language did protect rights to fish and hunt and that those rights were not repealed when the tribe was terminated.

Even after termination was reversed, issues of access remained. The decision meant that the Menominee could again legally fish on lands and waters that had once been part of their reservation, but there were other obstacles. Sturgeon (Box 7.1), which have an important role in the culture and economy of the Menominee, are prevented from reaching Menominee waters by a dam on the Wolf River (Beck, 1995). The Menominee argued in court that they were entitled to fish for sturgeon in waters outside the reservation because of long uninterrupted use of those waters, but the courts did not agree that the right to fish on non-reservation lands was implied by the treaty.

Taking fish at all usual and accustomed grounds … in common with all citizens. In 1854, several tribes and bands of Indians living near the southern end of Puget Sound in Washington State (Figure 9.1) signed a treaty with the USA in which they ceded their land but reserved the "right of taking fish at all usual and accustomed grounds and stations … in common with all citizens of the Territory."

Salmon were abundant when the treaty was signed. But subsequently they declined precipitously because of habitat degradation and overfishing. The clean beds of gravel that salmon need for spawning were degraded by sediment from streams that were used to transport logs downstream, and the exploitation increased when commercial processing made it possible to ship canned salmon far from the point of capture.

Washington State required Indians to obtain permits even though they had treaty rights to fish. The regulations also prohibited the use of traditional devices such as nets and traps. To protest the situation, Indian activists organized fish-ins, similar to the sit-ins of the Civil Rights movement in the 1960s, in which they protested the situation.

By 1970, tribes were taking just 5% of the total salmon harvest. In response, the federal government, acting as trustee for the treaty tribes, sued the state of Washington. The case went to the Supreme Court, which upheld the decision of Judge G. H. Boldt, the district court judge, who ruled that the tribes had treaty rights to fish at "usual and accustomed places" and that they were entitled to 50% of the harvestable salmon runs. The Court noted that the treaty tribes "shared a vital and unifying dependence on anadromous fish," which influenced their religious rites, seasonal movements, diet, food preservation, trade, and methods of fishing.

The Boldt decision was a major disruption for non-Indian fishers, and it was met with widespread defiance. When the State of Washington refused to carry out the court order, Judge Boldt ordered the US Coast Guard and federal agents to enforce it. Subsequently, the Supreme Court endorsed Boldt's actions (Echo-Hawk, 2013:163).

Even where national governments never signed treaties with Aboriginal peoples, there is increasing legal recognition of customary use rights in some contexts. In 2020, the High Court of Sweden unanimously ruled that Indigenous Sámi reindeer herders in part of Arctic Sweden have the exclusive right to hunt and fish in their territory on the basis of their presence there from time immemorial. In making its decision, the court considered a general principle of international law regarding the rights of Indigenous people to the resources in their traditional territories (Carstens, 2020).

9.6.2 International Declarations

In 1948, the United Nations General Assembly adopted a Universal Declaration of Human Rights (UN Declaration). This document affirmed that human rights belong to all human beings, and it expanded the scope of human rights beyond civil and political rights such as freedom of religion and the right to free assembly to include economic, social, and cultural rights of all people to have their basic needs met.

In 2007, 144 member states of the UN adopted a Declaration on the Rights of Indigenous Peoples (Indigenous Declaration). Australia, Canada, New Zealand, and the USA voted against this declaration at that time but subsequently reversed their positions. These declarations on human rights and the rights of Indigenous peoples are not legally binding, but they set standards for all peoples and all nations regarding their responsibility to respect those rights. In addition, compliance with many of the provisions of the declarations is required by international treaties (also termed *covenants*) that many nations have agreed to. For example, over 170 nations have signed the International Covenant on Civil and Political Rights. That agreement includes the right to practice one's culture, which is also addressed in Article 11 of the Indigenous Declaration. Some provisions of the UN Declaration and the Indigenous Declaration have also become part of customary international law, which provides a legal framework for implementing those declarations (Echo-Hawk, 2013).

Several of the articles in the Indigenous Declaration address potential impacts of conservation on the rights of Indigenous peoples. These include rights not to be forcibly removed from one's homeland, right to religious freedom, right to control profits from commercialization of traditional ecological knowledge, right to participate in decisions about commercial development of natural resources, and the right to conservation and protection of the environment and productive capacity of lands or territories.

National governments that do not recognize native peoples within their borders as Indigenous refuse to grant them rights recognized under international law. For example, this is the situation of the Bedouin of the Negev Desert in southern Israel (Box 11.3).

9.7 Diagnosing the Problem

The diverse perspectives and issues that we have reviewed in this chapter reveal complex and interrelated problems. The challenges we now confront are unprecedented in their complexity and scale. Whether we look at contemporary problems in conservation from a perspective of the natural sciences, the social sciences and humanities, or traditional cultures, we encounter situations that involve multiple stakeholders with different backgrounds, interests, values, and dilemmas.

Insights from utilitarian and preservationist conservation have provided a foundation for conserving living natural resources in varied contexts. But those approaches tend to focus on a narrow set of goals: production or preservation, respectively, within the context of a mindset that sees nature as tending toward balance and predictability. By the late twentieth century, it was becoming clear that business as usual in conservation was not necessarily good at preventing or coping with unpredictable and sometimes irreversible change. This set the stage for the emergence of a new approach.

9.8 Response: Stewardship Conservation

> Human activities affect Earth's life support systems so profoundly as to threaten many of the ecological services that are essential to society. To address this challenge, a new science agenda is needed that integrates people with the rest of nature to help chart a more sustainable trajectory for the relationship between society and the biosphere.
>
> *(Chapin et al., 2011:1)*

By the late twentieth century, increasing awareness of the interdependence between human and natural systems and of the rapidity and extent of anthropogenic changes led conservationists to focus on how to manage for complex ecosystems and their human dimensions. This stewardship approach to conservation strives to maintain variable and resilient ecosystems and cultures.

A stewardship approach requires collaboration between multiple disciplines. Questions about whether people are, or should be, considered part of nature belong in the domains of ethics and philosophy, and questions about what social policies are needed to remedy ecological problems belong to the domain of politics and sociology. Science provides relevant information to help us make these decisions.

The underlying concepts and strategies of the stewardship approach, as well as some of the challenges it faces, are described in the next three chapters. Chapter 10 covers how ecologists' ideas about the natural world and our place in it are changing and how different perspectives can point the way to strategies for conservation that meet the challenges we have been discussing. Chapters 11 and 12 describe the accomplishments of some methods of doing that.

10

Central Concepts

Complexity and Change

> Wherever we seek to find constancy, we discover change …. The
> old ideas of a static landscape, like a single musical chord sounded
> forever, must be abandoned, for such a landscape never existed
> except in our imagination. Nature undisturbed by human influence
> seems more like a symphony whose harmonies arise from variation
> and change over every interval of time. We see a landscape that
> is always in flux, changing over many scales of time and space,
> changing with individual births and deaths, local disruptions and
> recoveries, larger scale responses to climate from one glacial age
> to another, and to the slower alterations of soils, and yet larger
> variations between glacial ages.
>
> *(Botkin, 1990:62)*

We have seen that utilitarian and preservationist conservation tend to envision nature as moving toward equilibrium. In this view, ecological systems are seen as closed and self-regulating, returning like a pendulum to their original state if they are altered. This balance-of-nature view led ecologists to concentrate on certain phenomena and generated important insights in several areas of inquiry. First, when biologists looked at populations from this perspective, they focused on those that reached equilibrium with their resources, that is, populations near carrying capacity (Section 2.1.1.3). Density-dependent population growth and competition within populations were explored in this context. Second, ecologists viewed communities as proceeding toward a stable climax and focused their attention on vegetation sequences that occur after a disturbance (Section 2.2). This led them to distinguish between the existing plant community on a site and what was thought to be the potential vegetation at that location under typical circumstances. Third, the theory of island biogeography turned the attention of conservation biologists toward the equilibrium that occurs if island extinction rates and colonization rates are equal. Insights from this theory are used to examine the dynamics of extinction and colonization among populations in fragmented habitats.

Balance, stability, equilibrium – this theme comes up repeatedly in the writings of resource managers and conservation biologists. But by the middle of the twentieth century, some scientists were questioning the view that most of nature is in equilibrium or on its way to a state of equilibrium most of the time. This does not mean that the insights generated by equilibrium theories are always incorrect or insignificant, but it does suggest that important non-equilibrium processes in the natural world were overlooked because scientists were not interested in them or were biased against seeing them. Perhaps this was because at the time scientists wanted definitive answers, and the absence of a steady state meant that conditions would be unpredictable, uncertain, and less able to be controlled.

These problems led to the development of an alternative view. But, before considering that view, let us reexamine some ideas about equilibrium.

10.1 Revisiting Equilibrium Theories

10.1.1 Density-Dependent Population Regulation

10.1.1.1 Northern Fur Seals in the Pribilof Islands

In some cases, the idea that populations are regulated by density-dependent processes works well as a guide for managing harvests of game and commercially exploited species. But not always.

Until the 1950s, the northern fur seal harvest in the Pribilof Islands appeared to be a well-managed and successful hunt that was based upon conventional harvest theory (Box 3.3). But after that, there were some surprises. The rookeries began to show signs of overcrowding. Pups were crushed by bulls, and diseases and parasite loads increased. Managers believed that these were density-dependent problems that would be corrected when the harvest was increased and population density was reduced. This was accomplished by killing females (because the killing of polygynous males would have little effect on a population's reproductive rate (Section 4.2). These measures succeeded in decreasing the seal population density, but the population did not stabilize. Surprisingly, except for a brief increase in the early 1970s, the downward trend in fur seal numbers continued. Something was going on that conventional thinking had failed to grasp. Between 1800 and 1950, northern fur seals had twice recovered from precipitous declines, yet in spite of the fact that for more than a century, fur seal populations exhibited classic, density-dependent responses, fur seals continued to decline after 1950.

Clearly, the seals' ecology is more complex than managers had assumed. Evidently, the marine environment changed in ways that led to increased mortality or decreased reproduction, or both, and these effects were independent of population density. One possibility is that the prey base in the North Pacific has changed. The decline of seabird populations at the same time that the fur seal populations on the Pribilof Islands fell is consistent with this interpretation (Gentry, 1998). As of 2021, the Pribilof Islands fur seal population was still low (NOAA, 2021). Thus, in spite of over a century of successful management and several decades of apparently sustainable harvest, equilibrium models of seal population dynamics ultimately proved inadequate. Rather than being a textbook case of sustainable yield achieved by applying the concept of density-dependent population dynamics, it appeared that management of northern fur seals may need to consider both density-dependent and density-independent processes.

Considerable evidence suggests that this situation is not unusual. Population regulation is not necessarily a matter of density-dependent *or* density-independent processes. Rather, the two types of phenomena may occur in the same population.

Managing for sustained yield (and usually for maximum sustained yield) has been the central goal of utilitarian conservation since the days of Gifford Pinchot (Section 1.5.1.2). This approach assumes that density-dependent regulation determines the size of a harvested population, an assumption that is not always justified. It may miss changing environmental conditions or unintended consequences of the harvest.

The objective of maximum sustained yield might not be achieved when estimates of allowable harvest are too optimistic, perhaps because recommendations made by scientists are ignored, or because the underlying assumption of density-dependent regulation of the harvested population is incorrect. Nevertheless, the philosophy of conservation as a means of ensuring continuous production pervaded the disciplines of wildlife management, forestry, and range management in the Western world for many decades. More recently, that concept was also applied in international development programs.

10.1.1.2 Livestock in Arid and Semiarid Rangelands

In arid and semiarid rangelands, animal numbers fluctuate in response to frequent droughts. After massive livestock die-offs during droughts in East Africa in the 1950s and 1960s, international donor agencies invested millions in development projects aimed at improving the welfare of nomadic pastoralists and their livestock.

Western range managers attributed the problem to overgrazing. To remedy the situation, they applied recommendations based on equilibrium models of rangeland dynamics developed for wetter environments in the USA and Australia. According to those models, an environment can support a set number of grazing animals, its carrying capacity (Section 2.1.1.3). In these models, if a population is close to the environment's carrying capacity, it is limited in a density-dependent fashion by competition for limited food. As a population approaches carrying capacity, limitations in the quantity and quality of its food plants lead to lowered reproduction and a decrease in the population growth rate. In that view, if livestock were allowed to exceed a range's carrying capacity, overgrazing would occur, and vegetation would be damaged. On the basis of those models, managers concluded that livestock populations should be kept at fixed levels rather than being allowed to fluctuate widely.

However, livestock numbers in semiarid and arid rangelands are not necessarily regulated by food supply in the way that equilibrium models predict. In those environments, the interacting physiological and environmental effects of drought – such as dehydration, heat stress, and the death of plant foods – cause livestock losses that are independent of herd size. Although this outcome is distressing to outsiders, it has the advantage of maximizing productivity during years when resources are available. Thus, it is advantageous in the variable and unpredictable environment of East Africa. Proponents of the pastoralists' approach point out that with traditional management, mortality in response to periods of low rainfall prevents livestock populations from becoming high enough to harm vegetation (Homewood and Rodgers, 1987).

The objectives of Western range managers were to maximize the growth rates of animals in their first few years of life, in order to produce high-quality meat for commercial markets. Pastoralists, however, had different objectives. They sought to support high numbers of animals in years of favorable weather. This arrangement produced meat that was not as profitable but provided subsistence for more people.

The debate about the relevance of equilibrium models for arid and semiarid rangelands stimulated more research. The resulting studies suggest that both equilibrium processes and non-equilibrium processes play a part in the ecology of semiarid and arid rangelands, depending on local vegetation characteristics and other factors (Illius and O'Connor, 1999). This continues to be an area of active research.

10.1.2 Competition

In equilibrium models, competition between members of the same species is the mechanism responsible for density-dependent population regulation. The concept of *interspecific competition* (competition between different species) also figures prominently in ecological theory.

Competition between species is difficult to observe, so ecologists often infer competition from species' patterns of resource use. The American evolutionary ecologist John Thompson compares this to a situation in *Through the Looking Glass*, Lewis Carroll's sequel to *Alice's Adventures in Wonderland*.

"But you've got a bee hive—or something like one—fastened to the saddle," said Alice.

"Yes, it's a very good beehive," the Knight said in a discontented tone. "One of the best kind. But not a single bee has come near it yet. And the other thing is a mouse trap. I suppose the mice keep the bees out—or the bees keep the mice out, I don't know which."

(Thompson, 1982:39)

In this amusing example, the Knight interprets the failure of bees and mice to use potential resources (the hive and the bait in the mouse trap) as evidence that they are excluding each other. Similarly, ecologists have often assumed that the failure of a species to use a potential resource is due to competition with another species. On the other hand, when two similar species occur together and use the same resource but in different ways, this has often been attributed to past competition as well, even if specific evidence of competition is lacking, a kind of "ghost of competition past" (Connell, 1980:137).

If we see competition in either of these situations – when similar species overlap or when they do not overlap – how could we ever falsify the hypothesis that competition is responsible?

Ecology, like other sciences, is influenced by the cultural context within which it is practiced. Bias in favor of seeing competition in nature is consistent with the assumptions of a capitalist economic system, which is based on competition between individuals (Rozzi, 1999). Perhaps this has influenced ecologists' tendency to attribute contrasting phenomena to competition.

The idea that competition is common in nature is closely tied to the idea that people are naturally competitive and motivated by self-interest. We saw in Section 5.3.1 that Garrett

Hardin saw human population growth as the inevitable outcome of self-interested exploitation of commonly held resources. This interpretation has been challenged.

10.1.3 The Commons

In *The Tragedy of the Commons*, Garrett Hardin argued that communally owned lands are inevitably overused and degraded, and that therefore the only way to avoid catastrophe is through government regulation or privatization. Hardin assumed that the common grazing lands he described were a *common-pool resource*, a resource that is limited and cannot easily be defended. If one person uses a common-pool resource, that reduces what is available to others. Rangelands and fisheries are classic examples of common-pool resources (Jensen, 2000). If I eat a fish, it is no longer available for anyone else to use it. Breathable air is not a common-pool resource. No matter how much air I breathe, there is always enough to go around. Because common-pool resources are limited and access to them cannot be controlled easily they are vulnerable to overexploitation. Overexploitation of common-pool resources is not inevitable, however.

In the scenario Hardin described, there were no agreements between participants about how to cooperate, and there were no obvious short-term negative consequences for seeking individual gain at the expense of the community. In his hypothetical situation, access to the commons is always unrestricted, users are selfish, their behavior is not influenced by social norms or taboos, they have perfect knowledge about the economic costs and benefits of their actions, and they always try to maximize their short-term gains (*see Q10.1*).

The American Nobel Prize-winning economist Elinor Ostrom considered other possible outcomes of social interactions involving resource use. She examined alternative arrangements in which group members could theoretically devise a mutually agreed upon cooperative strategy, and she found many examples of communities that have devised workable arrangements for preventing tragedies of the commons (Ostrom, 1990). Box 10.1 describes two examples of community arrangements for regulating fisheries. (See Section 12.7 for another example of a small community that is active in management to avoid overfishing.)

We will see another example of a common-pool resource in Box 12.3.

Box 10.1
Cooperative Fisheries Management Developed by Local Resource Users

Small-scale fishery at Alanya, Turkey: In the 1970s, fishers in Alanya, a small fishery in a community on the Mediterranean coast of Turkey, faced economic uncertainty due to competition for good fishing spots (Figure 10.1). To address this problem, members of a local cooperative began experimenting with ways to allot fishing sites to individual fishers in a fair and efficient manner. Sites were arranged so that nets at one site would not block fish from reaching other sites. The cooperative also devised a schedule for fishers to move regularly from site to site, so that everyone had a chance to fish at each site.

With this system, there was no incentive to cheat because anyone who tried to fish at a good site when they did not have a defined right to do so would be detected by the legitimate fisher at that site as well as other fishers who benefited if the system was respected. Although it took

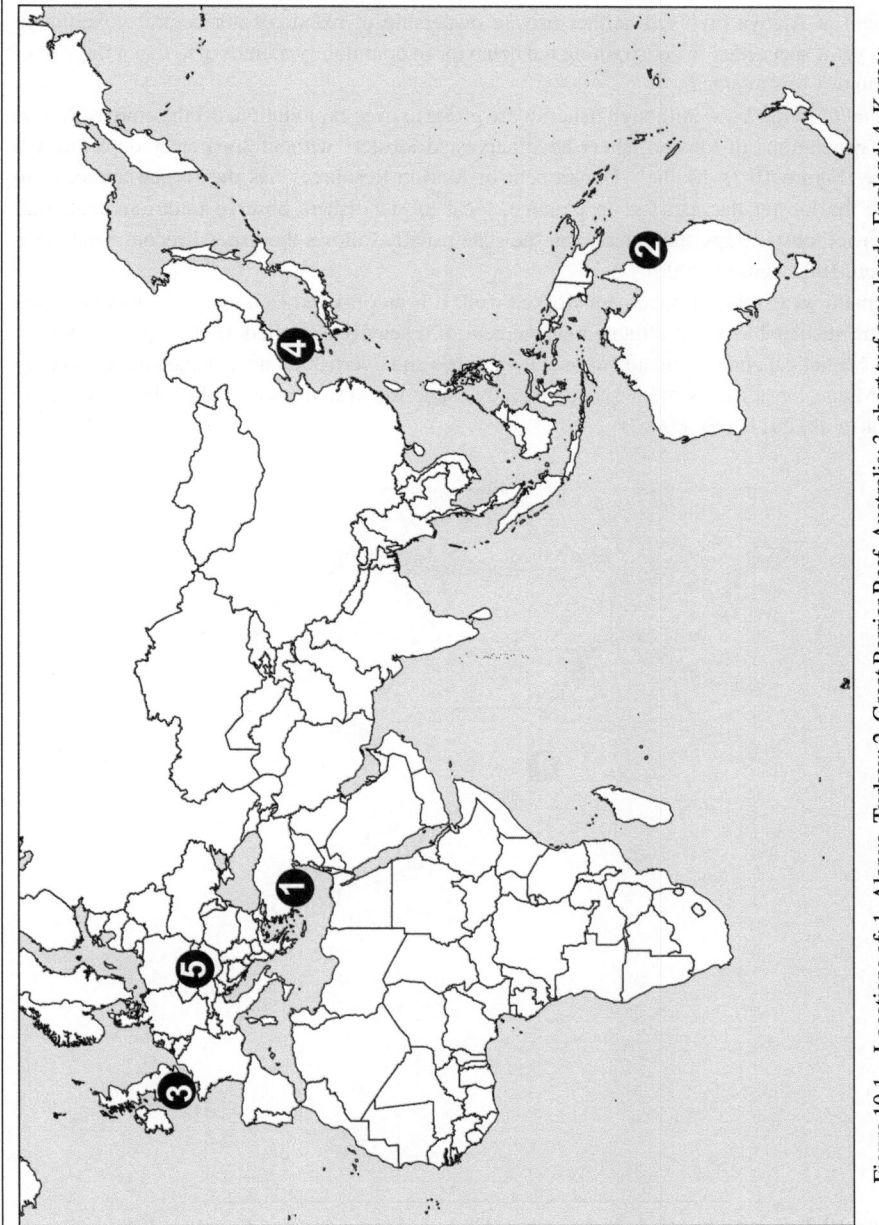

Figure 10.1 Locations of: 1, Alanya, Turkey; 2, Great Barrier Reef, Australia; 3, short turf grasslands, England; 4, Korea; 5, White Carpathians, Czech Republic. Map created by Eva Strand using Esri, DeLorme World Countries Generalized Data & Maps for ArcGIS 2013, with permission.

trial-and-error experiments over several years for users who knew the area to work out an effective system, it could not have been developed by central government officials without spending even more time and money to map the productivity of the different sites (Ostrom, 1990). The arrangement at Alanya involved neither private ownership of resources nor central government control, yet it succeeded in constraining the behavior of community members so that a tragedy of the commons was avoided.

Maine lobster fishery: Although fisheries are prone to over exploitation, off the coast of Maine several generations of lobster fishers have harvested lobsters without apparently depleting the resource (Figure 10.2). Maine's Department of Marine Resources has the formal authority to regulate the lobster harvest. But in practice, local lobster fishers observe traditions that limit who can set lobster traps and where, and the state usually follows the fishers' recommendations (Acheson, 1987; Jensen, 2000).

For many years, this arrangement worked well. It is an oft-cited example of collective action achieving sustainable and profitable management of a heavily exploited marine species. Nevertheless, ecological changes related to predatory fishes and overfishing have depleted the diversity of the Maine lobster ecosystem and potentially made the system more vulnerable to collapse (Steneck et al., 2011) (*see Q10.2*).

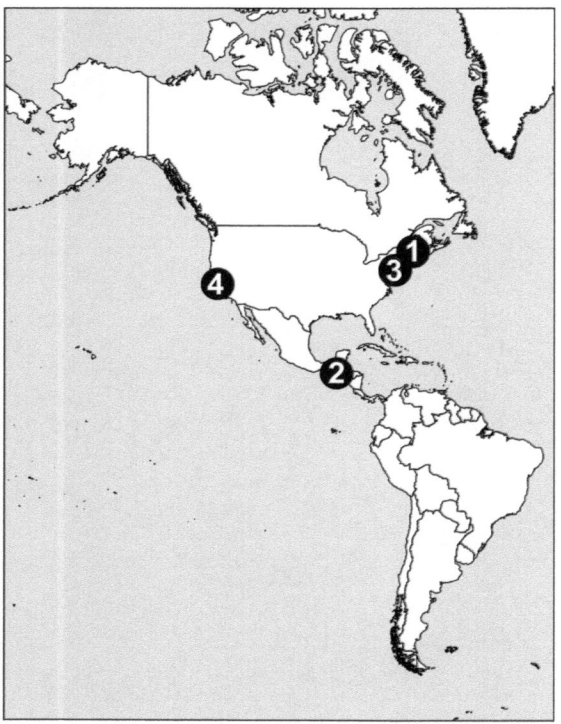

Figure 10.2   Locations of: 1, Maine coast, USA; 2, Morán and Las Cebollas, Guatemala; 3, Mettler's Woods, New Jersey; 4, Jasper Ridge Biological Preserve, California. Map created by Eva Strand using Esri, DeLorme World Countries Generalized Data & Maps for ArcGIS 2013, with permission.

These examples raise the question: what conditions promote the development of cooperative arrangements like the ones in Alanya and Maine? To address that question, scientists at the Indiana University Bloomington, USA, in collaboration with two universities in Guatemala compared forest management by two similar rural farming communities (Box 10.2).

Box 10.2
Local Forest Management in Two Communities in Guatemala

The small communities of Morán and Las Cebollas (each with between 40 and 70 households at the time of the study) are located on steep slopes in mixed pine and hardwood forests in Guatemala (Figure 10.2). In both places, *campesinos* (small-scale farmers) raise corn and black beans for local consumption, plus a few other products that are traded in local markets. The local forests provide timber, fuelwood, and grazing for cattle and horses.

Because these communities are far from urban areas and are often inaccessible even with four-wheel drive vehicles, they have been largely left alone by regional governments. As a result, community members have developed their own institutions for self-government. The community of Las Cebollas has set aside part of its forest as a Protective Forest within which harvesting of forest resources is prohibited. Because the community of Morán has no such forest designation the two communities allow for a comparative study.

Researchers conducted interviews and field studies to obtain data on ecological and socioeconomic variables that might shed light on the causes and consequences of this difference between the two communities. Interviews revealed that residents in both communities believed the supply of good farmland was limited. Each community regulated the use of agricultural land through explicit rules about rights to farmland and local enforcement of those rules. In contrast to this, the people who were interviewed did not express concern about the supply of forest resources, and, with an exception described below, they did not regulate use of those resources: "Community members at both sites repeatedly said that if a community member wanted firewood or timber, all he or she had to do was cut or collect it" (Gibson, 2001:81). However, in Las Cebollas the community prohibited the collection of resources from the Protective Forest and enforced those rules by marking and patrolling its boundaries. People in Las Cebollas explained that it was important to do this because that part of the forest safeguarded the water supply needed for farming during the dry season.

Researchers studied the effects of this policy on forest condition at Las Cebollas by comparing several ecological variables within and outside of the Protective Forest. They used species diversity, tree size, tree density, and ground cover as indicators of forest condition. At Las Cebollas, each of those variables was higher in the Protective Forest than in community forests that were not protected (Table 10.1). For example, the average diameter (measured at breast height) of pine tree trunks sampled in the Protective Forest was more than twice the average diameter (58.5 cm compared to 24.7 cm) of pine trees at sample sites in other community forests at Las Cebollas.

This study demonstrated the ability of a small farming community to regulate use of its forest under certain conditions. In particular, the study results suggest that perceptions of scarcity influenced community members' motivation to regulate use of their resources. Forest resources, which were generally not considered scarce, were not regulated, except where the forest could influence the value of agricultural land by protecting the supply of water, a scarce resource (Gibson, 2001) (*see Q10.3*).

Table 10.1   *Forest conditions in Protective Forests and other community forests of Las Cebollas.* Diameter at breast height (DBH) is a measure of tree size.

Type of forest	Tree density, all species (stems/m^2)	Sapling density, all species (stems/m^2)	Average DBH (cm), pine species
Community	0.062	0.101	24.7
Protective	0.087	0.184	58.5

On the basis of studies like the ones described above, researchers have identified common characteristics of successful regulation of common-pool resources (Ostrom, 1990; Jensen, 2000). These include the following:

- There are clear rules about who has the right to use resources.
- Users believe that the rules are fair.
- Most of the people who are affected by the rules participate in making those rules.
- Uses of resources and compliance with the rules are monitored.
- There are penalties for violating the rules.
- There are clear local procedures for resolving conflicts about use.

Garrett Hardin underestimated the potential for local groups to cooperatively manage their resources for the common good. He also minimized the problems from privatization or state regulation. Privatization can lead to commodification and overexploitation, and state regulation does not guarantee prudent resource use either. Resources can be depleted under state control where some or all of the following are true: (1) bureaucracies are inefficient, (2) administrators are vulnerable to political pressures from special interest groups, (3) people do not trust the state and do not comply with regulations, (4) resource conservation is not as high a priority as military or economic growth, or (5) the state does not take advantage of the expertise of local people (McCay and Acheson, 1987).

Although a considerable body of research has highlighted problems with Hardin's original argument, the idea that communal ownership leads to tragedy continues to be very influential.

10.1.4 The Theory of Island Biogeography

Conservation biologists have applied the theory of island biogeography to many kinds of communities in habitat fragments, ranging from small mammals on mountaintops to insects on patches of a single plant species. Data from a field experiment by Simberloff and Wilson (Box 6.12) with arthropods on mangrove islands support the model, and Karr took advantage of a natural experiment when Barro Colorado Island was created (Box 8.2). But additional experimental tests of the model are rare. Simberloff himself lamented the fact that the theory became "so widely accepted as an accurate description of nature that failure of an experiment to yield the result deduced from the theory leads *not* to rejection of the theory but rather to attempts to fault the deductive logic or experimental procedure, or simply to willful suspension of belief in the experimental result" (Simberloff, 1976:578; emphasis in original).

Clearly, local extinction and colonization occur. The theory of island biogeography states that the two processes will eventually occur at equal rates, at which point they will be at equilibrium. But local factors may continually boost or reduce either the colonization rate or the extinction rate (Section 6.5.4). In those cases, colonization will not balance out extinction. In cases like the mountaintops in the Great Basin, the number of species with limited ability to disperse continually declines (Section 8.3.4) because extinctions continue to occur while colonization in that group is non-existent. As stresses from climate change mount, this trend is likely to become even more pronounced.

The theory is useful because it focuses ecologists' attention on the processes of local extinction and colonization in islands and isolated habitats. The idea that the relationship between the two processes will determine the long-term fates of populations is an important insight. On the other hand, the specific management recommendations derived from the theory are controversial. The argument for larger reserves, with its all-other-things-being-equal caveat (Section 8.3.2.1) is not always helpful, because all the other relevant factors are rarely equal.

There is no substitute for detailed knowledge about the specifics of each situation, including the habitat requirements and dispersal abilities of the species in question, their reproductive potentials and sources of mortality in different habitats, the amounts and sizes of habitat patches and disturbance frequencies in a proposed reserve, and so on. For this reason, the debate over whether a single large reserve will conserve more species than several small ones of equivalent area has subsided. More and more ecologists are turning their attention to other pursuits, such as studying the dynamics of sources and sinks (Section 8.3.3.1) in habitats of varying quality, as a way of getting the information they need to make recommendations for managers (Caughley and Gunn, 1996; Simberloff, 1997).

10.1.5 Succession

10.1.5.1 The Clementsian Model: Predictable Stages and Stable Endpoints

The influential ecologist Frederic Clements believed that a predictable sequence of plant communities on a site inevitably led to a stable climax community at equilibrium. The climax was expected to reproduce itself indefinitely unless it was disturbed. Clementsian ecologists didn't pay a lot of attention to disturbances. Biotic disturbances – such as grazing, digging, wallowing, and trampling – were often overlooked, and abiotic disturbances – such as fires, volcanic eruptions, hurricanes, and floods – were seen only as catastrophic agents that set back a plant community to an earlier stage in succession.

The story of Mettler's Woods (Box 10.3) illustrates how the concepts of predictable stages and stable endpoints influenced scientists' expectations.

Box 10.3
Mettler's Woods: Surprises from a "Truly Virgin" Forest

In 1701, a Dutch settler purchased a tract of mature forest in central New Jersey from the Lenape Indians. Subsequently he cleared all except 26 ha, which became known as Mettler's Woods (Figure 10.2). When a botanist described the woods in 1749, he described them as consisting of oaks, hickories, and chestnuts with very little underbrush. He noted that the understory of the forest was so open that one could easily drive a carriage through it.

By the middle of the twentieth century, Mettler's Woods had not experienced any large anthropogenic disturbance for centuries (although the chestnut blight (Box 2.3) had wiped out the chestnuts, and hurricanes had toppled some trees). In an article in *Life* magazine in 1954, the woods were described as a

truly virgin forest .... an unbroken succession dating back to the ice ages .... Its preservation in primeval state derives from the fact that one family has maintained continuous ownership since 1701, consistently refusing, generation after generation, to allow it to be cut for timber, cleared for farming or scarred by fire.

Mettler's woods stands as a "climax" forest community, which means that it has approached a state of equilibrium with its environment, perpetuating itself year after year essentially without change, secure against the invasions of all other forest types that might seek to displace it.

(Barnett, 1954:78,99)

Around the time that article was written, Rutgers University acquired the woodlot and added it to a larger reserve which became known as the Hutcheson Memorial Forest. Since it was thought to be in a climax state, all that was deemed necessary to preserve the forest community was to exclude people and disturbances.

But the absence of disturbance did not lead to the expected result. Surprisingly, the existing community did not perpetuate itself, even though it had been rigorously protected from most disturbances that would have set back succession. The oaks produced large acorn crops periodically, yet virtually no oak seedlings survived. Old oaks died, but the young saplings beneath them were sugar maples, not oaks. The forest floor was no longer clear of underbrush, as it had been in the eighteenth century. Instead, dense shrubs and saplings had grown up. These changes were unexpected, since, by definition, a climax community is capable of perpetuating itself under the conditions it creates.

It seemed that if these trends continued, the oak forest would give way to a community dominated by maples. When scientists from Rutgers examined trunk sections from trees that had died during a hurricane, they found that before 1701 the forest had burned about once every 10 years. They hypothesized that oak regeneration requires frequent fires. Historical records revealed that the Lenape had often burned the woods. The oaks had persisted in the face of frequent fires because they require fire to regenerate on a site, and they are more resistant to fire than maples. The suppression of fire had degraded a community that was thought to be in a climax state (Buell, 1954; Botkin, 1990, 2012.

If the Clementsian ecologists had been right in their characterization of change, then the oak-dominated community at Mettler's Woods should have reached a point at which the plant species perpetuated themselves in proportion to their original abundance in the community as long as the area was protected from fire and other disturbances. But eventually it became clear that the oaks at Mettler's Woods require disturbance in order to reproduce.

By the 1970s, most ecologists recognized the importance of natural disturbances for a variety of species and communities. In addition to the oak forest of Mettler's Woods, frequent disturbances are characteristic of northern boreal forests, chaparral, tropical forests, temperate forests, savanna, and many kinds of wetlands (Appendix). In these communities, "natural disturbance is so common that it keeps the system from ever reaching a stable state, so it is unrealistic to assume that a climax is the 'normal' condition for ecosystems to be in" (Sprugel, 1991:3).

Field studies in the tropics shed additional light on the consequences of disturbance for species assemblages. For many decades, ecologists thought communities with high species diversity were due to low rates of disturbance. In tropical environments, species diversity is very high in forests and coral reefs, which were thought to be at equilibrium. But when the ecologist Joseph Connell obtained field data on the population dynamics of tropical corals and trees in Australia, he found that communities with infrequent disturbances were not the most diverse. He suggested that intermediate levels of disturbances promoted species diversity, and that in the absence of disturbance, communities lost some species. For example, in the Great Barrier Reef off the coast of Australia (Figure 10.1), Connell's observations of permanently marked plots over several years revealed that the highest number of coral species occurred in places exposed to hurricanes. These disturbed areas were recolonized by many species of corals. At sites that were not exposed to hurricanes, old colonies of a few species such as staghorn coral covered much of the reef (Connell, 1978).

10.1.5.2 Challenges to Clements' Ideas

Unique Settings; Unique Histories: Every site occurs in a unique setting. No two sites are identical. Similar plant communities at different sites differ in many respects. Even if they are similar in climate, geology, topography, and soils, they may differ in microclimates, water availability, seed sources, pathogens, interacting species, and many other abiotic or biotic characteristics.

Although there are general patterns in vegetation that tend to repeat across landscapes, each site also has a unique history. Every disturbance is unique in its size, timing, patchiness, and severity. Sites may also differ in how often they experience insect outbreaks or floods, or in whether some of their products were harvested or burned by people. Consequently, even similar sites differ in the availability of resources such as water and soil nutrients, as well as in the prevalence of seeds, spores, and so on.

In addition, the environment is not static. Continents continue to drift; mountains are uplifted and eroded; glaciers form and recede; sea level rises and falls; populations of predators, pests, and mutualists emerge and disappear. When Margaret Davis, an ecologist at the University of Minnesota, examined fossil pollen in lake sediment cores from the northeastern USA and adjacent Canada, she found evidence that about 4,800 years ago something happened which caused a lasting change in the region's vegetation. Hemlock pollen declined abruptly in the pollen cores for that period, probably because of a novel pathogen. She concluded that a single, large-scale event at that time caused hemlock populations to plummet throughout the region. In some areas, hemlock pollen regained its former abundance after 2,000 years, but in other places it never returned to its pre-disturbance level (Davis, 1981). The landscape certainly was not in equilibrium during that entire period.

Figure 10.3   Acacias in the Serengeti. Credit: SeppFriedhuber / E+ / Getty Images.

The events in the Serengeti-Mara Ecosystem during the past century and a half provide another illustration of unique historical circumstances that had an overriding influence on the landscape. Most of us are familiar with images of the Serengeti savanna (Box 9.2) from nature shows. Mention of the Serengeti brings to mind grassy plains with widely spaced, umbrella-shaped acacia trees (Figure 10.3). But recent work suggests that rather than being the climax vegetation of the region, this savanna is the product of unique historical circumstances.

Prior to the late nineteenth century, a combination of grazing by large herbivores and burning inhibited tree reproduction throughout most of the region. But in 1889, an Italian military expedition introduced rinderpest, a viral disease of cattle and their relatives, to Somalia (Sinclair, 1977). This exotic disease spread rapidly throughout East Africa, devastating populations of livestock, wildebeest, and buffalo. The human population of the region also declined because the pastoral tribes starved when their cattle died. In addition, during this period, elephant hunting for the ivory trade increased. The absence of cattle, native ungulates, and elephants allowed scattered clumps of trees to become established, leading to the development of savanna vegetation (Dublin et al., 1990; Sinclair, 1995).

Eventually, a vaccine for rinderpest was developed, and native herbivores acquired some resistance to the disease as well. Cattle and native ungulates again inhabit the plains. The mature acacia trees that grew up during a historically unique period, when there were few herbivores to prevent trees from becoming established, are now dying. But young saplings are not taking their place. The landscape we see photos of is not reproducing itself.

The acacias that we think of as the hallmark of the "natural" Serengeti landscape may be "due to an extraordinary outbreak of trees precipitated by the rinderpest epizootic late in the nineteenth century" (Sinclair, 1995:109). What we see as typical is actually unique.

This situation has practical implications. Since the tree-dotted savanna is a tourist attraction, land managers are in a difficult position:

[V]isitors' to East Africa (who provide the money that justifies the parks' existence) expect to see big, umbrella-shaped trees, and are disappointed if they do not. Regardless of the fact that the plains without trees may be more 'natural' than plains with trees, our image of the savannas has become fixed in a static early 20th century mold.

(Sprugel, 1991:7–8)

Viewing Ecosystems at Different Scales: Some parts of an ecosystem may appear to be at a stable equilibrium when examined at some spatial or temporal scales but not others. As early as 1965, S. A. Schumm and R. W. Lighty, two geologists with the US Geological Survey, pointed out that although a landform might appear to be in equilibrium if viewed over a short period of time and a small area, when larger time spans and larger areas are taken into consideration, dynamic change could become evident. The amount of water entering and the amount of sediment leaving a small stream reach could briefly balance each other out, but over a longer time span and a larger area it might be evident that erosional processes are changing the landscape's topography (Schumm and Lichty, 1965).

The Changing Stage: Multiple Stable States: If plant community composition has changed because of a missing disturbance, it is sometimes possible to return it to its prior state by reintroducing the disturbance. But this is not always the case. Sometimes a plant community that has been in what Clements would have called a stable climax has gone through a transition to an alternative stable state that is very unlikely to return to its previous condition. This is the case for rangelands in many parts of the western USA (Laycock, 1991). (See Box 11.1 for an example.)

The American wildlife biologist Holly Dublin and her coworkers tested several alternative hypotheses about stable states in the vegetation of the Serengeti. To reconstruct the history of Serengeti vegetation since the late nineteenth century, they compared the predictions of mathematical models to information from historical written records, recent interviews, repeated aerial photographs, and field data on tree seedling mortality. This allowed them to construct models of what would happen under a variety of conditions. The models predicted that rather than a single climax state in the Serengeti, two stable states are possible: grassland/savanna or woodland. Fire and elephant herbivory affect which state prevails. Fire could initially convert woodland to grassland by killing trees. Elephant browsing could then maintain that stable state by preventing trees from regenerating and re-establishing woodland (Dublin et al., 1990; Sinclair, 1995). Although other factors, including variation in rainfall and grazing by wildebeest, affected the vegetation, elephants and fire were the principal drivers affecting whether grassland/savanna or woodland would prevail.

If succession led to a predictable community of plants and animals, then it would be fairly simple to decide what the "natural" state of a community was. In that case, a hands-off approach to preserving it might be appropriate. But communities are on "trajectories of change from the past, through the present, and into the future" (Tausch, 1996:99). People tend to overlook these changes, because on the time scale of a human lifespan they are difficult to perceive. Often we manage for "some past magical time of supposed ecosystem perfection when things were still 'natural'" (Tausch, 1996:99). Even when "we recognize succession as a dynamic play, [we] regard the stage on which it is played out as static and unchanging" (Sprugel, 1991:6).

To summarize, on the basis of new observations and theoretical developments, ecologists modified their ideas about succession. Instead of expecting a predictable march toward a stable climax in the absence of disturbance, ecologists now recognize that because of unique histories, variable settings, complex responses to natural and human-caused disturbances, diverse interspecific interactions, and long-term environmental changes, it is often difficult to predict how a community will change over time.

10.2 Before the Wilderness: Anthropogenic Landscapes

Before the Wilderness: Environmental Management by Native Californians is a collection of papers edited by Thomas Blackburn and Kat Anderson. The anthropological material in this work challenges the blind spot of missionaries, miners, and settlers who thought that California's Indigenous peoples were not capable of managing their environment (Blackburn and Anderson, 1993). The new arrivals thought they had come to wilderness – a pristine, primeval, virgin, natural, untouched, original state of nature. If that was the case, then there was never a time before the wilderness. By using the phrase "before the wilderness" in their title, Blackburn and Anderson were being intentionally ironic, suggesting that instead of a wilderness, the world in which the newcomers arrived was actually managed by the Indians of California to provide the resources they needed.

Scholars in ecology, history, archaeology, anthropology, and other disciplines are realizing that many productive and diverse ecosystems which were previously thought to be natural are really the result of human intervention. Box 10.4 describes some of these anthropogenic landscapes.

Box 10.4

People behind the Scenes

Butterflies, livestock, and ants in England's hot pastures: Populations of the large blue butterfly began to decline in the grasslands of southern England in the 1880s (Figure 10.1). This butterfly has very specific habitat requirements and depends on a complex web of abiotic factors and interspecific interactions. It needs short grass that provides its principal foods: wild thyme and ant larvae. At one stage in their development, large blue caterpillars feed on the thyme plants. At a later stage, they secrete a sticky fluid that attracts ants, which carry the butterfly larvae to their underground nests. Once they arrive in the nests, the butterfly larvae eat the ant larvae in the nests for nearly a year.

The butterfly's decline was attributed to collecting, so in the 1930s a reserve was created. Collectors were excluded, as well as livestock. After grazing ended, the height of the grass in the reserve increased by about 2 cm. This was enough to cause the previously warm microclimate at ground level to become cooler.

In the cool soil beneath the taller turf, a different species of ant replaced the one whose larvae had been a good source of food for the large blue caterpillars. Their survival plummeted, and the population in the reserve fell to such a low level that females could not find mates. The large blue disappeared because a change in human use (the cessation of grazing) interacted with complex but subtle (at least to our eyes) changes in an abiotic factor (ground temperature due to grass height), combined with changes in interspecific interactions (livestock herbivory on grass, butterfly predation on ants) and population structure (an unbalanced sex ratio).

After suitable habitat was re-established when burning and grazing were reinstituted, large blues from northern Europe were successfully reintroduced to southern England (Barkham, 2022).

Korean red pine forests: the human touch: Forests of the Korean, or Japanese, red pine grow throughout eastern Asia. Red pine forests are a product of traditional Korean agriculture. When nomadic hunter-gatherers in the Korean Peninsula settled and began farming, temperate deciduous forests dominated by hardwood trees such as oaks covered much of the landscape (Figure 10.1). Because farming depleted the soil, villagers fertilized it, using human and animal manure plus plant material collected from the forest around the villages. When they removed fallen leaves from beneath the oaks, the trees declined because they needed the nutrients in the leaf litter. However, pines, which do well on nutrient-poor soils, were successful and gradually replaced the oaks. Thus, as villagers enhanced the soil within their farm plots, they converted the deciduous hardwood forest to pine forest. Chun Young-woo, a professor of forest resources at Kookmin University in Seoul, South Korea, urges Koreans to "restore the human touch" to red pine forests by pursuing traditional forest management (Chun, 2009:47).

Grasslands in the White Carpathians of the Czech Republic: record high plant species diversity: World records of high plant species richness occur in two community types: temperate grasslands and tropical moist forests. Some meadows in the White Carpathians, a mountain range in the Czech Republic and Slovakia (Figure 10.1), hold the record for grasslands with the highest species diversity. Up to 67 species (including many rare endemics) have been recorded in plots of just 1 m^2). The White Carpathian grassland ecosystem is a conservation priority for Europe.

The landscape in the White Carpathians consists of a mosaic of grassland, cropland, and woody vegetation. It has been farmed for thousands of years, using traditional farming practices of regular mowing and livestock grazing that prevented nearby forest from expanding into the grassland. Meadows were mowed in summer to obtain hay, and livestock grazed common pastures for part of the year. During the Communist era (1948–1989), small private farms were combined in large cooperative or state-owned farms. Traditional mowing and grazing declined on these large farms, tall grasses increased, and species richness declined. However, in small fields that were inaccessible to modern machinery, mowing continued through the Communist period, and species richness did not decline.

When mowing was reinstated as part of post-communist restoration projects, conditions recovered quickly (Merunková et al., 2012; Michalcová et al., 2014).

These examples describe the results of some unplanned before-and-after studies in different parts of the world and periods of time. In each case, the cessation of traditional management (burning, livestock grazing, forest product collection, mowing, haying) was followed by unanticipated and unwelcome (at least for some people) changes in plant community structure and composition. These scenarios do not prove that the changes in customary patterns of human intervention caused the communities to change. Other things changed at the same time, and it is likely that they influenced the outcomes. The changes described in Box 10.4 probably resulted from anthropogenic influences combined with biotic and abiotic factors. But they do suggest some intriguing relationships between anthropogenic effects and plant communities. Researchers in historical ecology continue to work on teasing out those different factors and how they interact.

So far in this chapter we have seen that (1) ecosystems can change in ways which often do not conform to equilibrium theories and (2) people often have a pivotal influence on the direction of those changes. Changes in the Earth's climate illustrate both these points.

10.3 Climate Change

10.3.1 Past Climate Change

The Earth's climate has changed many times. Plants and animals that live in the tropics today once lived north of the Arctic Circle, and the poles were once free of ice caps. At other times, ice sheets covered much of the planet. During the Pleistocene Epoch, glaciers advanced and retreated many times, and sea level fell and rose as the amount of water bound up in ice increased or decreased.

Over the past 2,000 years, the global climate has been relatively stable and mild compared to much of the Earth's history. Climate change can occur quite rapidly, however, with serious consequences for people. During a brief, regional warming in Greenland around 900 CE, settlers from Iceland and Norway grew crops and raised livestock, but a few centuries later another climate shift, this time to a colder climate, contributed to the collapse of European farming in Greenland.

We noted in Section 5.2.1.4 that as far back as the nineteenth century, a few scientists recognized that the concentration of carbon dioxide (CO_2) in the Earth's atmosphere was increasing. By the middle of the twentieth century, some scientists concluded on the basis of information from varied scientific disciplines that this change in the atmosphere was causing the Earth to become warmer and that this phenomenon could have far-reaching effects.

10.3.2 Climate Science Background: What We Know

From physics, chemistry, and biology (Weddell et al., 2012) we know that:

- Some of the energy from the sun that penetrates the atmosphere and warms the Earth is radiated back into space as heat.
- Several gases (termed *greenhouse gases*), including CO_2, water vapor, nitrous oxide, *methane* (a greenhouse gas that traps heat in the atmosphere about 20 times more efficiently than CO_2), and some other gases, trap some of the Earth's heat and act like a blanket, warming the land, air, and ocean in a phenomenon known as the *greenhouse effect*.
- Life can exist only in a narrow range of temperatures that is made possible by greenhouse gases which trap enough heat to allow life on Earth but not so much that the planet becomes uninhabitable.
- CO_2 is produced by the burning of fossil fuels (petroleum, coal, and natural gas), great amounts of which were formed millions of years ago when high-energy molecules in the tissues of plants were exposed to heat and pressure in the Earth's crust.

- CO_2 is removed from the atmosphere when photosynthesis incorporates carbon dioxide into plant biomass or when it is stored in soil or the ocean.
- The net difference between the amount of CO_2 released into the atmosphere and the amount removed or stored affects the Earth's climate.
- The amount of CO_2 in the Earth's atmosphere began increasing markedly around 1750, at the start of the Industrial Revolution.
- During the Industrial Revolution, the amount of fossil fuel that was burned for electricity, heat, and transportation increased.
- Ice sheets and glaciers expand in the polar and temperate zones when the Earth is relatively cool and shrink when the Earth is relatively warm.
- Because ice sheets and glaciers contain a lot of water, land along the sloping continental shelves at the margins of continents was exposed when the Earth's climate was cool. Conversely, coastal areas and some oceanic islands are submerged during warm periods because less ocean water is locked up in glaciers and ice sheets.
- When CO_2 dissolves in water it forms *carbonic acid*.
- Carbonic acid interferes with the formation of shells in marine organisms and can cause them to weaken or dissolve.
- Within limits, photosynthesis, and therefore plant productivity, increases as the level of carbon dioxide in the air increases.
- Air and water temperatures influence the timing of many seasonal cycles, such as migration, breeding, flowering, and autumn leaf fall, and the growth and development of plants and some animals.
- As the Earth's temperature increases, *permafrost*, ground that remains frozen for at least two years (and often much longer), begins to thaw, releasing CO_2 and methane.

10.3.3 Predicted Changes if Climate Change Is Already Occurring

Given this information, what consequences should we expect if the greenhouse effect was already occurring by the late twentieth and early twenty-first century? Table 10.2 summarizes some relevant physical and biological changes that have been reported for the period from the nineteenth century through the first decade of the twenty-first century. The information in the table draws from several sources of information about climate trends and their effects. Reliable temperature measurements go back to the late nineteenth century. Long-term annual observations of the dates of many seasonal phenomena – such as flowering, spring ice melt, and migration – are available, as are long-term natural records in rocks, fossils, crystals, sediments, reefs, tree trunks, and ice. In addition, scientists conduct experiments to test specific hypotheses about many of these processes. For instance, plants are grown under various controlled temperatures, and some invertebrates are raised in seawater of different acidities.

The information presented in Table 10.2 shows that many physical and biological changes that we would expect to see if the Earth is warming have indeed been observed, and some are accelerating. Assessing the impacts of these changes is challenging because of complex interactions between many variables.

Table 10.2 *Evidence of physical (A) and biological (B) changes that would be expected to occur if the Earth's climate was warming by the late twentieth century and early 2000s.* The left column summarizes some predicted consequences of warming. Findings that are consistent with the predictions are shown in the column on the right. (After Weddell et al., 2012 with permission.)

Predictions	Observations
(A) Physical consequences	
The ocean's surface will become warmer.	The global average temperature of the upper ocean increased by 0.10° C between 1961 and 2003 (Bindorff et al., 2007).
Glaciers will melt.	Glaciers in the Tien Shan mountains of Central Asia decreased by an estimated 18% from 1961 to 2012 (Farinotti et al., 2015).
Continental ice sheets will melt.	The Greenland and Antarctic ice sheets declined between 2002 and 2009, and the rate of loss accelerated (Velicogna, 2009).
Sea level will rise.	A global network of tide gauges showed that sea level rose about 21 cm from 1880 to 2009, and the rate of increase sped up (Church and White, 2011).
The ocean will become more acidic.	During the past 200 years, ocean pH, a measure of acidity, changed by 0.1 unit (Haugan and Drange, 1996). Because the pH scale is exponential (Section 5.2.1.5), this corresponds to about a 30% increase in acidity.
Lakes that freeze in winter will thaw earlier than they used to.	Lake ice has been melting earlier in New England since the 1800s (Hodgkins et al., 2002).
The extent of permafrost will shrink.	The area of permafrost in the Tibetan Plateau has been shrinking, and the lower altitudinal boundary has moved upslope (Cheng and Wu, 2007).
(B) Biological consequences	
Some species will shift their geographic ranges north or upslope of their current ranges.	The ranges of many wild plants and animals shifted northward and/or to higher elevations in the twentieth century (Parmesan et al., 1999; Parmesan and Yohe, 2003; Hickling et al., 2006; La Sorte and Jetz, 2010).
Extinctions of local populations of some animal species will occur in the southernmost and lowest parts of their geographic ranges.	Extinctions of Edith's checkerspot butterfly were most likely to occur at low elevations and southern locations than at high elevations and northern locations (Parmesan, 1996).
The amount of calcium in the skeletons of corals will decline.	Calcium in the skeletons of a major-reef-forming coral found in the Great Barrier Reef of Australia declined by 14% between 1990 and 2008, an unprecedented change in the past 400 years (De'ath et al., 2009).

Predictions	Observations
Spring migrations, breeding seasons, and other events that occur in spring will begin earlier.	Earlier migrations in birds and earlier breeding in birds and frogs have been documented (Sanz, 2002; Parmesan and Yohe, 2003). Many plants in temperate zones have begun to develop new leaves, flowers, or fruits earlier and lose their leaves later in autumn than they used to (Schwartz and Reiter, 2000; Peñuelas et al., 2002; Wolfe et al., 2005).
Some insects that can reproduce more than once in a year will have more breeding cycles per year.	In 44 European moth and butterfly species, the number of generations per year increased between 1980 and 2009 (Altermatt, 2010).
Some disease-causing organisms will expand or otherwise change their geographic ranges.	A fungal disease has spread among many species of Central American frogs, causing some to become extinct (Pounds et al., 2006).
Forests will burn more often, with greater intensity, and with more serious impacts on human-well-being.	In the mid-1980s, large wildfires in forests of the western USA became more frequent, and the length of the wildfire season increased (Westerling et al., 2006).

Some changes have the potential for both beneficial and harmful effects. In temperate climates, warmer winters can mean lower heating bills. But this change can also mean improved overwinter pest survival and therefore increased crop damage, loss of agricultural income, and escalating hunger. Warmer summers can result in more deaths from heat waves and damage to crops from heat stress (changes that have already occurred in parts of North America, Europe, and South Asia). The increase in concentrations of atmospheric CO_2 associated with global warming can increase crop yield, but the resulting crops have lower levels of mineral nutrients and are therefore less nutritious (Loladze, 2014).

Other factors besides the Earth's temperatures affect the observations listed in Table 10.2. Air temperature is affected by cloud cover, which in turn depends partly on volcanic activity and pollution. Habitat changes and exploitation affect the distributions of populations; forest management affects fire frequency and intensity; life span and reproductive potential affect the vulnerability of species to extinction.

10.3.4 Modeling the Climate System

Collecting data on the many variables that affect climate on a planet-wide scale and the consequences of those effects on the climate is very complex. To deal with this complexity, climate scientists use detailed models that link simpler models of changes in the atmosphere, ocean, land, and ice as well as models of economic, social, and health variables. Experts not involved in the original studies review and critique researchers' methods, results, and interpretations. They test model predictions against measurements and then revise, test again, and update the models with new information as necessary.

On the basis of observations, experiments, and models from hundreds of countries, the vast majority of climate scientists agree that the Earth is getting warmer, that human activity is the principal cause of this warming, that we are already seeing its effects, and that the warming climate has had far-reaching consequences and will have even more serious consequences in the future if levels of greenhouse gases in our atmosphere continue to increase.

Although scientists have amassed a lot of data about current climate trends and their impacts, questions remain. Ongoing research is addressing these questions, and new models are continually developed and tested.

Some critics suggest that these questions and discussions mean there is no scientific consensus about climate change. That reflects a misunderstanding about how science works. Science progresses when people challenge conventional wisdom by considering alternative explanations, making new observations, formulating new hypotheses, and testing them. Errors can be corrected, and our understanding of the natural world increases. Science is not static. It is not a panacea, but it does have a built-in, self-correcting mechanism that can advance our understanding of natural phenomena. Dialogue among scientists is part of the process.

Models allow scientists to forecast probable outcomes given specific assumptions. One topic that has received a lot of attention among scientists and the public alike is the implications of climate change for extinction rates.

10.3.5 Climate Change and Extinction Risk

From what we know about past extinctions, we can make some generalizations about what characteristics are associated with a high risk of extinction and how a changing climate could influence a species' vulnerability.

We have seen that a species' risk of becoming extinct depends on ultimate and proximate factors (Section 6.5.1.1). Climate change may contribute to the extinction of species that are closely tied to habitats which are likely to change as the climate changes. This is the case for species with ultimate risks that limit their ability to survive except in specific environments or that require one or a few critical interacting species.

In addition, ultimate limitations on the ability to move between patches of suitable habitat add to vulnerability. This ability depends on a species' range of tolerance and its capacity for movement. Species that cannot tolerate conditions in the matrix between suitable habitat patches are vulnerable. Species that can cross over or through unfavorable habitat may be able to escape warming by colonizing cooler habitats that are closer to the poles or upslope, while others will not and will be left behind. Plants such as dandelions, which have seeds that are spread far and wide by wind, can generally colonize new habitat more readily than plants with heavy seeds such as acorns, which are spread by animals.

Ultimate factors such as poor dispersal ability or specific habitat requirements that predispose a species or subspecies to extinction under relatively stable conditions become especially critical in a rapidly changing climate, which influences proximate factors that affect a population's ability to persist. Since the Industrial Revolution, many of the Earth's landscapes have become inhospitable to many kinds of plants and animals. Habitats have also become fragmented, separated by landscape matrices that some plants and animals cannot cross. In such landscapes, many organisms, even those that are protected in preserves, are often unable to shift their geographic ranges to environments where they could survive.

The plight of the Bay checkerspot butterfly illustrates the effects of interrelated biological and environmental characteristics on extinction risk (Box 10.5).

The case of the Bay checkerspot butterfly reveals the interplay of a species' attributes and features of its habitat. Certain characteristics of the Bay checkerspot's biology – such as its limited ability to move to new sites and its specific habitat requirements – make this species vulnerable. But, until relatively recently, environmental heterogeneity – particularly variation in the direction of slopes – allowed some populations to survive in each year.

Box 10.5

Habitat Complexity and Survival of the Bay Checkerspot Butterfly

For six decades Paul Ehrlich and his colleagues at Stanford University have studied the dynamics of a metapopulation (Section 8.3.3.1) of the Bay checkerspot butterfly, a threatened subspecies of Edith's checkerspot butterfly (Table 10.2) that is endemic to hillsides at the Jasper Ridge Biological Preserve near San Francisco (Figure 10.2). Within this limited range, the Bay checkerspot inhabits a distinctive plant community that occurs on patchily distributed soils that have high concentrations of the mineral serpentine.

The area has mild, wet winters and dry summers. In this climate, herbaceous annual plants begin growth after the onset of the fall rains, grow throughout the winter and spring, and then dry up when summer arrives. Female checkerspots lay their eggs in spring, and when the larvae hatch, they must develop to the point where they can enter summer *diapause* (dormancy) before their host plants complete their annual cycle. If they don't make that deadline, the larval butterflies run out of food and die. Since the larvae can move only short distances, they are at the mercy of the microclimate in the patch of habitat where they hatch.

During unusually dry summers, the host plants dry up early on south-facing slopes, which are warmer and dry out sooner than slopes facing in other directions. Many larvae die in this microhabitat because their food plants dry up before the larvae complete their development. But some survive elsewhere, on the cooler slopes facing north, northeast, or northwest. In wetter years the reverse is true. Because hatching is delayed on the cooler slopes, few larvae have time to complete their development there. In those years, survival is better on the warmer, south-facing slopes, because the larvae get a head start on development early in the spring (Murphy and Weiss, 1992).

If the Bay checkerspot's habitat did not provide slopes facing in a variety of directions, it would not be able to cope with year-to-year weather variations. The complex topography produces a variety of microhabitats in this heterogeneous habitat. Environmental variability provides refuges that allow checkerspots to survive despite variable weather conditions.

Their ability to persist in the face of long-term, global climate change is questionable, however. The Stanford research group has documented extinctions of several local checkerspot populations related to extreme weather conditions, such as prolonged droughts or unusually wet winters. Even a population in a relatively large habitat patch was not secure because the patch failed to provide enough microclimate options. The Coyote Reservoir site, which consists mainly of an east-facing slope, supported a dense butterfly population in 1971, but by 1976 the population was gone (Ehrlich and Murphy, 1987).

Different species move at different rates; this differential movement disrupts interspecific interactions. The loss of an interacting species such as a pollinator or a food source may interfere with survival in an otherwise acceptable habitat. A good disperser may be able to shift its range as the climate changes but nevertheless may fail to survive because a food plant or a pollinator on which it depends cannot relocate.

Plants and animals have adapted in the past, and genetic change is ongoing in the present. Can this adaptative ability prevent extinction? Some species that have short generation times and high reproductive rates, such as insects, are already adapting to the changing climate. However, for many species evolutionary adaptation to a warming environment probably occurs too slowly to improve their chances of avoiding extinction. (Keep in mind the difference between the geological and ecological time scales. The evolution of an adaptive change that appears suddenly in the fossil record actually takes hundreds of thousands of years.)

In addition to its effect on habitat availability, climate change influences some environmental processes in ways that are likely to abruptly produce surprising, and sometimes alarming, results

10.3.6 Positive Feedback Loops and Tipping Points

Some processes that contribute to global warming are self-reinforcing; they set up positive feedback mechanisms (Preface: Overview). Unlike negative feedbacks, which decrease the direct effects of a process, positive feedbacks increase it. In the Earth's atmosphere, several positive feedback loops – including one that involves water vapor and another that involves methane – reinforce the effects of greenhouse gases.

Water vapor is a greenhouse gas; it lessens the escape of heat from the atmosphere. If the atmosphere becomes warmer, the amount of water vapor in the atmosphere increases because warm air holds more water than cold air. The resulting increase in water vapor will then reduce the amount of heat able to escape from the Earth. As the atmosphere warms further, it will hold even more water vapor, which will cause further warming of the Earth. Thus, an initial increase in the amount of water vapor in the atmosphere due to warming causes the atmosphere to trap more heat, which causes the amount of water vapor in the atmosphere to increase even further, and so on.

In a similar positive feedback loop, as the climate warms, permafrost melts, releasing the greenhouse gas methane, which contributes to further warming, which causes more permafrost to thaw, which releases more methane, which speeds up climate change and the loss of more permafrost, and so on in a continually reinforcing positive feedback loop.

There is no brake on these processes. With positive feedback, a small change leads to continuously accelerating change. Positive feedbacks can suddenly build up to irreversible *tipping points* – changes that can convert a stable system to a new state that cannot be returned to its original state. Crossing a tipping point is like going through a one-way door. It has been compared to a wine glass that tips over. If wine spills from a tipping glass, standing the glass back up will not put the wine back in the glass. The state of a full wine glass becomes a new stable, in fact, virtually irreversible, state – an empty glass. Tipping points are sometimes referred to as *sleeping giants* because often they begin with small changes that are hard to recognize initially, and their large potential for harm is not recognized until it is too late to do much about them.

Climate change is affecting planet-wide patterns of carbon dioxide storage in ways that could reach tipping points. Forests and the ocean can act as carbon sinks. They remove carbon-containing compounds such as carbon dioxide from the atmosphere and store it in a process known as *carbon sequestration*.

Trees and other organisms that contain chlorophyll take up carbon dioxide from the atmosphere through photosynthesis, which uses CO_2 and converts it to biomass. In forests, trees are the main photosynthesizers; in the ocean, plankton, corals, algae, and some bacteria carry out photosynthesis. When the capacity of the ocean to take up carbon dioxide in this way is exceeded, additional CO_2 from the air dissolves in water and forms carbonic acid. Without these processes, there would be a lot more carbon dioxide in the air, and the greenhouse effect would be stronger than it is now.

A community will sometimes reach a *threshold*, a point where its current state cannot persist. If a forested ecosystem such as the tropical forest of the Amazon reaches a tipping point at which it loses its ability to act as a carbon sink, the consequences for the region's biodiversity and human well-being will be profound. We do not know at what point the capacity of the Amazon's forests to absorb carbon dioxide might be exceeded, and no one is deliberately going to try to find out because "there is no sense in discovering the precise tipping point by tipping it." However, models suggest that climate change combined with deforestation and burning (which is done in the Amazon to eliminate downed trees and weedy vegetation) could reach a tipping point at which much of the region's tropical rainforest would become savanna, and the CO_2-removing service of a major carbon sink would be lost (Lovejoy and Noble, 2018:1).

10.3.7 Evaluating Climate Reporting

In spite of the scientific consensus about climate change, there is debate about what to do about it. Much of the argument about climate change is a political one about whether the situation is dire enough to warrant action. In other words, are the costs and risks of not taking action greater than the costs and risks of acting (*see Q10.6*).

When evaluating material about climate change, it is important to evaluate both the content that is presented and the way it is presented. Review the discussion in the Introduction on critically evaluating evidence. This is especially important with the topic of climate change, because some groups have strong economic interests in preserving our dependence on fossil fuels.

Climate scientists and experts in many related disciplines continue to extend our understanding of environmental change in the Anthropocene. In addition to the changing climate, we are confronted with other escalating changes to the biosphere and unpredictable, non-equilibrium processes. These challenges have set the stage for new conceptual developments.

10.4 An Alternative Perspective: Nature in Flux

10.4.1 Background

The nineteenth-century writer George Perkins Marsh complained that wherever "man ... plants his foot, the harmonies of nature are turned to discords" (Marsh, 1874:34). We noted in the Preface that to Marsh, people were "everywhere a disturbing agent," upsetting the natural balance of nature. The metaphor of nature in balance is strong in the popular imagination, but many ecologists today see nature in a different light, as the quote that opened this chapter indicates. Scientists have criticized the balance-of-nature metaphor as being "unproductively vague, value-laden and not representative of ecological systems in most of its incarnations" (Zimmerman and Cuddington, 2007). Equilibrium models do not satisfactorily explain current trends in northern fur seals of the Pribilof Islands, oak forests of New Jersey, acacias of the Serengeti, and a good many other cases. And we have seen that many ecosystems thought to be pristine really depend on human intervention.

In addition to these accumulated observations from a variety of organisms in a variety of settings, equilibrium theories face another problem. Many ecologists have an uneasy sense that all too often their colleagues have molded their interpretations to fit equilibrium models. In a paper on *The Shifting Paradigm in Ecology*, the ecologists Steward Pickett and Richard Ostfeld suggested that by "idealizing and simplifying the ecological world the tenets of the classical [equilibrium] paradigm have blinded ecologists and managers to critical factors and events that can govern ecosystems. The assumptions have also caused scientists and managers to neglect important dynamical pathways and states, and to disregard important connections among different systems" (Pickett and Ostfeld, 1995:265).

These misgivings suggest an alternative perspective: the idea that the natural world is more often in flux than in balance. Proponents of this view suggest that "flux of nature" is a better metaphor than "balance of nature" for describing how the natural world operates (Pickett and Ostfeld, 1995). This viewpoint does not deny that density-dependent population growth sometimes occurs or that plant communities sometimes follow each other in a predictable succession of stages leading to a stable climax. But it does suggest that many populations are sometimes regulated in ways that are independent of population density and that plant communities do not always follow an expected pattern. Equilibrium models are useful in certain situations, but they have limitations.

This flux-of-nature perspective is sometimes referred to as the non-equilibrium paradigm or the flux-of-nature paradigm. (A *paradigm* is a central concept that organizes a body of knowledge; a paradigm shift occurs when a prevailing theory is toppled by the accumulated weight of evidence it cannot explain (Kuhn, 1970)). There is some debate about whether the flux-of-nature viewpoint is a new paradigm or simply a useful metaphor, but clearly it represents a different way of looking at natural phenomena.

10.4.2 Key Points

Key points that follow from a non-equilibrium perspective (Fiedler et al., 1997; Ostfeld et al., 1997) are presented below. Chapters 11 and 12 cover strategies for putting these concepts into practice.

- Equilibrium is not the usual state for nature.
- Phenomena that proceed toward equilibrium are the exception rather than the rule in nature. In addition, sometimes a system is at equilibrium when considered at one scale but not at another.
- Disturbances are widespread and common.
- Virtually, all ecosystems experience disturbances, although they differ in how often and how intensely disturbances occur. Whereas formerly "managers viewed disturbance as having mostly negative impacts, … currently, evidence suggests nearly the opposite: preservation of natural disturbance regimes is essential to promote healthy, dynamic ecosystems" (Rogers, 1996:13).
- Ecosystems are open and interconnected across a landscape.
- The movements of organisms across ecosystem boundaries make it necessary to manage landscapes including the matrix between reserves, rather than discrete ecosystems. Matter, energy, and organisms don't stop at political boundaries; they move between private and public lands, across ecosystem boundaries, between states and nations, and through international waters.
- Heterogeneity has a pivotal influence on ecosystems.
- Most wild species are more abundant and have higher survival rates in structurally complex habitats than in homogeneous environments. Temporal heterogeneity is as important as spatial variation. In streams, for example, some species thrive in wet years, while others do best in dry years, so overall biological diversity benefits from year-to-year variations in stream flow.
- Many states previously considered natural are influenced by the activities of people.

In the balance-of-nature view, management should strive to exclude people from systems that are thought to be natural. Research grounded in this perspective has often overlooked or failed to appreciate the effects of people on ecosystems. Management and conservation grounded in a non-equilibrium perspective explicitly recognize the role of people in ecosystems.

To summarize, "natural systems … have many states or 'ways to be' and many ways to arrive at those states" (Pickett et al., 1992:70). These many possibilities happen because ecosystems are influenced by disturbances, anthropogenic actions, chance events, and long-term changes in geology and climate.

10.5 Implications: Stewardship

The application of insights from the balance-of-nature perspective to conservation has been successful in achieving certain objectives, but in other instances strategies based upon the idea that nature tends toward balance backfired. One consequence of Clements' view of equilibrium was fire suppression, which had many unintended and far-reaching consequences. In dry climates with high biomass production, the suppression of forest fires allowed flammable fuels to build up, and the big fires that resulted destroyed valuable timber – an outcome that was exactly the reverse of what was intended by utilitarian managers. Similarly, the protection of Mettler's Woods resulted in a decline in the old-growth oak community, the opposite of the intended preservationist outcome. In these instances, utilitarian and preservationist goals were not well served by an equilibrium perspective because important ecological interactions and processes were missed.

A non-equilibrium (flux-of-nature) perspective sheds light on why these policies sometimes failed to achieve their objectives and suggests an alternative approach based on stewardship. It provides a different theoretical basis for conservation as well as new strategies for the challenging task of conservation in the twenty-first century. The word "stewardship" implies the careful and responsible management of something entrusted to one's care. A stewardship approach recognizes the importance of processes that do not necessarily involve equilibrium, and it seeks to maintain ecosystems that are variable, diverse, and resilient. Because those ecosystems are diverse, the stewardship approach is inclusive – of species regardless of their economic value and of cultures and groups regardless of their participation in the dominant culture. This approach is compatible with the interests of many of the new players in conservation who raise social, economic, and political considerations as well as ethical and philosophical concerns (Chapin et al., 2011; Rozzi, 2015).

Stewardship conservation incorporates the dimension of history. Histories of both human and biological communities are relevant. It is important to compare current and past rates of change in biological, environmental, and cultural phenomena so that we can understand what regulates those rates and how they change over time. A stewardship approach involves research on questions such as: (1) How do current rates of change compare to changes in the recent and in the distant past? (2) What were the effects of management by different societies? (3) How can we support resilience and avoid tipping points? (4) What role do cultural and biological legacies play in resilience? Answering these questions involves diverse branches of knowledge, including ecology, anthropology, archaeology, genetics, environmental science, landscape dynamics, sociology, and more.

Thus, to implement a stewardship approach we need to consider the recent past and the distant past. Similarly, we need to bear in mind a range of spatial scales, from DNA molecules to the atmosphere as we study how conservation may help us to foster resilience and avoid irreversible tipping points.

We now turn our attention to how these ideas can put into practice a stewardship approach to conserving ecosystems that are complex, diverse, and resilient. Chapter 11 covers the conservation of key processes, structures, legacies, and interactions. In Chapter 12, we look at ways of integrating the conservation of biological and cultural diversity.

11

Strategies

Stewardship to Conserve Complex, Resilient Ecosystems

As we understand more about how complex and unpredictable the natural world is, and how profoundly people have changed it, the challenges of stewardship stand out. We need to confront problems created by past actions such as fire suppression, displacement of people from preserves, management to reduce environmental variability, and reliance on fossil fuels, which have had serious ecological and social consequences. We also need to plan for an uncertain future in a rapidly changing world.

In this chapter, we consider ways that we can use insights from studying the ecology of disturbances to promote the conservation of complex ecosystems and minimize the chance of triggering undesirable tipping points. We will see how processes, interactions, material legacies, and key species play crucial roles in addressing that objective.

To begin, we will consider what we can learn from studying responses to some natural disturbances (Section 11.1) and look the conservation of some disturbance regimes (Section 11.2). We will then turn our attention to maintaining or restoring key interactions (Section 11.3) and structures (Section 11.4). Finally, we will discuss some examples of management to address the critical challenge of promoting carbon storage (Section 11.5).

11.1 Responses to Disturbances

Understanding responses to disturbance is important for conserving complex and diverse ecosystems. We may want to create disturbances that mimic natural processes (for instance, by designing timber harvests that promote post-disturbance recovery). Or we may want to restore natural disturbances (for example, by returning fire, floods, or grazing to a landscape). In other cases, we may want to prescribe small-scale disturbances such as deliberately set fires in order to decrease the risk of large, frequent, intense disturbances (such as wildfires) that result from climate change and past management.

Understanding disturbance dynamics is also important if we want to prevent anthropogenic disturbances that threaten biodiversity (as in the case of fire in tropical dry forest), or if we want to identify situations where assisting with recovery is important (for example, if we want to minimize the opportunities that disturbances create for invasions by introduced species).

Or we might want to reduce the likelihood that an ecosystem will reach a tipping point (Section 10.3.6). The likelihood of tipping depends on many factors, some of which we cannot control. In terrestrial ecosystems, climate, geology, volcanic activity, and topography may influence ecological resilience. So will things that we can sometimes control, such as the intensity and scale of logging, grazing, farming, or herbivory, and sometimes fires and floods. If we understand what causes transformations between steady states, we can use that knowledge to modify the factors over which we have control in ways that help to conserve complex and diverse ecosystems.

11.1.1 Small, Common Disturbances

In temperate and tropical forests, small gaps created by the death of one or a few trees or branches are common. The openings that result are sites for recolonization by early successional species, which do not grow in shade. Seeds in the soil germinate in these clearings, and remnant trees may resprout. Those remnants, along with trees on the gap edges provide food and perches for birds, which assist recovery by depositing seeds in their droppings. For example, in a forest that was cut and burned in the Amazon territory of Venezuela (Figure 11.1), forbs dominated during the first year of succession, followed by other pioneer species and, after five years, a closed canopy.

This process creates a landscape mosaic of patches of different-aged forest. The size of the disturbance, the amount and kinds of materials that can promote vegetative growth by germinating or sprouting, and the frequency with which disturbances occur are critical factors affecting the community that develops in a such a gap (Chazdon and Arroyo, 2013).

Evidence from soil science, archaeology, and anthropology reveals a long history of anthropogenic disturbances in tropical forests. For hundreds to thousands of years, tropical forests in Latin America, Africa, and South and Southeast Asia have been used for shifting cultivation. Typically, a family clears a forest plot, farms it for a few years, and then leaves it fallow for several to many years. However, there are many variations on this widespread and common practice (Box 12.5). In some systems, households simultaneously farm permanent plots and also practice shifting cultivation (Fox et al., 2000). If intermittently cultivated plots are left for long enough to recover their fertility this type of agricultural system can be productive for many years and can contribute to biodiversity (van Vliet et al., 2012).

This can only happen, however, if people are able to move from site to site several times before returning to a plot. Because of growth of the human population in the tropics and increased settlement and commercial exploitation of the forest, shifting cultivators have fewer places where they can farm their temporary plots. Consequently, they must return sooner, and therefore the fallow periods between cultivations are now shorter.

The situation is different in tropical cattle pastures that have been cleared. They are larger than plots that are used for shifting cultivation, and they are subject to treatments that inhibit their recovery. Pastures are burned and weeded repeatedly, seeds are destroyed, nutrients are lost, and there is not enough time for vegetation to return to pre-disturbance conditions. Consequently, tropical pastures may require active restoration if the goal is to restore forest vegetation (Chazdon, 2003).

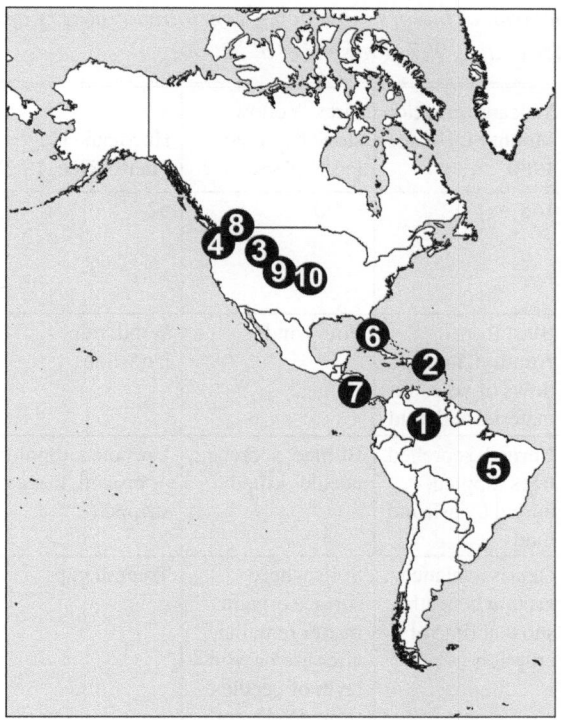

Figure 11.1 Locations of: 1, San Carlos de Rio Negro, Venezuela; 2, Luquillo Experimental Forest, Puerto Rico; 3, Greater Yellowstone Area, Wyoming; 4, Mount St. Helen's, Washington State; 5, two Indigenous Territories in the Cerrado of Brazil; 6, Everglades National Park, Florida; 7, Guanacaste National Park, Costa Rica; 8, Nelson, British Columbia, Canada; 9, Rocky Mountain National Park, Colorado; 10, perennial wheat study in tallgrass prairie, Kansas. Map created by Eva Strand using Esri, DeLorme World Countries Generalized Data & Maps for ArcGIS 2013, with permission.

Forest fragmentation hinders recovery from disturbances. Even small treefall gaps may not return to pre-disturbance conditions if they are in a fragmented landscape where sources of seeds and seed dispersers are not nearby.

11.1.2 Large, Infrequent Disturbances

A large, infrequent disturbance can change the trajectory of an ecosystem for decades to centuries by modifying successional pathways and the availability of the physical and biological materials that are available following disturbance. Because some kinds of large disturbances are becoming more frequent, understanding how ecosystems respond to large disturbances is increasingly important.

Table 11.1 compares four large disturbances in regions that had not experienced disturbances of similar magnitude for decades or centuries: (1) a flood of the riparian zone in a semiarid savanna along the Sabie River, part of severe flooding that took place in

Table 11.1   *Comparison of factors influencing vegetation recovery after four large disturbances (Turner et al., 1997; Parsons et al., 2006)*

	Volcanic eruption, Mount St. Helens, 1980	Fires, Yellowstone National Park, 1988	Hurricane Hugo, Puerto Rico, 1989	Flooding, Kruger National Park, 2000
Time (years) since last similar disturbance nearby	163	~290	57	75
Physical causes of destruction	Blast from eruption, fire, heat, flows of volcanic material and mud	Fire, wind	Wind, rain, landslides	Flowing water
Types of damage to vegetation	Burned, scorched, trees toppled, buried by ash and mud	Burned, scorched, needles killed	Uprooted, toppled or broken, leaves stripped	Uprooted, toppled, submerged, buried
Sites suitable for recovery	Debris avalanche, ground beneath snow at time of eruption	Soils where surface organic matter remained after fire or with layer of needles from dead trees	Treefall gaps	Wood piles, exposed mud
Recovery mechanisms	Survival of buried vegetation, germination, and seedling establishment (enhanced by burrowing mammals), sprouting of roots or stems	Germination and seedling establishment (especially seeds released by cones exposed to heat)	Germination, seedling establishment, growth of new tree foliage	Resprouting of toppled trees, seed germination, seedling establishment
Rate of vegetation recovery	Slow	Intermediate	Rapid	Rapid

South Africa, Zimbabwe, and Mozambique in September, 2000; (2) a Category 4 hurricane that struck tropical moist forest at the Luquillo Experimental Forest in Puerto Rico in September, 1989; (3) a complex of 31 fires that burned in the Greater Yellowstone Area, Wyoming, during a severe drought in the summer of 1988; and (4) the eruption of Mount St. Helens in Washington State, in a coniferous forest landscape in 1980 (Figures 11.1 and 11.2). During these events, wind, water, and fire moved across the landscape with great force. The resulting changes in physical and biological conditions favored some species over others, causing post-disturbance shifts in species dominance and richness.

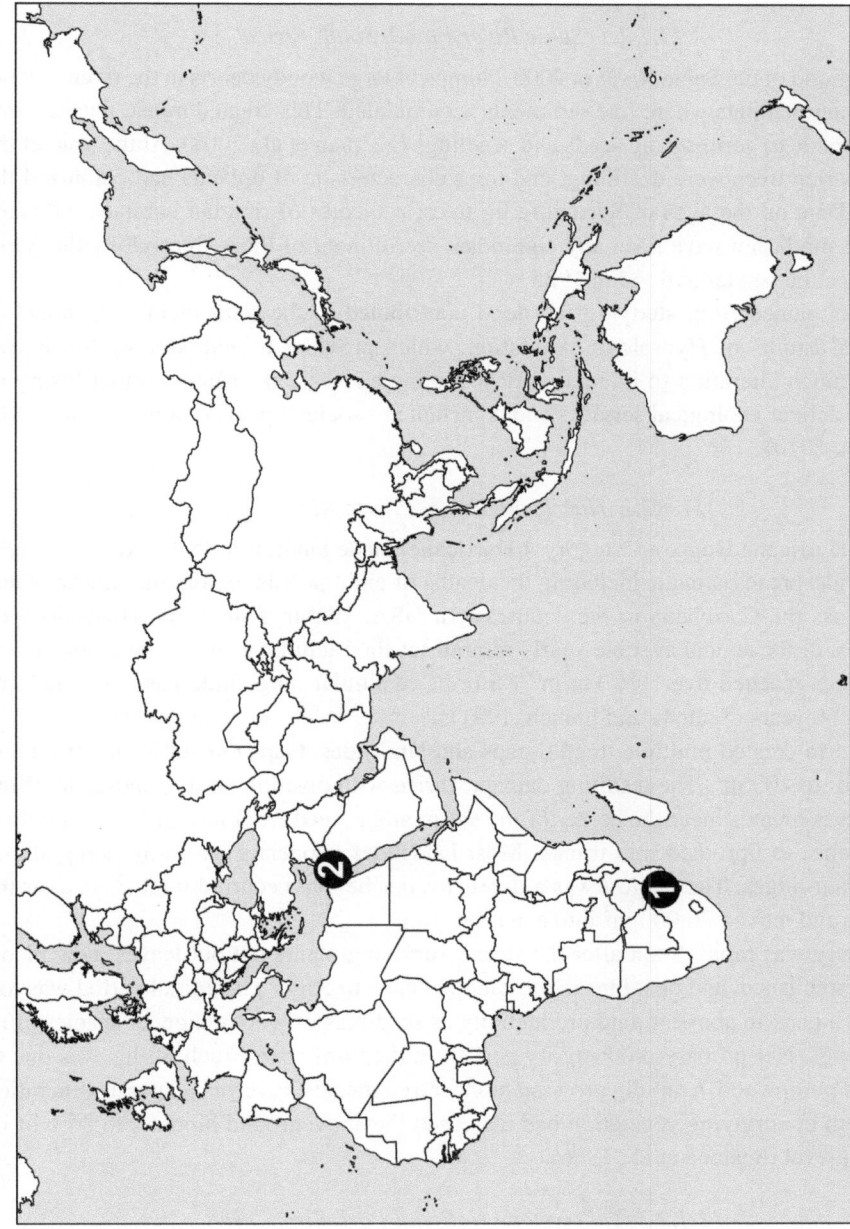

Figure 11.2 Locations of: 1, Sabie River, South Africa; 2, Negev desert, Israel. Map created by Eva Strand using Esri, DeLorme World Countries Generalized Data & Maps for ArcGIS 2013, with permission.

By destroying and damaging vegetation, these disturbances created sites favorable for the establishment and growth of seedlings, saplings, and sprouts from pre-existing woody plants and provided or removed resources favorable to recruitment.

11.1.2.1 Sabie River Flood, South Africa

After the flood of the Sabie River in 2000, clumps of large woody debris in the river channel formed microhabitats where fine sediments accumulated. This created moist, nutrient-rich sites favorable to germinating seeds and seedlings (Naiman et al., 2008). At the time of the flood, riparian trees were declining, and trees characteristic of uplands had colonized the channel. Data on the ages of Sycamore fig trees, a species of riparian habitats, indicated that there might not have been any significant recruitment of those trees along the Sabie River since the last large flood in 1925.

Insights gained from studying the flood contributed to the refinement of a model of Ecological Limits of Hydrologic Alteration, which provides a methodology for assessing how much alteration to its natural flow regime a river can sustain without losing its ability to deliver ecological services and contribute to societal goals (Naiman et al., 2008; Poff et al., 2010).

11.1.2.2 Hurricane Hugo, Puerto Rico, 1989

In 1989, Hurricane Hugo, a Category 4 Hurricane, made landfall in Puerto Rico, where it caused widespread damage, including the deaths of eight people, before moving west and north across the Caribbean to the southeastern USA. Within four hours, Hugo dropped about 15% of the annual average yearly rainfall for the Luquillo Experimental Forest, and wind speeds reached over 166 km hr^{-2}. Storms of similar magnitude pass over the area every 50–60 years (Scatena and Larsen, 1991).

The storm created multiple treefall gaps and landslides. Gaps ranged in size from less than 25 m^2 to 400 m^2. The resulting damage varied with position on the landscape. Wind damage was greatest near the center of the storm and ranged from loss of leaves and broken branches to uprooted tree trunks. Most landslides happened on north-facing slopes of mountain ridges. The biggest was a debris avalanche that occurred three days after the hurricane and moved 30,000 m^3 into a nearby river.

Recovery was rapid. Soon after the storm, surviving plants put out leaves, new plants became established, and openings in the canopy began to close. This initial period was followed by a peak in aboveground productivity as seedlings of woody pioneer species grew into saplings. Net primary productivity peaked in the third year. Much of this was due to growth of pumpwood, a rapidly growing pioneer tree species. Five years later, regeneration and growth of surviving vegetation had increased the aboveground biomass to 86% of its pre-storm level (Scatena et al., 1996).

11.1.2.3 Yellowstone National Park Fires, 1988

The fires in Yellowstone National Park in 1988, which received national and international coverage, likewise spurred the development of new, and sometimes surprising, insights about large disturbances. When I began writing this section, I expected it to be a story about how catastrophic fires led to recognition of the negative effects of fire suppression on forests. In other words, I thought the fires of 1988 were abnormal, an artifact of an unwise policy that had allowed excessive levels of fuel to build up.

Despite my efforts to find support for that interpretation, I found a different story. Yes, after 1910, when a single large fire complex burned 3 million acres and killed nearly 90 people in the West, the US government adopted a policy of fire suppression (Section 3.3.3). However, that policy did not change much. Attempts to put out wildfires were not effective until the middle of the twentieth century when fires could be fought from the air.

A few decades later, fire policy on public lands in the USA changed again. Beginning in 1972, fires caused by lightning in some parts of Yellowstone National Park were allowed to burn if they did not threaten human life, property, cultural sites, or endangered species. Thus, effective fire suppression throughout the park was only in place for a few decades.

So I was wrong. The fierce fires in Yellowstone in 1988 were not a result of fire suppression.

When fire ecologists analyzed the fire history of the park's lodgepole-pine-dominated forests using a combination of information from tree-ring cores, aerial photographs, and field studies, they found that the region had experienced large fires every 200 to 400 years. They concluded that the fires of 1988 were similar in terms of severity and total area burned to fires in the same region around the year 1700 and that drought and wind conditions, rather than fire suppression, were the major influences on the behavior of the 1988 fires (Romme and Despain, 1989). In its report on the ecological consequences of the fires, a scientific committee concluded that "Rather than ecological disasters or catastrophes, high-intensity fires in ecosystems such as those of the GYA [Greater Yellowstone Area] are virtually inevitable and are even essential for the reproduction of some species" (Christensen et al., No date:iv).

Lodgepole pine is one such species. Some populations of this common conifer have cones that do not open to release their seeds unless they are exposed to high temperatures such as those caused by a forest fire. In lodgepole pine populations with this adaptation, fire promotes reproduction.

The Yellowstone fires and information on their historical context raised difficult questions about the role of fire management in public lands. Media coverage of the fires, which was likened to a "celebrity scandal," emphasized the destruction of scenery and wildlife. Public acceptance of the National Park Service's let-burn policy was not high. Some groups suggested that the agency should take an active role in mitigating the effects of the fires by promoting post-fire recovery by seeding to promote forest regeneration and by feeding wildlife. However, the Service instead adopted a policy of preserving natural processes, combined with enhanced monitoring and research, rather than more active management (Christensen et al., 1989).

11.1.2.4 Mount St. Helens Eruption, Washington State, 1980

During the eruption of Mount St. Helens, vast areas were covered with mud and ash, yet in some places, recovery proceeded much more rapidly than scientists had expected. The effects of the eruption were heterogeneous. A host of physical and biological factors affected the severity of local effects. Where the impacts of the eruption were most severe, flows of volcanic material, debris, and mud obliterated all plants. By the end of the first post-disturbance growing season, hardly any plants grew in those areas. In areas where some vegetation survived, however, recovery was rapid. Where snowpack afforded protection, plants were able to resprout after the eruption and emerge through volcanic deposits up to 9 cm thick. Individual trees or clumps of vegetation also acted as foci for regeneration.

11.1.3 Implications from the Study of Disturbances

Recovery from large and small disturbances is patchy. Because they vary in extent and intensity, disturbances contribute to heterogeneity across a landscape. The mosaic that remains after a disturbance contains openings with plants of different ages and growth stages. They provide refuges for some organisms as well as sites for recruitment of young plants. They maintain species diversity and structural heterogeneity by creating sites for early successional species.

After the large disturbances at the Sabie River, Mount St. Helens, Yellowstone, and in Puerto Rico, "the effects on plants were not as catastrophic as suggested by first impressions" (Turner et al., 1997:765). This was in part because of biological legacies. Just as in human society we are aided by legacies that we receive from our past, *biological legacies* – organisms and organic materials that persist after a disturbance – boost a site's return to a pre-disturbance state (Franklin, 1990). In the small and large disturbances described above, legacies such as dead trees, partially decayed matter, sprouts, seeds, and live plants hastened the return to pre-disturbance conditions. For example, plants that remained on a cleared site enhanced the potential for revegetation because they provided shade, moderated temperatures, and offered seeds and habitat for seed-dispersing animals to repopulate a disturbed patch.

On a longer timescale, evolutionary history influences responses to disturbance. Natural selection has favored adaptations – such as the ability to sprout or to germinate after a fire – that allow organisms to reproduce after the kinds of disturbances they were exposed to during their evolution.

Insights from studying these and other large, infrequent disturbances can be used to develop practices that promote rather than impede post-disturbance regeneration. Return to pre-disturbance conditions is slower and less complete if the intensity and frequency of those disturbances is outside the range of conditions to which organisms are adapted. Knowledge of past disturbance regimes along with the legacies and interactions that promote recovery can help managers and scientists anticipate responses to disturbances. For instance, regeneration of a harvested forest can be facilitated by leaving standing green trees and other biological legacies (Sections 11.4.2 and 11.4.3). In this way, it may be possible to maintain a "safe operating space" for ecosystem recovery and reduce the chances of triggering abrupt shifts in "ecosystem composition" (Johnstone et al., 2016:369).

Although biological legacies, adaptations, and interactions sometimes allow ecosystems to recover rather rapidly from disturbances that seem, to us, to be very destructive, sometimes ecosystems are not so resilient. Some reasons for that situation are discussed next.

11.1.4 Why Are Some Ecosystems Slower to Return to Pre-disturbance Conditions than Others?

11.1.4.1 Environmental Limitations

Ecosystems with low productivity tend to take a long time to return to pre-disturbance conditions (Section A.4). When biomass in an environment of low productivity is removed, it is usually replaced slowly, because environments with low productivity, by definition, have slow rates of accumulating biomass. Usually this is due to climatic limitations or infertile soils. To make matters worse, during the time that vegetative cover is reduced, wind or water may erode exposed soils, creating a surface that is even less favorable for plant establishment. Consequently, dunes and deserts as well as alpine, arctic, subalpine, and subarctic communities return to pre-disturbance conditions relatively slowly after vegetation is removed by wind, water, fire, logging, grazing, trampling, burrowing, cultivation, or traffic. They also tend to be vulnerable to the effects of introduced species. Again, this stems in part from their limited plant cover.

11.1.4.2 Evolutionary Limitations

The rate at which an ecosystem returns to pre-disturbance conditions depends on the unique evolutionary histories of its species. Limited experience with certain kinds of antagonistic interactions can contribute to vulnerability (Box 11.1). Because of the circumstances of its evolutionary history, a species might not have come in contact with certain types of antagonists during the past. It might therefore be vulnerable to competition from introduced species that are better at capturing resources. Or it might lack defenses against them. As a result, such species are vulnerable when novel antagonists arrive.

Box 11.1
Vulnerability of Native Perennial Bunchgrasses to Livestock Grazing, a Novel Antagonistic Interaction

Native grasses of the steppes east and west of the Rocky Mountains evolved with different biotic disturbances. East of the Rockies, in the shortgrass prairie (Section A.1.5), the native perennial grasses form a dense sod that resembles a lawn. When the hooves of bison or cattle create an opening in this spreading mat of grass, the native grasses grow in from the edges of the disturbance. This minimizes the time when bare soil is exposed and vulnerable to erosion or to colonization by introduced grasses.

On the west side of the Rocky Mountains, in the *Intermountain West* (the region between the Rocky Mountains to the east and the Cascade-Sierra ranges to the west), native steppe vegetation is dominated by *bunchgrasses*, which form tufts. In drier parts of the Intermountain West, a thin crust formed from bacteria, algae, mosses, and lichens covers most of the soil surface between bunches of grass. These *soil crust*s are common in deserts and some steppe ecosystems (Section 11.4.1).

Bunchgrasses such as Idaho fescue are more erect than sod-forming grasses. Actively growing buds are higher on the stem in bunchgrasses than in grasses that form sod. The low position of the buds in sod-forming grasses is an adaptation to grazing by large herbivores, which cannot easily replace them. The bunchgrasses of the Intermountain West, with the higher placement of their buds, are more susceptible to grazing by large mammals because the grazers can reach the buds more easily.

Some bison inhabited the Intermountain steppe, but they never reached the enormous numbers that occurred in the bison herds of the shortgrass prairie. Thus, in the Intermountain West, there had been little selective pressure on the bunchgrasses for adaptations that would have made them less susceptible to new kinds of disturbance.

When settler agriculture and livestock grazing came to the Intermountain West in the nineteenth century, the native bunchgrasses were unable to reestablish themselves rapidly after grazing by cattle and sheep. That left them vulnerable not just to disturbances caused by livestock, but also to cultivation or other land uses that removed vegetation. The bunchgrasses grew back only slowly if at all under the new disturbance regime (Mack, 1981; Mack and Thompson, 1982).

That situation also left them susceptible to invasion by non-native grasses that were good at colonizing bare soil. Downy brome, also known as cheatgrass, is such a grass. This non-native annual grass evolved with centuries of tillage in the steppes of Eurasia. When it arrived in western North America in the nineteenth century, it rapidly colonized ground that had been cultivated or grazed. Its numbers and distribution increased rapidly.

The invasion of cheatgrass set in motion a change in the disturbance regime of the Intermountain West that reinforced its spread. Because cheatgrass is an annual, it dies after a single growing season. When it that happens, it becomes fine, dry fuel that burns easily. The resulting fires create bare ground on which cheatgrass germinates readily. This sets in motion a positive feedback loop Cheatgrass promotes fire, which favors cheatgrass, which favors more fire, a conversion that is almost impossible to reverse. It appears to have caused steppe in some places west of the Rockies to reach an alternative stable state.

11.2 Restoring Natural Disturbance Agents: Fire and Water

11.2.1 Restoring Fire in Brazil's Fire-Dependent Savanna: A Conservation Dilemma

Ecosystems can be classified according to their responses to fire as fire-sensitive, fire-dependent, or fire-independent depending on the *fire regime* (fire frequency and intensity) that prevailed during their evolutionary history. In *fire-sensitive* ecosystems, fire damages or kills many organisms and disrupts key processes. Fire is rare in *fire-independent* ecosystems. On the other hand, the fire regimes of *fire-dependent* ecosystems involve periodic fires. In fire-dependent ecosystems, selective pressures favor adaptations that allow survival, reproduction, and, in the case of animals, escape from fire. (Note that the ecosystems themselves did not evolve in response to fire; the species within ecosystems responded to selective pressures created by the fires to which they were exposed.) Alterations in a fire regime can create conditions that threaten the persistence of native plant and animal populations associated with that fire regime.

The *Cerrado* (the vast savanna of Brazil and neighboring nations) is a fire-dependent ecosystem (Figure 11.1). Plants of the Cerrado are adapted to fire, and some require fire in order to survive and reproduce. Like many other ecosystems in which fire is prevalent, the Cerrado has a climate that is marked by pronounced seasonal variation in rainfall. High rainfall during the wet season promotes vegetative growth. At the end of the dry season (the beginning of the rainy season), this herbaceous biomass dries out and provides fuel that is readily ignited by lightning. But because the rains are returning and green biomass is regenerating, fires at that time remain small and cool.

Fires started by lightning have been modifying the vegetation of the Cerrado since long before people arrived. Indigenous peoples of this biological hotspot (Box 8.3) began using fire to manage the Cerrado at least 4,000 years ago. They set fires to stimulate the flowering and fruiting of certain plants, to control pests, to favor vegetation that attracted game, and for a variety of other purposes. Indigenous semi-nomadic hunter-gatherers and shifting cultivators used detailed, site-specific fire management to create patches of vegetation at different stages of succession.

After the Portuguese colonized Brazil, the fire regime changed. Cultivators became more sedentary and burned larger plots more often. During the nineteenth and twentieth centuries, the use of fire to clear land for agriculture increased further. Unlike fires ignited by lightning, which occur during the rainy season, these fires were set during the dry season. The result was fires that were more frequent and hotter.

Plants remove carbon dioxide from the air through photosynthesis and store some of that carbon as biomass. When plant biomass burns, carbon dioxide and other greenhouse gases are released into the atmosphere. For this reason, large fires can contribute to climate change.

Conversely, since they contain a lot of biomass, trees can play a major role in storing carbon (Section 10.3.2). For this reason, the IUCN (Section 7.2) and some other conservation organizations have supported planting trees and suppressing fires to promote the expansion of forest into savanna (Veldman et al., 2015b). This strategy makes sense in fire-sensitive ecosystems such as tropical rainforests. But it is problematic in fire-dependent ecosystems like savanna.

In places where trees have expanded into the Cerrado, the diversity of savanna plants has declined. When a team of researchers from South American universities and the University of North Carolina assessed species diversity in three sites where forest had replaced savanna, they found that plant species richness declined by 27%, and ant species richness declined by 35%, compared to nearby plots where forest was not encroaching into the savanna (Abreu et al., 2017).

However, when researchers estimated carbon stored in soil and vegetation at those sites, they found that the forest plots had four times as much carbon as the savanna sites. This is important because management that promotes the storage of carbon in aboveground biomass or in soil can help mitigate climate change by sequestering carbon (and thus preventing it from entering the atmosphere in the form of carbon dioxide or other greenhouse gases).

Thus, the spread of forest into the Cerrado savanna is bad for the diversity of savanna species but good for carbon storage. Fire prevents encroachment by trees, but fire also converts the carbon that is stored in wood into carbon dioxide that is released into the atmosphere.

Clearly, this situation poses a dilemma for conservation. How should this tradeoff between carbon storage and biodiversity be handled?

There is no easy answer to this question. However, fires vary in their effects. Those effects depend on many things, including the season when the fires occur – which in turn influences how large and hot they are – and how they are managed.

Can fire in the Cerrado be managed in a way that minimizes biodiversity loss and maximizes climate mitigation? For instance, can management promote relatively cool fires that foster species diversity in fire-dependent ecosystems while preventing hot fires that hinder post-fire recovery? The following example illustrates the potential for fire management to play a role in addressing this question.

From 2008 to 2013, the Brazilian government had a zero-fire policy. Dry fuels accumulated, and large fires, including *megafires* (wildfires that are very large and hard to put out) followed during the dry season in parts of the Cerrado.

By 2014, however, the government had moved from its zero-fire policy to a policy of integrated fire management that included deliberately set fires. A few years after the new program went into effect, scientists from Brazil, Portugal, and Germany conducted an unplanned before-and-after study on the effects of integrated fire management in the Cerrado. With the new approach, *prescribed burns* (controlled burns that are set for the purpose of management) were conducted during managed fire seasons that extended from the rainy season to the middle of the dry season. Using data from satellite imagery, fire scars, and long-term environmental datasets, the team compared fires during the managed fire season to fires during the wildfire season (the end of the dry season) in two Indigenous territories (Section 12.2.1) that had adopted integrated fire management.

Wildfires occur during the dry season. With prescribed burning, large wildfires occurred less often, were less extensive and less intense, and emitted less air pollution than wildfires that had occurred during the period when fires were suppressed (but occurred in spite of the zero-fire policy) (Santos et al., 2021).

The researchers concluded that fire management can contribute to reductions in large, severe wildfires during the Cerrado's dry season. This finding has important implications because fires that occur late in the dry season are more severe and have more negative effects on biodiversity and higher greenhouse gas emissions than the cooler fires that are set earlier in the fire year.

Additional examples of managing ecosystems to store carbon are described in Section 11.5.

11.2.2 *Restoring Water: Dynamic Sheet Flow in the Florida Everglades*

The Everglades ecosystem in South Florida is a large freshwater wetland in a landscape dominated by intensive agriculture (principally sugar production) and rapid urban growth (Figure 11.1). Prior to development of the region, a huge, shallow sheet of water from Lake Okeechobee in the north slowly moved over much of the land for several months during most years in a process of *dynamic sheet flow*. This unusual hydrology is a key feature of the South Florida landscape. Because the topography of South Florida is relatively flat, water flows across it quite slowly. Slight differences in topography provide habitat diversity (Harwell, 1997).

Although much of it might appear homogeneous, topographic relief in this vast, gently sloping plain is significant. Sawgrass (a species in the sedge family that gets its name from saw-like serrations on the edges of its grass-like leaves) dominates the vegetation of the Everglades. Patches of ground that are slightly higher or lower than the landscape matrix provide habitat for distinct assemblages of organisms. Elevated mounds known as *hummock*s dry out sooner than the surrounding area and support palms and other trees. Depressions made by alligators hold water after it has receded from the rest of the landscape. During the dry season and in dry years, these wet spots provide important refuges for many species of invertebrates along with animals that feed upon them.

Because of these distinctive features, the landscape of South Florida is a heterogeneous landscape that historically was utilized by a variety of animals, including the Florida panther, American alligator, wading birds, fish, crayfish, snails, turtles, and the locally endangered snail kite, a hawk that feeds exclusively on apple snails.

Major disturbances periodically affected this landscape. In years of low rainfall, vegetation dried out sooner than usual and was subject to fires. Hurricanes with extreme storm tides and high winds, occasional freezes, and droughts also damaged local patches of vegetation. Recovery from these disturbances created patches of vegetation in different stages of succession.

The Everglades ecosystem faces many threats. In the 1970s and 1980s, it became apparent that cattails, which grow in nutrient-rich water, were replacing sawgrass in areas adjacent to agriculture. Historically, the water was low in nutrients, but agricultural runoff increased levels of phosphorus, resulting in the spread of cattails and a decline in structural diversity and species richness. Around the same time, levels of the heavy metal mercury that were toxic to people and wildlife were detected in fish. Habitat loss due to intensive real estate development, sea level rise from climate change, and invasions of exotic species also caused serious problems.

Projects aimed at controlling water in the Everglades began in the mid-twentieth century, with the goals of preventing floods, supplying water for urban and agricultural needs, and draining land for development. Although this management was successful in terms of its objectives, it disrupted dynamic sheet flow and fragmented and reduced habitat for wildlife.

Since then, the scope of water management has expanded, and its goals have changed. A massive restoration effort is currently underway to mimic some aspects of historical hydrology in the Everglades. This restoration is the focus of one of the largest and most expensive environmental restoration efforts in the world. It is plagued by political and funding difficulties, however, including competing visions for the region.

11.3 Restoring Interactions

In some situations where a missing interaction cannot be restored, it is possible to devise creative ways to mimic it. Restoration of tropical dry forest in Costa Rica provides an intriguing example (Box 11.2).

Box 11.2

Substituting for Extinct Seed Dispersers in Costa Rica's Tropical Dry Forest

Species diversity is high in *tropical dry forest*, the characteristic ecosystem in tropical regions with a pronounced dry season (Section A.1.8.1). Much of the formerly extensive tropical dry forest of Central America has been converted to savanna, however.

Enhancing seed dispersal in this ecosystem is important because mutualistic interactions between some tree species and their seed dispersers have been disrupted. Large native megafauna that once dispersed tree seeds in this ecosystem are now extinct (Section 6.4.1.1). Until the end of the last ice age, a variety of large herbivores – including horses, giant ground sloths, and relatives of the mastodon – roamed the region. Some of the ecosystem's trees and shrubs produce large fruits with tough seeds that were eaten by these large, now-extinct, herbivorous megafauna (Janzen and Martin, 1982) (Figure 11.3). The disappearance of these fruit-eating mammals means that more of the fruits, and the seeds they contain, are likely to remain where they fall. If they germinate there, they are unlikely to survive because of density-dependent mortality beneath the parent tree.

Today, domestic horses and cattle disperse those seeds. Daniel Janzen (Section 8.3.3.3) showed experimentally that when seeds of the guanacaste tree, a conspicuous element of Costa Rica's present-day tropical dry forest (Figure 11.1), are consumed by a large mammal, some seeds survive passage through the digestive tract. Guanacaste seeds cannot germinate unless their hard seed coat is broken. This happens when horses or cattle digest them. The herbivore also carries some of the seeds in its gut as it moves around the forest and eventually deposits them in its dung (Janzen, 1981).

Thus, there is a mutualistic relationship between large herbivores of the dry tropical forest and trees with fruits that contain large, hard seeds. When the herbivore eats the fruit, it benefits from the nourishment it receives from digesting some seeds, while the tree benefits when some of its seeds with digested coats are deposited away from the parent tree.

Figure 11.3   Model of an extinct gomphothere, a prehistoric seed disperser of tropical dry forests. Credit: Nobumichi Tamura / Stocktrek Images / Getty Images.

As part of Costa Rica's dry tropical forest restoration in Guanacaste National Park (*Parque Nacional Guanacaste*), horses and cattle are fed seeds of the guanacaste and other important forest trees, and then allowed to roam through the area that is being restored. Local children collect the seeds from dung, and older residents plant them where young trees are needed (Allen, 1988).

The extinction of the megafaunal seed dispersers raises an interesting question. How were the guanacaste and the other trees that were involved in this mutualism able to persist in the millennia after the megafauna went extinct and before people with cattle and horses got there? Researchers in Brazil have shown that gravity, runoff, flooding, and the activities of medium to large mammals such as monkeys and some birds disperse some fruits of the affected trees. However, they hypothesize that these substitute dispersal mechanisms may affect how far from the parent tree the fruits are dispersed, which might also affect the trees' genetic diversity (Guimarães et al., 2008).

11.4 Managing Ecosystem Structures

11.4.1 Modifying Patch Structure to Convert Desert to Savanna

In the above examples, fires and flooding affect the landscape heterogeneity. This can happen at fine spatial scales. The next example describes the effects of a management intervention involving small artificial disturbances that range in size from centimeters (Box 11.3) to meters (Box 11.4).

Box 11.3

Increasing Structural Heterogeneity by Altering Surface Topography in the Negev Desert, Israel

The Negev Desert of Israel (Figure 11.2) is a land of only modest topographic relief. The terrain is hilly, with low soil mounds in a matrix of crusted soil (Shachak and Pickett, 1997). Water limits productivity in this hot, dry environment. Annual grasses and forbs make up most of the vegetation, along with scattered patches of shrubs (Boeken and Shachak, 1994). For several thousand years, the area has been inhabited by Bedouin who until recently practiced nomadic pastoralism.

The matrix soil and mound soils differ in texture. Mounds consist of loose soil particles, whereas in the matrix between the mounds, a soil crust formed by microorganisms covers the ground. Particles within the crust are bound together by chemicals that the microorganisms produce (Box 11.1). Water does not move easily between the particles of cemented soil in the crust. Instead, it runs across the crusted surface of the ground until it intercepts a pit or a mound, where it flows downward through the more porous soil it encounters there.

Scientists at the Jacob Blaustein Institute for Desert Research in Israel investigated the effects of topography on productivity and plant species diversity. To do this, they dug pits measuring 1 m by 30 cm to a depth of 20 cm, and used the soil they took from the pits to create mounds further downslope. In this way, they created 28 experimental units consisting of a pit, a mound, and a portion of the surrounding matrix.

Three and a half years after the experiment was set up, the researchers recorded the plants that had become established in their experimental units. Plant species diversity and biomass were greater in the pits and on the mounds than in the matrix. Plants with small seeds were more common in the matrix, while plants with seeds that were relatively large (but nowhere near the size of the seeds the gomphotheres ate in the previous example) were more abundant in the pits and mounds. Apparently, runoff and wind moved the larger seeds until they encountered a pit or a mound, where they became established. Species with tiny seeds lodged in small cracks within the crust.

The results of this experiment demonstrated that increasing the heterogeneity of a desert ecosystem by altering its topography changed the availability of resources and affected productivity, seedling recruitment, and species diversity. Scientists and managers involved in developing a park near the city of Be'er Sheba reasoned that water running across the crust could be captured and used to create a landscape with "parks consisting of clusters of [planted] trees integrated into a matrix of shrubs and herbaceous vegetation" (Shachak et al., 1998:475). In this way, the desert ecosystem could be converted to something more like a savanna. To do this, elongated soil terraces were created to direct water where it could be used to support vegetation, including planted trees.

The parks have aesthetic and recreational value, but they are controversial. Critics of the parks suggest that conclusions from the original research have been extrapolated to an inappropriate scale, in which earth-moving machinery creates disturbances many times larger than the original pits and mounds. This disrupts a fundamental component of the desert ecosystem, because although organisms that make up soil crusts are well adapted to extreme growing conditions, they grow back slowly after physical disturbance (Johnston, 1997). Detractors argue that the Negev ecosystem should be protected rather than converted to savanna. They point out that few examples of the Negev ecosystem are protected in reserves, that the Negev ecosystem supports several endemic plants and animals, and that the planting of trees in regions such as the Negev, where they were not present historically, can have unintended consequences (Rotem et al., 2014). Planted trees in the Negev provide perches for predatory birds that may have allowed predation on an endemic lizard to increase and contributed to its decline. There is also evidence that construction of the terraces degraded soil quality, although some recovery of the soil occurred subsequently (Stavi and Argaman, 2016).

Box 11.4
Retaining Legacies That Promote Reproduction and Regeneration

Suzanne Simard, a forest ecologist at the University of British Columbia, and her colleagues discovered that the threadlike filaments of mycorrhizal fungi connect the roots of many trees to each other in the Canadian Douglas-fir forest they studied. There are so many of these physical connections among individual trees and between trees and fungi that a map of them appears crowded (Figure 11.4). Sugars produced by photosynthesis move from the older trees through mycorrhizal filaments (Section 2.1.2.4) to seedlings that they are genetically related to. When researchers mapped the connections between fungi and individual trees, they found that larger (and therefore older) trees had the most connections in this *"wood-wide web"* (Beiler et al., 2010).

When a forest is clear-cut, all of the trees and most of the understory vegetation are removed from a site. This interferes with two kinds of biological legacies: (1) living trees and shrubs that could be sources of seeds on the cleared area after the harvest, and (2) other plants that harbor potential fungal mutualists. As a result, the potential for trees on the site to grow back declines after clear-cutting.

If, instead of a clear-cut, a timber harvest leaves some trees and other plants, these biological materials can be sources of seeds and fungi. However, the seeds of early successional plants that colonize a site after logging or any other disturbance need light. This situation creates a dilemma

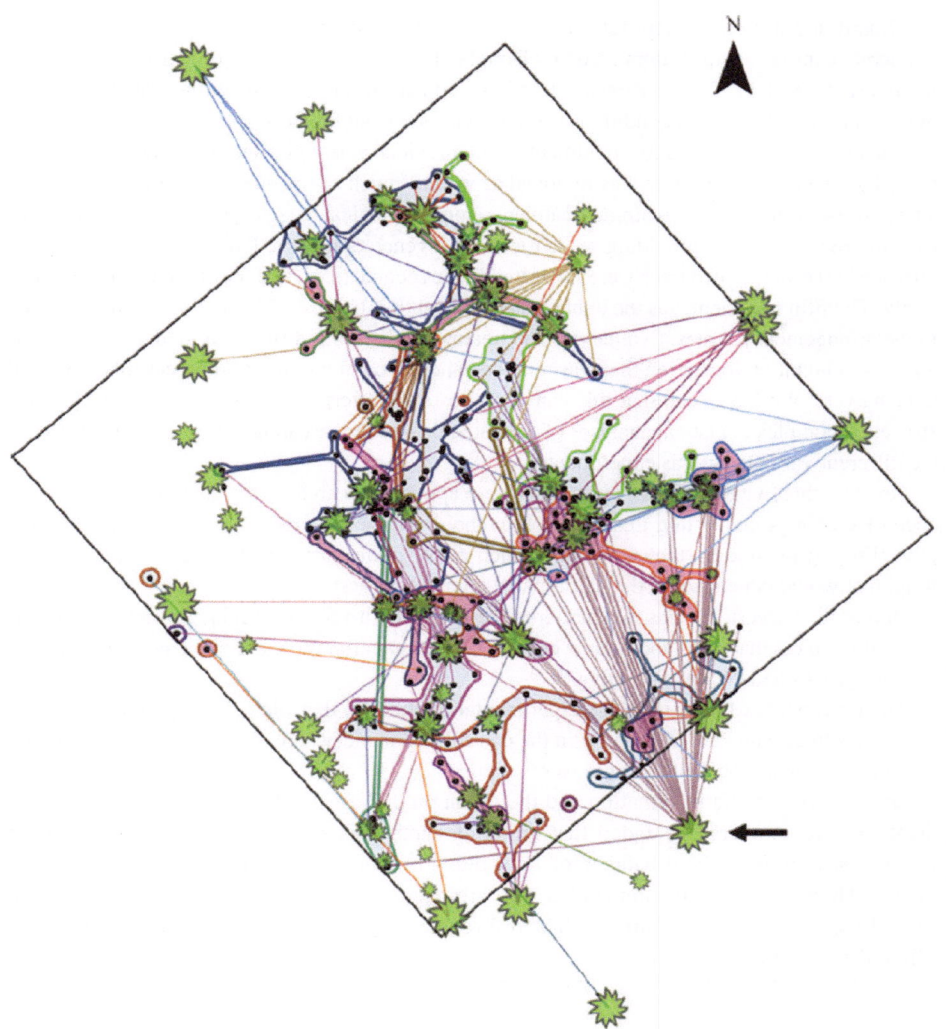

Figure 11.4   The wood-wide web showing connections between Douglas-fir trees and mutualistic underground fungi (mycorrhizae). Sunburst symbols indicate trees; size of the sunburst symbol indicates a tree's diameter and, therefore, age. Black dots indicate locations where mycorrhizae were sampled. Groups of genetically related mycorrhiza are outlined with curved lines. Arrow points to the most highly connected tree, which was linked to 47 other trees through mycorrhizae inside the 30-by-30-m plot. Used with permission of John Wiley & Sons Ltd., from Beiler et al. (2010).

for forest managers. If too many trees are removed, there won't be enough seeds and mycorrhizal fungi to promote good post-harvest germination and seedling establishment. But if logging is very selective and takes only a few trees, there may not be enough light to allow good regeneration of plants on the site. Hence, there is a tradeoff between the availability of seeds and fungal mutualists on one hand and access to light on the other.

Simard and her co-workers set out to discover the optimum solution to this dilemma in a temperate coniferous forest dominated by Douglas-fir trees near the town of Nelson, British Columbia (Figure 11.1). They designed a set of experiments to find the timber harvest arrangement that would provide optimal conditions for regrowth of a Douglas-fir forest.

The experiment involved five treatments, each of which was replicated four times. In the most intensive treatment, every tree was removed as well as much of the other vegetation. This mimicked clear-cutting. In the remaining treatments, some biological legacies in the form of *seed trees* (mature trees capable of providing seeds for forest regeneration) were left standing. Ten percent, 30%, and 60% of the seed trees were harvested in the second, third, and fourth treatments, respectively. The fifth treatment was the unharvested control (0% harvest). The treatments also differed in the arrangement of trees. (Notice that the treatments therefore differed in more than one variable: the amount of vegetation that was removed and the configuration of the trees that remained after harvest. We have seen that this can complicate the interpretation of experimental results. However, complex statistical analyses of the data allowed the researchers to assess the effects of the different variables in this experiment.)

One and three years after the treatments were applied, researchers recorded data on the presence of seedlings and young trees, as well as the diversity and coverage of other plants in the plots. From prior studies, the team knew which plant species harbored the types of mycorrhizal fungi that would benefit seeds that germinated after the harvest.

Before the harvest, species that harbored the beneficial mycorrhizal fungi covered about four fifths of the plots that were about to be clear-cut. One year after the clear-cut, coverage by plants in those groups had declined to only 6%.

Three years after the start of the experimental treatments, densities of Douglas-fir and other coniferous trees tended to be greatest in the control treatment (where all trees were retained) and the treatment where 60% of the trees were retained.

On the basis of this information and their data on the effects of the different tree arrangements, the researchers concluded that retaining seed trees spaced about 10 to 20 m apart would result in an optimal balance of light plus proximity to sources of seeds and mycorrhizae. Therefore, they recommended this arrangement for maximizing the potential for natural recovery of Douglas-fir-dominated forests in British Columbia after timber harvests (Simard et al., 2021).

11.4.2 Retaining Structural Legacies through Retention Forestry

Retention forestry is an approach to managing commercial forests in which specified kinds and amounts of forest structures are left following harvest. Biological legacies such as individual live or dead trees, patches of forest, and dead wood may be retained. Whether or not plants, animals, and fungi use the retained structures depends on variety of factors, including species composition of the forest community, site characteristics, and nearby land uses.

Currently, some forests in Western Europe, Australia, South America, Canada, and the USA are managed with this approach. Although the scale and methods of retention forestry depend on the specific ecological, social, and political context, all applications of this approach involve timber harvests that leave biological legacies.

Scientists from many nations have studied the use of retained structures by diverse groups of organisms in boreal and temperate forests. Retention is particularly effective at maintaining some groups of lichens, mycorrhizal fungi (Section 2.1.2.4), and small, ground-dwelling animals. Decaying wood in the form of snags and logs is especially important for some kinds of beetles and fungi (Gustafsson et al., 2020).

In boreal forests of northern Scandinavia and northwest Russia, European aspen is considered a keystone species. When dead aspen wood is retained after harvest, this hardwood tree provides refuges from which many species can recolonize cutover lands after timber harvest. Live remnant aspens host mosses and lichens, including some rare species. Fungi that specialize on dead wood tend to prefer retained aspens to other species. Dead aspens retained in sunny locations are important for the survival of many species and as sources of colonists (Gustaffson et al., 2020).

The evaluation of retention forestry is discussed in Section 11.6.

In addition to retaining structures, it is important to conserve interspecific interactions that facilitate responses to disturbance. Interactions between fungi and plants play a pivotal role in forest communities, as the next example illustrates.

11.4.3 Retaining Mutualistic Interactions and Structural Legacies: Mycorrhizal Fungi

Most plants have symbiotic relationships with soil fungi. In this type of relationship, which is known as a *mycorrhiza* (plural, mycorrhizae), extensive underground fungal filaments form around a plant's roots. The fungus obtains sugars and other products of photosynthesis from the plant, while the plant partner in the relationship receives soil nutrients from the fungus. The aboveground part of the fungus might be a familiar mushroom, but the belowground part is a vast network of thin filaments.

Mycorrhizae play pivotal roles in plant biology. In temperate forests, many tree species depend on these mutualistic interactions for their nutrition and reproduction. Recent research sheds light on far-reaching implications of this relationship for forest management.

The concept of a wood-wide-web has excited the popular imagination. Proponents of this idea hypothesize that mycorrhizal networks allow individual trees to "talk" to each other, signal danger, cooperate with other trees, and send resources to trees in need. The idea has inspired magazine articles, TV and radio programs, YouTube videos, and a Pulitzer Prize-winning novel about a woman scientist whose ideas about cooperating trees were initially ridiculed (*see Q11.1*).

This controversy is stimulating lots of discussion and research. Many questions remain. How widespread are underground networks between trees? How do they affect tree survival? Could trees be getting nutrients directly from soil rather than relying entirely on mycorrhizal networks? Do trees really help the needy members of their community? The topic tends to inspire strong feelings, which may be influenced in part by unrecognized biases about whether competition or cooperation is the dominant force in human society and in nature.

11.5 Managing Ecosystems for Carbon Storage

Simard's team also looked at how the different harvest treatments affected the amount of carbon stored in the forest plots. The more intensive harvest methods, such as clear-cutting, lost the most carbon from aboveground live trees and from the forest floor, whereas the treatments that promoted recovery of the forest after harvest also retained the most carbon.

The next example (Box 11.5) describes a different approach to carbon storage.

Box 11.5

Restoring a Keystone Species to Restore Hydrology and Store Soil Carbon

An estimated 60–400 million North American beaver inhabited North America before commercial fur trapping for beaver pelts began in the sixteenth century. The Eurasian beaver (a different species but a member of the same genus) was similarly abundant. Both species were widespread in their former range, but in both cases exploitation for pelts, and to a lesser extent for meat and *castoreum* (an oily secretion produced by beaver and used in perfume), led to pronounced declines. The beaver in North America disappeared from much of its former range, and the Eurasian beaver dwindled to about 1,200 individuals in eight isolated populations.

When beaver were still abundant, early European explorers in the USA complained that it was difficult to navigate the rivers because branching networks around log jams made it hard to find the main channel. Historical descriptions and photos, along with fossils and ancient sediments, also suggest that river channels in North America (as well as Europe) used to be more complex. These intricate channels resulted from water backing up behind beaver dams.

In the twentieth century, however, river managers in North America and Europe focused on straightening channels and constructing levees. As a result, the channels were simplified and became isolated from their adjacent floodplains. This approach succeeded in regulating flows, but in the process of accomplishing that objective it simplified landscapes.

As concern about those changes mounted, interest in reintroducing beavers to their former range increased. After they were introduced in North America and Europe, both species increased in range and numbers due to protection, regulated exploitation, and natural spread (Wohl, 2019).

Beavers are considered keystone species. They modify the functions and structures of ecosystems through their effects on stream channels. Wetlands created by beavers purify water, store floodwaters, enhance habitat and species diversity, cycle nutrients, and promote carbon sequestration. Beaver dams cause water to collect and form ponds. When the waters back up behind a dam and flow over channel banks, wet meadows form. These meadows are particularly important in mountainous regions, where most water flows swiftly because of the steep topography. Consequently, and there are few places where sediments, soil, and organic matter accumulate. However, in more level stretches of rivers and streams, beaver dams impede the flow of water and allow sediments and nutrients to accumulate at the bottoms of ponds and wet meadows. This material is high in organic carbon. The resulting carbon sinks are important sites for carbon storage.

Methane and carbon dioxide are both greenhouse gases, but methane is more potent than carbon dioxide. Because the water table is elevated behind an intact beaver dam, the soil, sediments, and organic matter (such as pieces of wood) stored at the bottom of and adjacent to an active beaver pond and meadow are saturated with water. Since the wood fragments and bits of other organic matter buried there are not in contact with oxygen, *aerobic decomposition*, the process by which organic matter is broken down in the presence of oxygen and releases carbon dioxide, cannot occur. In those conditions, the saturated soil, sediment, and organic matter undergo *anaerobic* (*an*, not; *aerobic*, air) *decomposition*. In the absence of oxygen, this process produces carbon dioxide, methane, and sulfur-containing compounds, which smell like rotten eggs. (You may have encountered this odor if you have spent time around permanently flooded marshes. Or if, like me, you have ever allowed the decomposing vegetables in your compost to remain too wet for too long.)

Anaerobic decomposition is much slower than decomposition that occurs in the presence of air. According to Ellen Wohl, an American scientist who studies physical and biological processes that affect rivers, wood buried in a soggy beaver meadow can last about 600 years!

As long as materials that are decomposing in the absence of oxygen remain buried, anaerobic decomposition continues. Methane and carbon dioxide are produced, but for the most part, these greenhouse gases do not escape into the atmosphere, and therefore they do not contribute to climate change. The carbon in the fragments and particles of organic matter that are stored in this way are sequestered. Across a landscape that once had a lot of beaver activity, this submerged organic matter tied up a substantial amount of carbon. But that changed after the beavers disappeared, and the soils dried out.

When Wohl and her colleagues compared the sediments in active and abandoned beaver sites within mountain streams in Rocky Mountain National Park in Colorado (Figure 11.1), they found that abandoned beaver meadows had lost some of their carbon. (The actual amount depended on several things including how long they had been dry.) Once the sediments were no longer saturated, exposure to air allowed the trapped methane and carbon dioxide to escape (Wohl, 2013). The organic carbon stored in the beaver meadows was no longer protected from air, and so aerobic decomposition took place. The beaver meadows that had been carbon sinks became carbon sources when they dried out. Some scientists think the amount of carbon that is lost in this way is a fraction of the amount that is stored in beaver meadows, but this question continues to be debated.

Because beaver activity increases landscape heterogeneity, promotes biodiversity, and provides many ecosystem services including carbon storage, it makes sense to reintroduce beavers to parts of their former ranges where they are compatible with other land uses or to restore landscape conditions that encourage recolonization by beavers. Another approach involves putting in *beaver dam analogs*, artificial structures that mimic beaver dams.

There are many places where people appreciate the return of beavers. On the other hand, they are not welcome everywhere. Their propensity to destroy trees and cause water to go where it is not wanted makes them unpopular in some places. For that reason, beaver reintroduction, like most other forms of resource management, requires dialog among stakeholders.

11.5.1 Planting Perennial Crops to Protect Soil Carbon

In several ancient settled agricultural systems, annual plants were the principal crops: corn in Mesoamerica, rice in Southeast Asia, wheat in the Middle East, and potatoes in South America. In most cases, harvesting annual crops involves pulling up or digging out the entire plant, leaving the soil bare for several months. During this time, the soil lacks a stabilizing cover of vegetation which leaves it vulnerable to wind and water. (Wetland rice cultivation is an exception because it is flooded for much of the year.) Because annual plants live for only one year, they must be planted annually. Usually this involves disturbing the soil with the aid of an implement – perhaps a digging stick or a plow. Therefore, the transition to settled agriculture brought about a big change in the duration and extent of soil disturbance.

In the middle of the twentieth century, interest in reducing the negative consequences of mechanized farming grew. Strategies for addressing those issues include practices that (1) maximize the time when soil is covered and minimize disturbance to soil and (2) maximize the addition of on-farm inputs such as manure and crop residues to soil and minimize the use of synthetic fertilizers and pesticides. A variety of practices – such as (1) minimizing tillage by seeding directly into crop stubble, (2) integrating livestock with cropping systems so that domestic animals recycle nutrients, (3) using integrated pest management (Section 4.4.1), (4) planting cover crops that stabilize soil and are subsequently incorporated into the soil, and (5) planting perennials – are used to address those objectives.

Perennial crops have several potential advantages over annual crops. Unlike annuals, perennial roots can grow for years. Because perennials have deeper roots than annuals, they can also reach more soil nutrients and water, and they have access to those resources for a longer time. When they die, the extensive roots of perennials decay and contribute to soil organic matter. Perennials also shield soil from erosive wind and water throughout the year.

In addition, perennials protect organic carbon that is stored in soil. When dead roots decompose, aerobic decomposition releases most of their carbon as carbon dioxide, but some is stored as soil organic matter. Soil microorganisms produce substances that glue particles together, forming *soil aggregates*. Tillage breaks open soil aggregates and exposes soil organic matter in the aggregates to air, releasing carbon dioxide. Because perennial crops are not tilled their soil organic matter stays in soil aggregates.

At several locations in Kansas, tallgrass prairie vegetation has been harvested annually for hay for 75 years or more (Glover et al., 2010) (Figure 11.1). When those perennial grasslands were compared to nearby annual croplands used for annual wheat production, the nitrogen content of aboveground biomass was similar in the two ecosystems, even though the cropland but not the perennial grassland had received additional fertilizer. Below ground, the perennial grassland had more soil carbon and more soil nitrogen than the cropland. The food web of soil invertebrates was more complex, and insects were more abundant and diverse in the perennial grassland soils than in the cropland.

However, perennial grain crops have lower yields than annuals. An annual plant can put a lot of energy into making seeds (for example, edible grain) each year because it is going to die after it makes those seeds, so it doesn't need to put energy into building any other kind of tissue. It needs only enough roots, stems, and leaves to keep it alive until it has made seeds for the following year. Natural selection acting on annuals has operated against wasting energy on unneeded tissue, and it has favored annuals with adaptations that allow them to put most of their energy into seeds.

In contrast to annuals, perennial plants must support tissue that will allow them to survive from year to year. That leaves them less energy for grain or other parts of the perennial plants that we would like to eat. This is a problem for us if we want to grow perennials for food. Although perennial crops can have several ecological and economic benefits, transitioning to large-scale agricultural production of perennial crops is hampered by the lack of high-yielding perennial food crops. Nevertheless, as of 2022, a perennial wheat is being incorporated into some commercial foods.

11.6 Evaluating Management for Complex, Resilient Ecosystems

The examples in this chapter illustrate a stewardship approach that strives to conserve complex, resilient ecosystems by maintaining, restoring, or manipulating material legacies (trees, snags), key species (beavers), interactions (seed dispersal, mycorrhizae), disturbance regimes (fire, dynamic sheet flow), or key structures (beaver dams, soil crusts, perennial vegetation, pits, and mounds). On the basis of our understanding of ecosystem dynamics, this approach makes sense. But how do we know if that kind of conservation is succeeding? That depends on our objectives.

Evaluating these endeavors is complicated by the fact that stewardship involves thinking about the connections between local and global scales. We would like to know how a beaver dam, a clear-cut, a perennial crop, or a prescribed fire affects the diversity of species and habitats on a local or perhaps a regional scale. But we are also interested in how it contributes to the accumulation of greenhouse gases and other processes in the atmosphere on a global scale.

Consider the assessment of forest management. One way to evaluate evidence from scientific studies about an approach is *meta-analysis*. A meta-analysis is a comprehensive synthesis of the results of many independent studies of the same topic. It uses methods that take into consideration the statistical characteristics of the individual studies. (The study of shifting cultivation that is discussed in Section 5.2.2.3 is an example of a meta-analysis.) Akira S. Mori and Ryo Kitagawa at Yokohama University in Japan conducted a meta-analysis of studies of the effects of retention forestry in three biomes (boreal, temperate, and tropical forest) on species diversity. They searched the scientific literature for relevant studies and found 23 studies that compared species diversity in logged and unlogged forests. In this meta-analysis, the species diversity of forests in which biological legacies were retained was equivalent to that of forests that were not harvested (Mori and Kitagawa, 2014). This was the case in each of the biomes they considered. In a different meta-analysis, a review of over 900 comparisons of retention cuts and either clear-cuts or unharvested boreal or temperate forests in Europe or North America found that many species of uncut forests used habitats provided by the retained structures (Fedrowitz et al., 2014).

These are important findings. But in addition to these assessments of the effects of retention forestry on one objective, biodiversity conservation, we would like to know how retaining legacy materials affects greenhouse gas emissions. This is more complicated. The retention of live and dead wood has effects on broader spatial scales and over longer time frames that raise important questions. For example, dead trees that are retained provide habitat for cavity-nesting birds, but as they decay, they release CO_2. How will this affect a forest's carbon budget? How long will the retained legacies last? Will they affect the frequency of fires? What about the effects of retention management on the availability of wood products? Will lowered supply of forest products in some regions lead to an increase in harvests elsewhere, and, if so, how will that affect greenhouse gas emissions? Clearly, answering these questions requires interdisciplinary and international collaboration. Even with that collaboration, the evaluation will depend on the objectives of those who ask the questions.

The examples in this chapter are not panaceas. I do not offer them as win-win situations. None of them would receive a positive evaluation on every variable we might evaluate. Some have definite drawbacks. Fires release carbon into the atmosphere but in some contexts, fires can promote biological diversity and reduce the probability of more severe fires. Creating a savanna-like park in the Negev desert is controversial. Beaver activity initially increases carbon storage but ultimately releases stored carbon. Perennial crops contribute to carbon storage but have low yields.

Furthermore, regardless of whether we are conserving processes or structures, disturbance regimes or interactions, ecosystems are never divorced from the activities of people. For this reason, we will now consider connections between cultural and biological diversity.

12

Strategies

Stewardship to Integrate Conservation of Biological and Cultural Diversity

12.1 Introduction

In this final chapter, we look at conservation that incorporates human histories and legacies as well as biological ones. The first parts cover conservation that sets out to address the needs of people. We begin with reserves where people live (extractive reserves, biosphere reserves) and then move to approaches that strive to create incentives for local participation in conservation (nontimber forest product harvest, ecotourism, bioprospecting). Those approaches attempt to structure programs that will allow, persuade, reward, encourage, or convince people to participate in conservation.

The last part of the chapter moves to conservation that explicitly seeks to conserve biocultural diversity. Examples of this approach involve mapping places where cultural diversity and biological diversity overlap (Section 12.4.1) and melding traditional knowledge and science to preserve local biota and traditional culture (Section 12.4.2). Finally, we consider cases where local people proactively advance their interests regarding conservation (Kuna, Inupiat, Yup'ik, small-scale fishers in Costa Rica (Sections 12.5 through 12.7)), at times working through institutional channels and at other times challenging conventional management.

By the 1990s, problems with fortress conservation (Section 9.4.2) were evident. Coercive, top-down conservation imposed on local people did not result in support or compliance. Environmental degradation and biodiversity loss persisted despite concerted actions to address them (Section 8.8).

This situation led conservationists to reassess concepts of community involvement (Western and Wright, 1994). The idea that people will overexploit resources in the absence of private ownership or state control was challenged in the light of research that demonstrated the importance of community norms for regulating common pool resources. New conservation strategies (often referred to with terms like *community-based conservation* or *integrated conservation and development*) emphasizing community participation, emerged. Many of these programs aim to benefit people who are required by conservation projects to give up some forms of resource use. One way to do this is through reserves that allow resource extraction.

12.2 Extractive Reserves

Reserves in or near places where people reside and are permitted to use resources are termed *extractive reserves*. Resource extraction within reserves is not new. The first nationwide reserve network in the USA was the system of national forests (Section 1.5.1.2), which was established to guarantee future supplies of commercial timber. The extractive reserves we are concerned with in this chapter differ from those reserves in that biodiversity protection, rather than commercial exploitation, is their principal goal. (In American forest reserves, biodiversity protection was not identified as a goal until long after the reserves were designated.)

12.2.1 Indigenous Reserves

In some parts of the world, particularly South and Central America, Indigenous peoples often live within lands that are designated as *Indigenous reserves* or territories. (These should not be confused with the Indian reservations that Europeans or their descendants established for the purpose of settling Indigenous peoples in North America, Australia, and Latin America.) Today many Indigenous reserves, especially in the Brazilian Amazon, encompass vast areas and substantial biodiversity, within which the inhabitants practice their traditional livelihoods.

Conservation biologists disagree about whether the inhabitants of Indigenous reserves are allies in biodiversity conservation. Some criticize Indigenous groups for responding to development pressures along their borders by granting concessions to logging and mining companies (Peres, 1994) (Section 9.5.2).

The legal status of the inhabitants of Indigenous reserves varies. Some own their land, but in other cases the reserves exist only on paper and have vague boundaries without clear guidelines about rights and responsibilities. Threats from commercial concessions and colonists abound.

12.2.2 Biosphere Reserves

A different strategy for integrating the protection of cultures and biota is used in the Man and the Biosphere Program that the United Nations established in 1971 to conserve representative examples of natural areas and their societies. *Biosphere reserves* contain lands and waters with long-established, presumably sustainable, uses rather than allegedly pristine ecosystems. The program uses an interdisciplinary focus that includes social and economic considerations.

Unlike conservation areas that emphasize outstanding or rare physical and biological phenomena, the international network of multipurpose biosphere reserves designates areas that are intended to showcase typical, rather than spectacular, examples of ecosystems. As of 2022, 134 nations had designated over 700 biosphere reserves.

Biosphere reserves usually involve *zoning*, the demarcation of areas for different levels of use. These hierarchies of protection are arranged so that resources are more strictly controlled in some areas, usually a central *core area*, than in others. This regulation of land uses is not an innovation; the idea of restrictions on land uses within habitats of special importance underlies the creation of any reserve. Zoning in order to group compatible activities – such as business, residential, and industrial uses – together is also the basis of land use planning in many cities and other jurisdictions. The new element in biosphere reserve zones is the degree to which ecological considerations are incorporated into the planning process.

The core area should protect ecologically significant sites within which resource extraction is prohibited or strictly regulated and few activities other than monitoring and research are allowed. Ideally, the core should be large enough to contain viable populations of wide-ranging species, but often this is not possible. However, previously designated wilderness areas, nature sanctuaries, and sacred lands are sometimes incorporated into biosphere reserve core areas. In regions that have been densely settled for centuries and where large natural areas no longer exist, biosphere reserves may consist of small nature reserves embedded in a landscape that supports traditional, local, or Indigenous land uses.

In addition to the core area, there are areas where more intensive resource use takes place. Activities that are compatible with protection of the core area – such as nature tourism, research, education, restoration, and some traditional land uses – are sometimes permitted in *buffer zones*. Adjacent to the buffer there should be a *transition zone*, a zone of cooperation that ties the reserve to the surrounding region. People live within the transition zones of many biosphere reserves and pursue economic activities such as forestry, farming, fishing, hunting, and gathering that are compatible with conservation of the core and buffer.

Biosphere reserves are often ethnically as well as biologically diverse. Some, such as the Maya Biosphere Reserve in Guatemala, include Indigenous reserves (Section 12.3.1.2).

In the USA, support for the Man and the Biosphere Program fluctuates in response to the political climate. When anti-government sentiment, opposition to participation in the United Nations, and concern about maintaining rights to private property are high, US participation in the program is minimal.

The examples in Box 12.1 of biosphere reserves in several biological and cultural contexts illustrate some of the opportunities and challenges of implementing the biosphere reserve concept.

Box 12.1
Examples of Biosphere Reserves (UNESCO, No date)

The *Dja Biosphere Reserve*, West Africa (Figure 12.1), is located in the nation of Cameroon, which is one of the most culturally and biological diverse countries in the world (Section 12.4). The reserve encompasses tropical moist forest that supports diverse plants (including birds and mammals and about 150 species of orchids). The reserve is one of the largest protected areas of this type of forest in Africa. The fauna of the reserve includes African forest elephants, chimpanzees, black-and-white colobus monkeys, leopards, and critically endangered western gorillas.

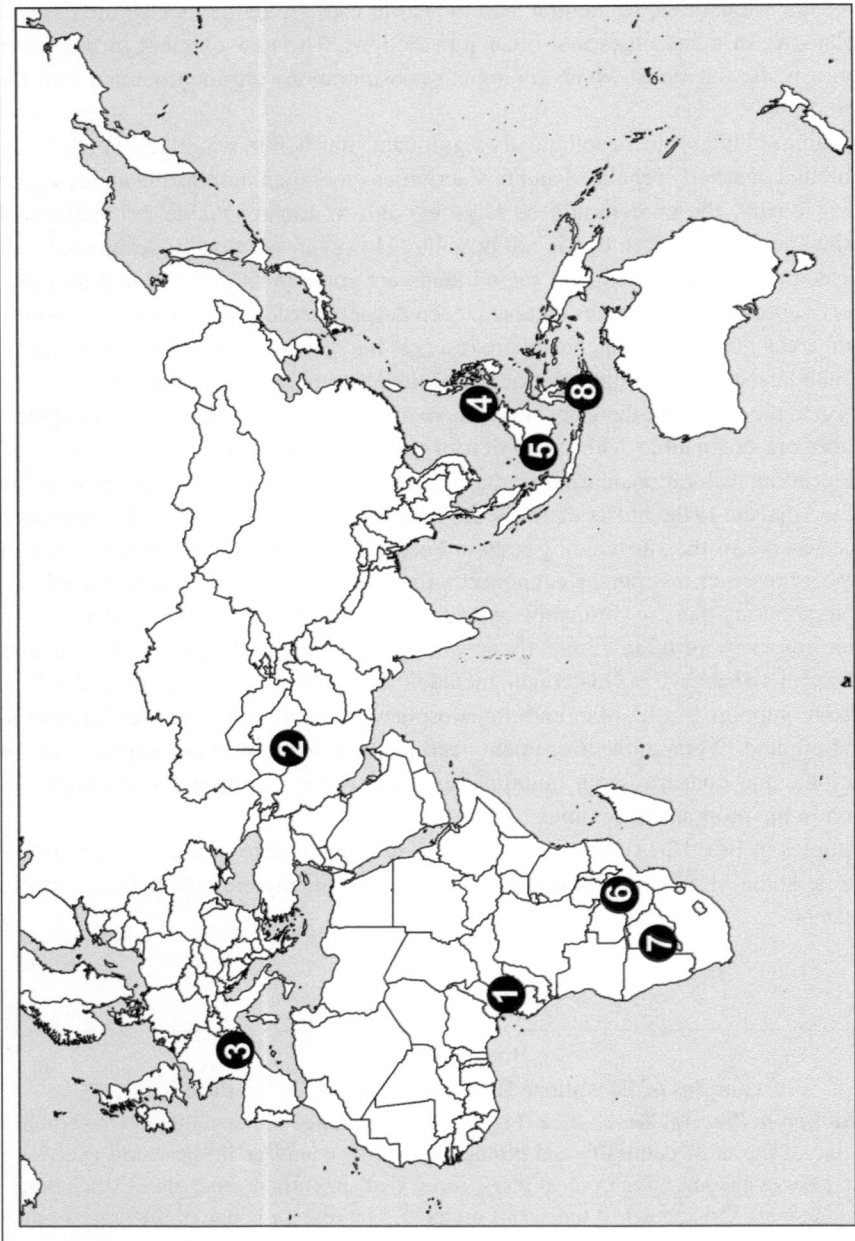

Figure 12.1   Locations of: 1, Dja Biosphere Reserve, Cameroon; 2, Golestan Biosphere Reserve, Iran; 3, Cévennes Biosphere Reserve, France; 4, Palawan Island, Philippines; 5, Gunung Palung National Park; Kalimantan, Indonesia; 6, Zimbabwe; 7, the Kalahari; 8, Flores Island, Indonesia. Map created by Eva Strand using Esri, DeLorme World Countries Generalized Data & Maps for ArcGIS 2013, with permission.

Thirty-seven villages are located within the reserve, where villagers practice shifting cultivation, fishing, mining, hunting, and gathering. Fruits, seeds, bark, and other nontimber products are collected for trade, sale, and household uses such as medicine, food, and construction. Evidence from laboratory experiments supports the medicinal effectiveness of some of the species that are used in traditional medicine.

The effects of harvesting vary by species and the methods that are used. Trees that supply valued nontimber products are often preserved when they occur in shifting cultivation plots.

About 4,000 people live within the core area of the Dja Biosphere Reserve. There are four main resident ethnic groups and two semi-nomadic groups. The latter category includes the Baka pygmies, who are allowed to hunt within the reserve. Bush meat from game obtained in the reserve is sold in markets in the capital. Logging around the reserve threatens plants and animals by fragmenting the forest habitat.

The *Golestan Biosphere Reserve*, Islamic Republic of Iran, is Iran's oldest protected area. It includes remote areas of temperate forest and steppe (Figure 12.1). The topography of this Central Asian reserve varies from level plains to foothills and mountains. Nearly 1,400 plant species, including 30 endemics, occur within the reserve. The vertebrate fauna include several species of threatened or endangered birds and mammals, such as the saker falcon, wild goat, and goitered gazelle. Several culturally significant historical sites also occur within the reserve.

The core area is uninhabited except for personnel at a monitoring station. In addition to Persians (the largest ethnic group in Iran), Turkmens and Kurds live in 43 villages within the transition zone. Agriculture, ecotourism (Section 12.3.2), and silk production are the principal economic endeavors. Villagers in and around the reserve receive revenue from tourism by renting accommodations and selling goods. However, farmers experience negative impacts from tourism because of regulations that prevent them from controlling wild animals that damage crops or threaten people and fines for illegally grazing their livestock. Poaching of wild ungulates is widespread (Ghoddousi, et al., 2018).

Silkworm culture has a long history in Iran, dating back to ancient trade along the Silk Road connecting Central Asia and China. It is well-suited to rural areas such as Golestan because it does not require a lot of space or investment, and the silk moth life cycle is only a little more than a month. The enterprise involves rearing the caterpillars on the leaves of mulberry trees and harvesting the cocoons for their tough, shimmering fibers.

The *Rio Plátano Biosphere Reserve* includes moist tropical forest, mangroves, and other coastal marine areas in the Central American nation of Honduras (Figure 12.2). The topography is mountainous, and habitat diversity, species diversity, and cultural diversity are all high. Burial sites and petroglyphs provide evidence of the region's long history of use by Indigenous peoples. Several Indigenous groups, as well as people of African descent and people of mixed European and Indigenous heritage, live within the reserve and in the surrounding area, supporting themselves with shifting agriculture and grazing supplemented by fishing, hunting, and gathering.

Rio Plátano provides habitat for nearly 400 species of birds, over 100 species of amphibians and reptiles, nearly 40 species of mammals, and over 500 species of plants, many of which are globally rare.

The *Cévennes Biosphere Reserve* in the mountains of southeastern France is a geologically and biologically diverse region supporting a mosaic of temperate deciduous forests, fields, pastures, and settlements (Figure 12.1). People have lived in the region continuously since Roman times. In the eighteenth and nineteenth centuries, overgrazing by sheep and the clearing of extensive

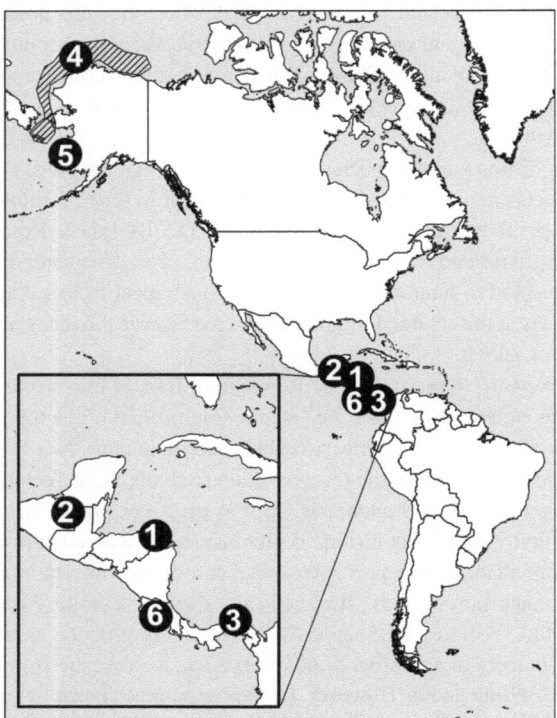

Figure 12.2   Locations of: 1, Rio Plátano Biosphere Reserve, Honduras; 2, Maya Biosphere Reserve, Guatemala; 3, Kuna Indigenous region, Panama; 4, distribution of the bowhead in the Beaufort Sea; 5, Yukon Delta National Wildlife Refuge; 6,Tárcoles, Costa Rica. Map created by Eva Strand using Esri, DeLorme World Countries Generalized Data & Maps for ArcGIS 2013, with permission.

areas of hardwood trees for agriculture and charcoal production resulted in widespread erosion. To counteract this problem, large areas of pines were later planted. When grazing declined, many previously open areas and grasslands were invaded by woody plants. These changes homogenized the landscape. Because of dwindling markets for traditional agricultural products such as chestnuts, sheepskins, and silk, Cévennes is now one of the most sparsely populated regions of France.

The Cévennes Biosphere Reserve includes the *Parc National de Cévennes* (which forms the core and buffer areas) plus an adjacent zone of cooperation. Management within the reserve supports rural activities by assisting farmers with contracts, restoring chestnut groves, and protecting traditional breeds of cattle and sheep. The quality of traditionally produced onions, some cheeses, and chestnuts from Cévennes is recognized through protected *denominations of origin*. These designations enhance a product's marketability by guaranteeing that it was produced in a specified area with traditional methods. Programs to preserve traditional music and other customs showcase the area's rural culture. Residents are offered incentives for using traditional architecture. These programs attempt to minimize migration out of the area and to attract tourists.

The Cévennes Biosphere Reserve cooperates closely with a similar biosphere reserve in the Catalonian region of northeastern Spain.

12.3 Economic Incentives for Conservation

12.3.1 Nontimber Forest Products

Collecting and marketing nontimber products such as nuts, rubber, resins, or wild game is permitted in many forests, especially in the tropics. This stems from the idea that alternatives to commercial timber may be harvested sustainably and generate profits that are at least as high as more destructive land uses (Peters et al., 1989). Many nontimber products can be obtained without cutting down trees, so they can potentially be harvested indefinitely (like maple syrup in temperate zone forests).

Historically, marketing nontimber products has not always been good for forests or for people. The rubber boom at the turn of the century generated profits for rubber companies but brought about environmental degradation and poverty in parts of the Amazon (Dove, 1995) and Southeast Asia. However, in situations where it is economically viable and ecologically sustainable, the collection and marketing of nontimber products has the potential to contribute to biodiversity conservation while providing economic gains for local users.

The economic benefits from extraction of nontimber products depend upon a host of ecological, socioeconomic, and political factors, however. The examples below illustrate some of those factors.

12.3.1.1 Forest Product Harvests by Indigenous Batak in the Philippines

Indigenous Batak inhabit coastal areas on Palawan Island in the Philippines (Figure 12.1). Prior to about 1960, they lived mainly at low elevations along the coast. Their traditional economy involves shifting cultivation with long fallow periods (7 to 18 years) as well as hunting and gathering.

The two most important commercially harvested forest products for the Batak are *rattan* and *almaciga*. Rattan refers to the tough stems of some climbing palms. It is used to make furniture, baskets, canes, and similar products, some of which – like a bed purchased by Brad Pitt – cost thousands of dollars. *Almaciga* is the resin from a large tropical coniferous tree. Sometimes known as Manila copal, *almaciga* is used in many industrial products including, lacquer, paint, and linoleum. Traditionally it is obtained from sap exuded by shallow cuts in the tree's bark. This harvest method is sustainable in contrast to deep cuts through the bark that kill the tree.

In the late twentieth century, several developments eroded Batak control over their ancestral lands. Migrants began settling along the coast, and politicians and other outsiders obtained concessions to extract *almaciga* and rattan. In addition, concessions for logging and mining in Batak lands went to outside companies. In response to these developments in the lowlands, the Batak shifted their activities to higher elevations.

Beginning in the 1980s, NGOs and government policies favored nontimber forest product harvests as part of integrated conservation and development in Palawan. However, stringent controls on extractive activities in sites that were set aside to protect rare and endemic species in high-elevation habitats limited the collection of forest products in many areas. Furthermore, where the Batak were allowed to harvest, they often ended up in debt to middlemen because they lacked skill at market-transactions. Consequently, trading forest products did not do much to improve Batak livelihoods.

The Batak faced two kinds of obstacles to successful harvesting and marketing of nontimber forest products. First, they were limited by technical problems such as lack of managerial and administrative skills, experience with bureaucracy, and handling credit. Second, strict protections of high elevation habitats exacerbated their difficulties. Problems in the first category could potentially be addressed with technical training and assistance. Problems in the second category would require a better understanding of land tenure and inter-group friction (Novellino, 2010).

12.3.1.2 Forest Product Harvests in Petén, Guatemala, and Kalimantan, Indonesia

In 1990, three researchers from the Center for Tropical Conservation at Duke University in North Carolina accompanied workers on harvest trips, conducted interviews, and spoke with researchers and people involved in the nontimber forest product supply chain in Petén, Guatemala (Figure 12.2) and West Kalimantan, Indonesia (Figure 12.1) (Salafsky et al., 1993).

In both those settings, Indigenous people have a long history of harvesting nontimber forest products for local use, and they probably also traded some surpluses at local markets. More recently, they have been harvesting such products for sale to international markets as well. In the Maya Biosphere Reserve in Petén, Guatemala, the principal harvested products were: (1) *chicle*, the sap of a long-lived, tropical evergreen tree that is used in making chewing gum and glue, (2) *xate*, leaves from palms that are used in floral arrangements, and (3) the dried fruit of the allspice plant, which is the source of the spices nutmeg and mace. These products formed the basis of an export-oriented industry that employed thousands of people as harvesters, contractors, or processors. In 1989, export income from these enterprises was estimated at US $4 million to $7 million (Reining and Heinzman, 1992; Salafsky et al., 1993).

The tropical forests of Petén have some characteristics that tend to favor sustainable harvest of these nontimber forest products. There are about 50 to 100 species of trees per hectare, a relatively low species richness for tropical forest (although still much higher than temperate zone forests!). Because there are relatively few tree species, many individuals of each species are present. This makes it easy for harvesters to get to trees of the desired species with relatively little travel through the forest. Furthermore, harvests of these species are spread out in space and time, so that overexploitation of any one species is avoided. *Chicle* must be harvested during the rainy season, from August to January; xate palm fronds are available throughout the year, but harvest peaks between March and June (partly because of the spring demand for wedding floral arrangements in Europe and North America); and allspice is available only in July and August. This staggered availability provides harvesters with year-round income. Because of the predictable supply of nontimber forest products, stable markets developed.

Chicle, *xate*, and allspice are usually harvested without killing reproductively mature individuals. *Chicle* is harvested by tapping trees to obtain sap. If properly tapped, individual trees can continue producing sap for decades. *Xate* harvest does not involve the removal of reproductive structures. Allspice fruits can be collected from the ground below the parent plant, where survival is low. In this density-dependent situation, harvesters can remove considerable numbers of fruits from beneath the parent allspice tree without appreciably inhibiting reproduction.

Chicle, xate palm fronds, and allspice are all easy to store and transport. Roads, airstrips, warehouses, and other components of the infrastructure necessary for processing and delivering these products have been available for some time because the export-based chicle industry dates back to the nineteenth century. Thus, a variety of biological and economic factors contribute to the relative sustainability and viability of nontimber forest product use in Petén.

In West Kalimantan, the situation is different. Species diversity is high (150–225 species per hectare) in Indonesian tropical forests. Because many species are packed into a given area, the number of individuals of any one species is low. Therefore, it takes a long time to get to each plant and a long time to transport the harvested products. Under these conditions, harvesting nontimber forest products is inefficient. In addition, obtaining some products kills the source tree.

Harvesters tend to concentrate on species that command a high price. *Gaharu*, the resin from diseased heartwood of agarwood trees, is one such product. Valued for its use in incense and perfumes, the resin from a single *gaharu* tree can be sold for thousands of dollars. With such high profits, there was an incentive to maximize short-term gains by killing a tree to obtain all the *gaharu* at once. When a market for this product developed, it was quickly exploited, often by professional harvesters, and the population declined rapidly. Today the main species of *gaharu* trees is considered virtually extinct in West Kalimantan.

To make matters worse, many of the species in Indonesian forests are mast trees. They produce fruit only once every three to five years, and many species produce fruit simultaneously. (Recall from Section 1.3.3.1 that the nut trees passenger pigeons fed on were mast species.) This unpredictable timing – an adaptation that prevents populations of seed-eating species from building up – also prevents stable markets from developing. When fruits do become available there is a glut on the market, and prices drop.

In both Kalimantan and Petén, forest products are common-pool resources (Section 10.1.3). But in Petén there are informal rules governing their exploitation. For example, it was reported that harvesters generally do not harvest *xate* leaves from trees that are within a 15-minute walk from camp because it is understood that those trees are reserved for harvesters who need some extra leaves to complete a bundle at the end of a day. In West Kalimantan, there do not seem to be such understandings. Furthermore, in Petén most of the harvesters work with each other and with contractors on a long-term basis. This seems to foster cooperation and some incentive for conservation. (However, this system is not permanent; a change in these relationships could result in different outcomes.)

The results of this comparative study suggest that the exploitation of nontimber forest products is more likely to be economically viable and ecologically sustainable in Petén than in West Kalimantan. Factors that bode well for the sustainability of the Petén enterprises include ecological characteristics such as the predictable availability of forest products and the relatively high density of harvested species, as well as social and economic factors that may put some constraints on how much is harvested but at the same time contribute to the economic viability of the harvests. However, in Kalimantan, the ecology of the harvested species as well as economic and social factors seem to work against the sustainable harvest of nontimber forest products.

12.3.2 Ecotourism

People living next to protected areas are not necessarily supportive of those areas. They may have been displaced when a preserve was designated, or they may have lost access to resources that they used to depend on. They may experience conflicts with wild animals and not be allowed to control them. In these cases, the externalities of conservation (Section 5.3.4) are borne by local people, whereas the benefits are enjoyed by outsiders.

Ecotourism, a type of tourism that attracts visitors who want to observe wildlife or experience special natural settings, can potentially address these problems by generating local benefits from preserves or private wildlife-related businesses. In addition to income from fees, ecotourism may provide jobs for local people as scouts, wardens, guides, maintenance workers, and service providers. If the money that tourists spend stays local, it may go to individual households or to community projects such as schools, clinics, or wells. In theory, benefits to local communities from conservation should compensate for any losses. But, as Iran's Golestan Biosphere Reserve (Box 12.1) illustrates, creating situations where that happens is challenging.

Cooperation with ecotourism projects is likely to be greatest where there is local participation in planning and implementation. In addition, ecosystems that are easily eroded, have slow-growing vegetation, are vulnerable to invasive species, or contain wildlife that is intolerant of humans may not be suitable for ecotourism. Wildlife viewing can interfere with normal behavior and reproduction and attract predators. The energy involved in fleeing from observers can disrupt wildlife energy budgets, resulting in diminished reproduction or survival. When wild animals are exposed to people, they may become less wary and be exposed to danger. In addition, care must be taken to avoid introducing weeds or diseases.

Inputs of foreign income from ecotourism are sensitive to weather and to political and military circumstances. When conditions are unfavorable, tourist revenue declines and those who depend on it suffer. Trophy hunting expeditions to China declined after the violent suppression of the Tiananmin Square uprising in 1989, and Rwanda's civil war interrupted tours to view gorillas in their natural habitat. During economic downturns, tourism declines, especially if gas prices are high. This problem may increase as concern about the high carbon footprint of international travel increases.

Because it occurs in places that are far from development, and because cultural and biological diversity often occur in the same areas (Section 12.4.1), ecotourism often highlights cultural practices and products such as locally made crafts in addition to natural features. This *cultural tourism* can lead to heightened awareness of Indigenous and local cultures and histories. However, it may perpetuate views of Indigenous and rural cultures as backward or romantic (or both) (Pi-Sunyer, 1982). Although this is not inevitable, it is a risk inherent in cultural tourism, which by its very nature involves wealthy people observing others in non-wealthy societies. Interactions between visitors and local people that occur in such settings can foster mutually beneficial rapport and respect, but they may instead reinforce stereotypes.

Boxes 12.2 and 12.3 illustrate some outcomes of ecotourism in Africa.

Box 12.2
Evaluating Evidence: How Did Ecotourism Affect Livelihoods in the Serengeti-Mara Ecosystem?

Precise information about the effects of ecotourism on local people isn't always available. Does ecotourism improve the material well-being of local households? How are revenues from ecotourism distributed?

Detailed data on some of the relevant economic issues are available for the Serengeti-Mara Ecosystem, the site of some of the world's most famous opportunities for ecotourism (Box 9.2). Tourists come to the region to see spectacular herds of wildebeest and other migratory ungulates and their predators. For decades, international donors have funded programs designed to link conservation to economic development in this region. This approach is based on the assumption that ecotourism replaces pastoralist livestock production with income from wildlife-based tourism.

To find out how revenues from tourism affected household incomes, Kathleen Homewood and her colleagues conducted a five-site study in Kenya and Tanzania involving over 1,000 household surveys of income and related variables. They found that the distribution of money from wildlife-based tourism varied greatly among households. Some received substantial revenue from those enterprises, but many households received no money from anything related to wildlife. For those that did, money from wildlife averaged less than 5% of their annual income. The researchers concluded that "few wildlife-derived [economic] benefits flow to pastoralists, while conservation restrictions constrain production and coping strategies, undermining potential for coexistence.... Revenues from wildlife rarely begin to compensate for loss of mobility, access to and control over important natural resources" due to conservation restrictions (Homewood et al., 2012:18). Most of the earnings related to wildlife went to tour operators, service industry workers, and the government. This was partly because of corruption; in some cases, revenue that was supposed to reach communities went to local elites or government officials.

Box 12.3
Zimbabwe's CAMPFIRE Program

The Communal Areas Management Programme for Indigenous Resources, or CAMPFIRE, in the African nation of Zimbabwe (Figure 12.1) is a well-known example of ecotourism. Most of Zimbabwe is savanna, parts of which support populations of large, charismatic, and sometimes dangerous wildlife. Elephants, leopards, lions, and buffaloes attract tourists but also damage crops and injure villagers, sometimes fatally.

Since the early 1980s, CAMPFIRE has sought to increase wildlife populations while raising the incomes of poor people living in the midst of wild animals. When the project began, most native people of Zimbabwe lived on "communal" lands owned by the state. Black people in Zimbabwe had virtually no rights to use wildlife, a holdover from colonial regimes that permitted hunting by white residents and international sport hunters but not native people (Section 1.2.2). In this context, people in local communities considered poaching a justifiable form of resistance to state regulations.

After independence from British colonial rule, changing government policies regarding wildlife management and land tenure set the stage for CAMPFIRE. The program introduced an innovative approach: wildlife management on lands that were common-pool resources rather than either state owned or individually owned land (Section 10.1.3). In 1982, it became legal for rural district councils in Zimbabwe to conduct safari hunts on communal land and return the resulting revenues to their communities.

The idea behind this policy change was that in Zimbabwe's drought-prone climate, wildlife utilization could be more profitable than farming. This was a shift away from the previous emphasis on preservationist wildlife management within parks. In less than a decade, dozens of rural district councils adopted CAMPFIRE (Murphree, 2005).

The CAMPFIRE approach assumed that creating a situation where wildlife had positive values for local communities would lead to wildlife conservation. That is how things worked in the successful districts. Wildlife management became a form of land use that made economic sense. The control of wildlife resources was transferred to local authorities, and villagers managed local wildlife as an asset, analogous to an agricultural crop. "We see now," said the chief of one village, "that these buffalo are our cattle. We are going to farm them" (Murphree, 2005:124).

Damage and injuries from wildlife decreased. Poaching came to be seen as theft from the community, and as a result it declined. Households benefited from wildlife-related revenue, and additional revenue was used for improvements to schools or other community infrastructure. By profiting from the international ecotourism market, community members reduced their vulnerability to the droughts that made farming uncertain.

Not all examples of the CAMPFIRE approach were success stories, however. CAMPFIRE worked best where wildlife densities were high and where the rural development councils fostered local empowerment. Those conditions were not always met. Wildlife densities varied according to habitat and land use, and in some communities the rural development councils appropriated revenues to themselves, and villagers received few or no benefits. In addition, support for wildlife management varied within communities. Some individuals, especially those who were relatively wealthy, preferred to manage for cattle.

If the success of CAMPFIRE varied among communities in Zimbabwe, efforts to import the CAMPFIRE model to other nations were even more varied, despite the fact that the program was hailed as a model for linking rural development and conservation. Many problems stemmed from the fact that CAMPFIRE was developed for specific political and economic conditions of Zimbabwe, such as the common-pool land tenure system, which were not duplicated elsewhere.

Although the experience of CAMPFIRE was disappointing in some respects, it was instructive. The varied experiences in the different villages demonstrated the pivotal importance of local control as well as the obstacles to its achievement. The villages where CAMPFIRE was most successful were the ones where communities achieved effective management at the local level.

12.3.3 Bioprospecting

Biodiversity tends to be greatest in the tropics, where most developing nations are located and where many people are poor. But technical expertise in research and product development is greatest in developed countries in the northern hemisphere.

Because species richness and therefore genetic diversity are high in tropical ecosystems, there have been many interspecific interactions and therefore many opportunities for coevolution. This led to the evolution of an extraordinary array of molecules that function in defense and counter-defense in tropical plants, animals, and microorganisms (Section 6.1.2). Because of their unique biological activity, those chemicals have the potential for applications in medicine, agriculture, and industry. Consequently, companies in developed countries are interested in controlling the commercial development of those species by patenting their genetic material.

In the 1980s, scientists began endorsing drug development from medicinal plants as an incentive for preserving tropical biodiversity. The rationale for this approach is similar to the argument for ecotourism or for harvesting nontimber forest products: the assumption that if people reap economic benefits from biodiversity where they live, they will have an incentive to save the habitats that provide that biodiversity. Might such arrangements lead to win-win-win situations in which northern companies, southern communities, and biodiversity all benefit?

The early record of the pharmaceutical industry's development of products from tropical species is troubling, however. In the 1960s, chemists at Lilly Research Laboratories developed two chemotherapy drugs from the leaves of the Madagascar rosy periwinkle. Between 1963 and 1985, the sale of those drugs saved many lives and grossed approximately $100 million, but the people whose "folkloric information" provided clues that led to the "discovery" and marketing of those multi-million-dollar products were never compensated (Ehrlich and Ehrlich, 1985; Farnsworth, 1988:95).

In 1991, a private organization in Costa Rica, the *Instituto Nacional de Biodiversidad (INBio)*, or National Institute of Biodiversity, and the American pharmaceutical giant Merck signed an agreement under which searching for molecules with useful biological activity in Costa Rica would be treated like prospecting for timber or minerals. Merck agreed to pay INBio $1 million to conduct *bioprospecting*, searching for substances with commercial potential. In addition, if the search yielded commercially viable products, the company would pay royalties on any products it marketed as a result of its investigations. Under this arrangement, Costa Rica not only would share in profits from the pharmaceutical company's discoveries, it also would receive payment up front for the right to search for sources of new products (Gershon, 1992).

The agreement between INBio and Merck was praised as a way to enhance a source country's control over uses of its biological diversity (Reid et al., 1993). However, in 2013 the arrangement was dissolved because of financial difficulties. Vivienne Solís Rivera, a biologist, and Patricia Madrigal Cordero, a lawyer, who had both participated in drafting Costa Rica's biodiversity law (which implemented the Convention on Biological Diversity, see below) pointed out that INBio lacked accountability. They suggested that international agreements are more likely than agreements among private companies to ensure equitable distribution of benefits from biodiversity utilization (Solís Rivera and Madrigal Cordero, 2013).

The Convention on Biological Diversity that was drafted in 1992 (Section 7.3.1) requires fair and equitable sharing of profits derived from utilizing biodiversity. Ideally, putting fair and equitable benefit sharing into practice "can help to build scientific and technological capacity within high-biodiversity countries, can promote legal and policy regimes that protect the rights of countries, individuals, communities and corporations, and can help promote sustainable development and the conservation of biological diversity" (ten Kate and Laird, 2000:264). In practice, doing so is complex for ethical and practical reasons (Box 12.4). Due to a basic cultural difference, some Indigenous advocates such as the Indian physicist and ecologist Vandana Shiva (2016) object to bioprospecting (also referred to as *biopiracy* and *scientific colonialism*) regardless of its terms. In developed countries, ideas and other intellectual products are treated as individual or corporate property that can be patented. In Indigenous societies, knowledge often belongs to a community, not individuals. Some are offended by the concept of patenting life and the suggestion that commercial value can be placed on a community's knowledge.

Box 12.4

Hoodia: A Cure for Obesity in the North and Poverty in the South?

For centuries the San – Indigenous peoples from the Kalahari of southern Africa (Figure 12.1) – reduced their need for food and water on long hunting trips by consuming the fleshy stems of plants they called *ghaap*. Later this succulent, spiny plant with large flowers that smell like rotting meat (an adaptation that attracts pollinating flies) was classified in the genus *Hoodia*, a member of the milkweed family.

The San are the oldest human inhabitants of the region. Traditionally they are hunters and gatherers, but colonial governments considered their way of life incompatible with civilization and sought to exterminate the San. Today about 100,000 San live in poverty in South Africa, Namibia, Botswana, and Angola on land to which they have no legal rights (Wynberg and Chennells, 2009). Books and movies (such as *The Gods Must Be Crazy I* and *II*) portray them as childlike people living a simple life in harmony with their environment.

In the 1960s, CSIR, a South African research institution, began confidential investigations into the potential of *Hoodia* as an appetite suppressant that might be suitable for military use. Eventually, they filed a patent application for the relevant active chemicals. Subsequently, CSIR sold the development rights for *Hoodia* to Phytopharm, a small British developer of plant-based medicines and supplements. A partnership with the US-based Pfizer (a company that later developed one of the first vaccines for Covid-19) followed.

In 2001, the results of a clinical trial on the effects of the resulting drug on 18 overweight men were reported. Although nearly four decades had passed since CSIR began investigating *Hoodia*, the San still had not been informed about its potential commercialization, nor had they been acknowledged as the source of the knowledge that made drug development possible. That changed only when two NGOs drew attention to this violation of the Convention on Biological Diversity.

After an article in a British newspaper highlighted the issue (Barnett, 2001), Phytopharm's executive director responded by saying that the San had not been notified because they were thought to be extinct.

Initially, the San were reluctant to trust CSIR because of its past association with the South African government's apartheid policy of forced racial segregation and its prolonged failure to involve the San in *Hoodia* development. Eventually, however, they opted to negotiate for a share of royalties from *Hoodia* products rather than to pursue a legal challenge. It was a slow process, but the parties eventually reached an agreement. Representatives of the San signed two benefit-sharing agreements, one with CSIR and one with the South Africa *Hoodia* Growers Association.

Nevertheless, several problems remained. Competition from legal and illegal quarters was rife. Commercial growers flooded the market with cultivated *Hoodia*. Unregulated and often illegal collection threatened it in the wild. Herbal supplements were marketed with false or unsubstantiated claims that they contained *Hoodia* and that they were developed from San knowledge. There were also disagreements about how to distribute benefits among San communities.

Reactions to the *Hoodia* agreements were mixed. They were praised as major steps toward fair and equitable sharing of benefits from commercialized traditional knowledge. They were also criticized as fundamentally illegitimate or as too little too late. In any case, the commercial exploitation of *Hoodia* highlighted the pitfalls and the potential of bioprospecting (Burrows, 2005).

The combination of past injustices and huge potential profits from bioprospecting is a volatile mix. Efforts to develop fair ways to share benefits from bioprospecting continue, as do non-commercial avenues for integrating the conservation of nature and cultures.

12.4 Conserving Biocultural Diversity

12.4.1 Overlap of Cultural and Biological Diversity

The Canadian anthropologist Wade Davis uses the term *ethnosphere* for the cultural web of life that is analogous to the biosphere – the "sum total of all thoughts and intuitions, myths and beliefs, ideas and inspirations brought into being by the human imagination" (Davis et al., 2007:8). As social scientists came to embrace the idea that people are part of nature, they turned their attention to links between cultural diversity and biological diversity, a concept expressed by the term *biocultural diversity*.

Places with high biological diversity also tend to have high cultural diversity (Figure 12.3). The information mapped in Figure 12.3 was developed from: (1) an index of cultural diversity (the number of languages, religions, and ethnic groups in a country) and (2) an index of some aspects of biological diversity (total plant, mammal, and bird species in a country). Using this method, Brazil, India, Indonesia, and Papua New Guinea ranked highest in biocultural diversity. However, this method of ranking biocultural diversity is biased in favor of countries that are large and populous because they are likely to have

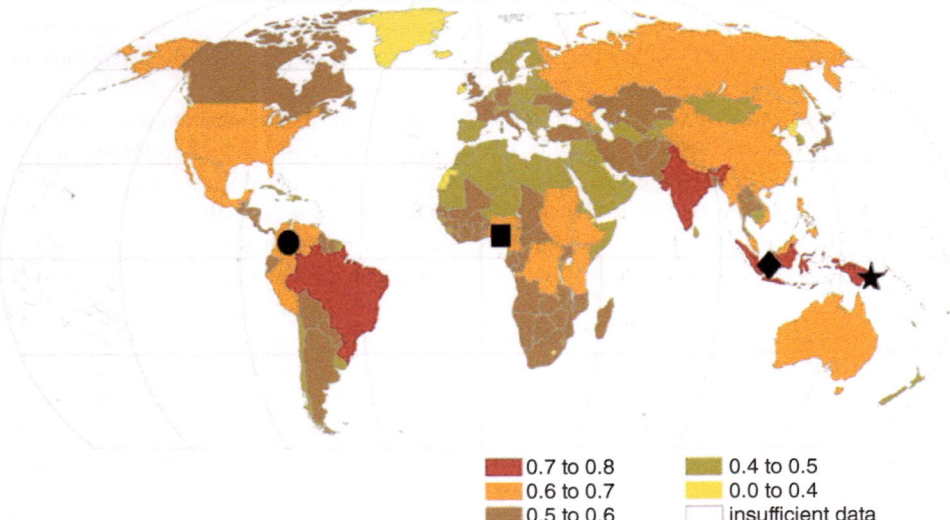

Figure 12.3   Worldwide distribution of estimated biocultural diversity. Numbers in legend are ranges of values for biocultural diversity index before being adjusted for area and population. Columbia (circle), Cameroon (square), Indonesia (diamond), and Papua New Guinea (star) are ranked the most bioculturally diverse nations when index is adjusted for country area and population. Note that regions where high linguistic and biological diversity coincide are mainly in the tropics. Adapted from Loh and Harmon, (2005), with permission from Elsevier.

more species and cultures than small, sparsely populated countries. When the data are adjusted to account for a country's area and population, the small nations of Colombia and Cameroon emerge as biocultural diversity hotspots.

Islands often have high concentrations of both languages and species, as do mountainous regions. On islands and mountains, cultures develop and species evolve in isolation, which favors both speciation and cultural adaptation (Section 8.4.1). The high habitat diversity in mountains may foster species diversity and diversity in the ways that people interact with their environment. Rugged high-elevation habitats also tend to isolate human communities, leading to the development of different languages and other aspects of culture over short distances. Perhaps the most striking example of this is the mountainous environment of Papua New Guinea, the eastern portion of the island of New Guinea, where over 800 languages are spoken, and scientists have described over at least 11,000 plant species.

Cultures and species face similar threats from development, habitat loss, and climate change. Both types of diversity are threatened by forces that homogenize ecosystems. Many languages have already disappeared, and more are endangered, often because they are not being passed on to future generations and their last remaining speakers are elderly. Each language expresses the knowledge and understandings of its culture. When a language disappears, unique vocabulary and concepts, including traditional knowledge about ways of interacting with the environment, disappear.

12.4.2 Applying Traditional Ecological Knowledge to Preserve Local Culture and Ecology

The classification systems of Indigenous cultures often recognize species, subspecies, and community types that are not known to Western scientists. Individuals in those cultures can frequently identify hundreds, and even thousands, of different kinds of plants and animals. Indigenous knowledge also includes information about ecology, including the life cycles of animals and plants, animal behavior, species' habitat requirements, interspecific interactions, and the uses and spiritual significance of species. (See Introduction: Oral traditions.)

For centuries, Indigenous informants have provided Western naturalists and scientists with information about local flora and fauna. In most cases, the studies that resulted came to an end when the visitors departed, and the sources of the information were never credited. Today those exploitative relationships are less common. Some projects use cross-cultural approaches that combine Western science and traditional ecological knowledge in cataloging biological diversity.

Parataxonomists are local individuals with skill in taxonomy, the science of classifying life forms. Their expertise is based on experience and observations rather than formal training. When trained in how to collect, preserve, and label specimens, parataxonomists can engage in mutually beneficial collaborations with Western scientists. They bring their knowledge of the diversity and uses of local species and ecosystems, while scientists provide financing as well as expertise in data processing, statistical analysis, mapping, and publication.

Working with parataxonomists improves not only the quantity but also the quality of data that is collected. In contrast to professional researchers, who often reside far from their study areas, parataxonomists are long-term residents. Because they are local, they can visit sites throughout the seasons and collect plants of interest during each phase of their life cycle. The resulting specimens, with flowers, fruits, and seeds as well as vegetative parts, are important for scientists who later examine them. Parataxonomists also have expertise in local environmental history. In some cases, ecologists have learned that community types they thought were natural are in fact products of human intervention.

Box 12.5 presents a case study of a collaborative investigation of the diversity of wild plants, animals, crops, rituals, and farming methods in an Indigenous tropical agricultural system.

In the next three examples, Indigenous groups have been proactive in maintaining their traditional modes of resource management in different settings.

Box 12.5
The Tado Cultural Ecology Conservation Program

Early in the nineteenth century, the English botanist Joseph Arnold in Indonesia learned of a plant with foul-smelling, reddish flowers up to a meter in diameter. Because it resembles rotting meat in color and smell and attracts flies that it digests, this plant has become known in the English-speaking world as the corpse plant. (Note that the corpse plant has adaptations that are very similar to those of *Hoodia*, although the two species are not closely related. This is an example of convergent evolution (Section 1.3.3.4)). Arnold was credited with discovering the corpse plant, although he learned of it from an Indigenous guide.

Today Indigenous farmers and Western scientists in Indonesia are working together to document local flora. The Indigenous Tado of Flores Island in eastern Indonesia (Figure 12.1) inhabit a mosaic of tropical forest and savanna, where they have historically cultivated upland (non-flooded) varieties of rice in a system of shifting cultivation. Traditional Tado agriculture uses many *landraces* (locally adapted, traditional varieties) of rice, which they plant along with other vegetables or cash crops. Landraces are chosen for their compatibility with the other crops and for their contribution to traditional pest control.

Beginning in the 1960s, the Indonesian government promoted the adoption of hybrid varieties of paddy rice (which must be irrigated) as part of the Green Revolution. These require expensive external inputs such as water, fertilizer, and pesticides. Hybrid rice varieties are uniform in color, yield, seed size, and other characteristics. This homogeneity increases the risk of crop failure in years of drought or pest attacks, which compromises food security. As a result of this policy, the number of landraces that Tado households maintain is declining, and knowledge of farming customs and local ecology is being lost. Few "residents under the age of 40 can describe ancestral cultivation rituals or sing the songs associated with traditional upland rice landraces" (Pfeiffer et al., 2006:615).

Tado rice cultivation is associated with stories, recipes, ceremonies, prayers, tools, baskets, and planting techniques, as well as detailed knowledge about the ecology of the landraces. Elders "repeatedly note the connection between the cultural survival of the Tado and the conservation of the ethnobotanical traditions and associated flora" (Pfeiffer and Uril, 2003:68).

Scientists associated with the University of California at Davis and the Tado Cultural Ecology Conservation Program along with the Tado parataxonomists jointly carried out a collaborative, interdisciplinary project aimed at protecting and maintaining Tado knowledge and culture. Tado leaders participated in all aspects of the program from design and implementation to analysis and review. Community members participated in sessions where traditional plant uses were described, and Tado elders also recounted their memories of the ceremonial and spiritual aspects of upland rice cultivation. Researchers and Tado farmers documented traits of landraces in traditional fields. Parataxonomists collected and processed plant specimens; recorded data on ecology, associated vegetation, stage of development, and uses; and collaborated with researchers on the preparation of publications and presentations for academic conferences.

In 2008, Tado residents began an ecotourism project in which visitors learning about Tado knowledge and resource management make traditional mats and oil lamps, prepare traditional foods and medicines, learn about using plants in jungle survival, and participate in traditional songs and games (Maffi and Woodley, 2012). Information about selected rituals (those that are not too sensitive to be shared with outsiders) is shared.

12.5 The Kuna-Yala's PEMASKY: An Uneasy Alliance

In 1925, the Indigenous Kuna-Yala of Panama's northern coast began a revolutionary uprising that led to their being granted territory for a *Comarca* (Indigenous region) within Panama. The Kuna homeland extends from the continental divide down to the Caribbean Sea (Figure 12.2) and encompasses wetlands, moist tropical forests, coral reefs, and mangroves. Most Kuna live on small coral islands a short distance from the coast. There are also a few settlements onshore and inland.

Because of their isolation and control over their territory, the Kuna retained a high degree of political and cultural autonomy. However, by the 1960s, threats from mining, tourism, and a planned military base became apparent. Cattle pastures appeared near the border of their territory, and squatters settled in and near Kuna lands.

By the 1980s, the threats had increased. The US Agency for International Development funded construction of a road through Kuna lands. Colonists who cleared and farmed a plot of land were given "ownership" by the Panamanian government because this was considered productive use, a policy rooted in European ideas about land ownership (Section 1.2.1.2). On paper, the Kuna had legal title to their lands, but they could not effectively defend those lands because they had never been surveyed. Wealthy land speculators also posed a threat. The mountains that had provided protection along the border no longer served that function.

In 1983, some Kuna leaders and technicians set up a protected area on the southern border of their territory. This Research Project for the Management of Wilderness Areas in Kuna (*Proyecto de Estudio para el Manejo des las Areas Silvestres de Kuna Yala*) quickly became known by its Spanish acronym PEMASKY. This was the first time Indigenous people in Latin America had established an internationally recognized nature reserve. (Subsequently, some of the funders suggested getting the project designated as an Indigenous-run Biosphere Reserve, but the Kuna feared that might mean they would have to deal with a corrupt government agency in Panama, so they decided against a biosphere designation.)

The Kuna started PEMASKY as a means of defending their territory from encroachment by colonists. At the same time, international conservationists saw the protected area as an important step toward biodiversity conservation in the region. The project expanded rapidly. Funds from international donors poured in. PEMASKY representatives signed collaborative agreements with several research and conservation organizations, and work on developing a management plan began. Environmentalists and Indigenous rights activists praised the project. PEMASKY seemed to be an ideal alliance between Indigenous people and conservationists (Chapin, 1998).

Mismatches between the priorities of outsiders and the Kuna soon became apparent, however. Conservationists were interested in studying the area's flora and fauna and in setting up ecotourism and environmental education, but those things were not priorities for the Kuna. Teacher modules developed by a consultant foundered when he realized that "the chiefs [did not] need lectures about the forest" because they knew more about the forest than he did (Chapin, 1998:257). Nor were the Kuna interested in scientific research that did not address their practical concerns: marking the boundary of their lands and developing sustainable agricultural production.

PEMASKY's funding was never tied to clear objectives. Because PEMASKY had been held up as an inspiration, there was pressure to live up to expectations and to impress funders. That meant that reports tended to put a positive spin on things, rather than to acknowledge problems. When the optimistic scenarios did not pan out, disappointment followed. Under those circumstances, PEMASKY's funding was not sustainable. By 1992, it had dried up.

Yet although PEMASKY did not live up to outsiders' (rather unrealistic) expectations, it accomplished much of the Kuna agenda. From the outset, the Kuna intended to protect their land and culture. Through the project they mapped and patrolled their border, repossessed some of their lands, and dealt with other external threats. They also formed environmental NGOs, networked internationally, produced scientific and popular publications, carried out fieldwork, facilitated transmission of elders' traditional ecological knowledge to the next generation, and created publications on their culture and environment.

Because of their long history of cultural cohesion combined with title to their lands, the Kuna have been able to exercise an unusual degree of control over their resources. But they face serious challenges. Besides dealing with interrelated external threats from consumer goods, expanding market incentives for overexploitation, and new roads, they face internal challenges from population growth, outmigration, Western education, and erosion of their culture. Signs of environmental degradation are increasing. Turtles, fish, lobsters, timber, firewood, and game are declining, and trash is increasing.

Although outside forces contribute to these problems, Kuna also acknowledge their own responsibility. A book by Kuna authors on *Plants and Animals in the Life of the Kuna*, addresses this point:

Overexploitation of marine fauna in Panama has reached alarming levels.... [t]he overfishing of marine resources occurs because of ignorance, complacency, and shared culpability of players both inside and outside of the Comarca.
It is ... up to the Kuna to protect the base of their existence.

(Ventocilla with Olaidi, 1995:54)

Unlike the Kuna, the Indigenous people in the next two cases achieved some influence over the management of resources in their communities by contesting the decisions of regulatory bodies. In both examples, hunters living in communities remote from regulators used their knowledge of wildlife to challenge conclusions by scientists regarding the management of animals central to their culture. In both cases, co-management of the resources resulted.

12.6 Co-management following Indigenous Challenges to Regulatory Agencies

12.6.1 Inupiat Bowhead Harvest

Meat, blubber, bones, skin, and baleen of the bowhead, a large baleen whale, are central to the cultures of people in the American, Canadian, and Russian Arctic. Even if modern substitutes are available, the traditions associated with obtaining, sharing, and preparing whale products are culturally significant. A report to the International Whaling Commission (IWC) on subsistence whaling found that nutritionally adequate substitute foods are not regarded as culturally satisfactory. The report concluded that

whaling is a focal point of Eskimo culture in which values are expressed and actualized, individual achievement is fulfilled, and social integration is manifested to its highest degree....

The north Alaskan Eskimos place an extremely high cultural value on their customary diet. Most people express a conviction that meals are incomplete without native food, and emphasize that they cannot remain strong, healthy, and satisfied when they rely on imported foods.

(International Whaling Commission, 1982:41)

Harvests of bowheads and other whales are regulated by the IWC (Section 7.3.1). Prior to 1977, Indigenous harvests of bowheads for subsistence were exempt from IWC restrictions. However, in the 1970s, the animal rights movement gained momentum. Activists objected especially strongly to killing whales because of their history of intense exploitation and because they are highly intelligent and form strong social bonds.

In 1977, a scientific committee estimated that the population of bowheads in waters near Alaska was down to only 600 to 2,000 individuals. On the basis of the committee's recommendations, and perhaps influenced by anti-harvest sentiment, the IWC canceled, without consulting with native whalers, the subsistence exemption for that bowhead population.

Reliable data on the sizes of whale populations are difficult to obtain. The *Inupiat* people of Alaska took issue with the claim that bowhead populations were too low to sustain a harvest. They argued that the scientists' estimate was biased because it omitted whales migrating under ice, where they were not visible from the census station. From their observations and their knowledge of whale behavior, the Inupiat estimated that there were 4,000 bowheads and that calving rates were high (Figure 12.2).

In response to these developments, Inupiat whalers formed the Alaskan Eskimo Whaling Commission (AEWC), to represent their whaling communities at the IWC. To back up their position on bowhead abundance, they verified the presence of whales under the ice using underwater earphones. When they presented the IWC with these data, it lifted its ban on native hunting of bowheads and instead allowed a limited take.

However, according to the AEWC, this quota was too low and jeopardized villagers' food security. Eventually better methods for estimating the population helped secure a higher quota that was closer to historic use by the Alaska whaling communities (Alaska Eskimo Whaling Commission, No date).

The Alaska Eskimo Whaling Commission currently manages its bowhead hunt through a Cooperative Agreement with the US federal government in a program that includes ice-based censuses, aerial surveys, photo identification of individual whales, genetic studies, bowhead health assessments, and movement tracking via satellite. These co-management efforts "blend western science techniques with local observation-based ecosystem understanding of the whale and its habitat" (IWC, No date). The result is an apparently sustainable harvest that supports Inupiat culture and food security.

12.6.2 Yup'ik Harvests of Caribou and Brown Bear in Western Alaska

Another example of power-sharing in resource management grew out of a crisis that occurred in western Alaska when the *Yup'ik* (Indigenous people of southwestern Alaska and Siberia) and state wildlife management agents reached a stalemate regarding caribou hunts. In this example, a tradition of passive resistance to conservation regulations ultimately developed into proactive engagement.

For most of the time since the USA acquired Alaska from Russia in 1867, Indigenous people in Alaska had no legal rights to land or resources. Until 1980, when a law granted them subsistence rights, they harvested game in violation of state laws, using secretive methods to avoid penalties or confrontations with authorities.

For Yup'ik in southwestern Alaska (Figure 12.2), the effects of the change to legal rights were mixed. It meant that they could legally harvest fish, game, and wild plants for the first time in over a century. But it also meant that Alaska Fish and Game, the state game agency, started to pay more attention to managing hunts within Yup'ik territory. This led to conflict.

Game agents searched the homes of suspected poachers, levied fines, and confiscated the equipment of hunters convicted of taking game out of season. Confrontations between hunters and game agents became contentious, verging on violence at times. In effect, access to game decreased, and Yup'ik meat supplies dwindled.

Alaska Fish and Game also began using aerial surveys to count caribou. From this information, they concluded that the caribou were declining, and they closed the caribou hunting season without consulting the Yup'ik. This policy was consistent with the equilibrium model in which hunting regulates population size in a density-dependent fashion. The reasoning behind this unpopular decision ran counter to the views held by the Yup'ik, who do not believe that animal populations respond to hunting pressure. The Yup'ik also distrusted results from aerial surveying, which they thought harassed wildlife and undercounted the caribou.

A group of villagers who were tired of the meat shortage and tense confrontations submitted a proposal for a permit to hunt more caribou. The request was denied, but the verdict in a lawsuit in federal court allowed the Yup'ik a one-time permit to take 50 caribou. This was a turning point. A Caribou Working Group composed of representatives from village councils plus Alaska Fish and Game and the federal government continued to meet and work out a mutually acceptable plan for co-managing the caribou. Village traditional councils participated in all phases of the negotiations. Yup'ik conferred in their own language and used traditional Yup'ik modes of making decisions, including consensus aimed at achieving unanimity.

The agreed-upon plan included a small caribou quota. Hunters were also included in the aerial surveys and were able to suggest ways that their knowledge of caribou behavior could be used to refine the aerial counts.

Yup'ik participation in wildlife management carried over into resolving a subsequent conflict over brown bear hunts. The Yup'ik revere bears. They attribute mystical powers to bears, including the ability to comprehend human motives and intentions. Treating bears with respect is of utmost importance. When researchers planned to place radio collars with transmitters on bears, the Yup'ik strenuously objected. The proposed project would involve chasing the bears with helicopters, injecting them with tranquilizers, and fitting them with collars. Hunters felt that the research would harass the bears, change their behavior, and perhaps make them more dangerous.

The events that occurred in response to Yup'ik objections to the bear research were similar to the caribou case. Research went ahead despite local opposition. Yup'ik again used legal challenges to advance their case. Ultimately, a higher jurisdiction got involved. Bruce Babbitt, Secretary of the US Department of Interior at the time, suspended the project for a year, which eventually led to discussion and compromise (Spaeder, 2005).

Co-management was well-suited to sparsely populated, remote ecosystems where communities like the Yup'ik and the Inupiat lacked formal rights to land and resources but asserted their rights to be represented in conservation. A different set of conditions set the stage for the development of local management in another context: small-scale fisheries in a tropical, coastal ecosystem.

12.7 Local Management of a Small-Scale Fishery in Costa Rica

Small-scale fishers are the world's largest group of ocean users, but their opinions and knowledge are often overlooked by managers and policy makers. These local users of marine resources live in coastal villages where they often face poverty and food insecurity along with threats from commercial fishing operations, pollution, and rising sea levels (Table 10.2).

Costa Rica has coastlines on both the Pacific Ocean and the Caribbean Sea (Figure 12.2). The Gulf of Nicoya on Costa Rica's Pacific coast is a productive estuary (Box 3.2) at the mouths of two rivers. Mangroves along parts of the gulf's coastline contribute to this productivity by acting as nurseries for young fish. Local fishers in the region catch fish for their own use and also sell some of their catch. Thus, they engage in both subsistence fishing and small-scale commercial fishing.

Until recently, fisheries in Costa Rica were managed in a top-down manner without local input. In the village of Tárcoles, located near the southern end of the Gulf of Nicoya, villagers fish with lines or nets from small boats propelled by outboard motors. In 1985, some of the fishers from Tárcoles formed Coope Tárcoles, R.L., a government-registered cooperative, with the objective of improving working conditions and market access. The coop provides members with gasoline, ice, bait, a place to store outboard motors, and help with getting fishing licenses and insurance. About half the fishers in Tárcoles belong to the cooperative.

Coope Tárcoles, R.L., handles all operations related to fishing and marketing including onsite processing and sale of the catch. Because it deals with buyers directly, members get better prices than in the past when they had to pay middlemen. Some buyers like Martec (a marine fish processing and farming company in Central America) export much of the product to North America.

The cooperative has a reputation for providing high-quality fish caught under fair working conditions. Ecotourism brings additional income to the coop while educating visitors about fishing traditions and the sustainable management of small-scale fisheries. Tourists meet people in the community, have a chance to fish from a boat, and share a snack prepared from their catch.

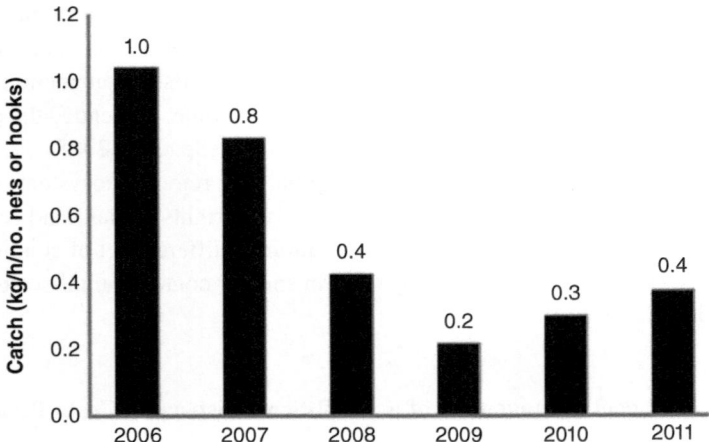

Figure 12.4   Total annual catch according to effort obtained from the Coope Tárcoles, R.L., central database for 2006–2011. From Solís Rivera et al. (2015). Reproduced with permission, © FAO 2015.

In 2001, Coope Tárcoles, R.L., formed an alliance with another cooperative, Coope SoliDar, R.L., a Costa Rican NGO that provides professional services related to integrating conservation, development, and protection of human rights such as land tenure, fair working conditions, and access to education and health care (Coope SoliDar, No date). Staff members assist with the integration of local knowledge and Western science in data management, participate in data analysis, and facilitate sharing of experiences with other fishing villages.

In 2006, Coope Tárcoles, R.L., members began recording information from their catches in a computerized database. When a fisherman or fisherwoman returns from a trip, their catch is immediately frozen, and they report their results. Date and duration of the trip, total catch of each species, time spent fishing, type of gear (nets or hooks), catch location, and phase of the moon are recorded. For selected species, the developmental stage (juvenile or adult) is also recorded. The resulting database is valuable for both scientific and social reasons. It ensures that each fisher receives fair payment for every trip, and it provides crucial information on trends in the harvested species.

The information from the database is then shared with fishers in the community, who contribute their knowledge and insight. Following that process, the information is analyzed, with technical support from Coope SoliDar, R.L., and used to formulate recommendations for managing the fishery (Figure 12.4).

The data for 2006–2011 on the relationship of the catch to fishers' efforts confirmed trends that the fishers had already observed. The reasoning behind taking effort into consideration is the assumption that if the abundance of a common harvested species does not change, then there will be no change in how easy it is to catch it. If that assumption is correct, then the results shown in Figure 12.4 suggest that the harvested species became less abundant after 2007. This convinced the fishers that conservation was needed (Solís Rivera et al., 2015).

In 2007, Coope Tárcoles, R.L., and Coope SoliDar, R.L., jointly asked the government to recognize an *Área Marina de Pesca Responsable* (Marine Area of Responsible Fishing) in Costa Rica's marine territory in part of the Gulf of Nicoya. The Marine Area of Responsible Fishing of Tárcoles that resulted from this effort was one of the first to be recognized by the Costa Rican government, which subsequently recognized nearly two dozen other similar designations. These areas are co-managed through shared *governance* (arrangements for governing) between the *Instituto Costarricense de Pesca y Acuicultura* (Costa Rican Institute of Fisheries and Aquaculture) and the local fishery. Fishers and governmental authorities in a Marine Responsible fishing areas collaborate on a management plan for the fishery.

The Tárcoles marine area included a zone within which shrimp fishing was banned for a year as recommended by the cooperatives. At the end of the ban, the shrimp stock appeared to be recovering. The state agency responsible for managing marine resources then opened the area to traditional (but not to industrial) shrimping for three months each year, again in accordance with the cooperatives' recommendations (Solís Rivera et al., 2015, 2017).

After the Marine Area of Responsible Fishing Area was recognized, the biomass of the catch increased, and ecotourism expanded. The increase in tourism provided additional income to the fishers, but also increased pressure on the fishery.

Coope Tárcoles, R.L., is a successful business and an advocate for fisheries conservation and the conditions of local fishers. However, its influence on policy matters is limited. In 2021, Costa Rica, Panama, Colombia, and Ecuador announced the creation of a marine corridor that consolidates and increases the size of the four nations' protected territorial waters in the Eastern Tropical Pacific Ocean. The designated area encompasses one of the world's most productive and biodiverse oceans. Many endemic species occur within it, and many migratory species, including sea turtles, whales, and sharks, pass through it.

Small-scale fishers were not consulted about this decision. The agreement is regarded as an encouraging example of regional cooperation leading to the creation of an international network of marine protected areas in Latin America. However, it will likely cause large shipping operations that had been fishing in the open ocean to move their operations closer to shore, where they will compete with Coope Tárcoles, R.L., and other small-scale fishers, affecting their livelihoods and way of life (Enright et al., 2021).

Unlike many of the cases discussed in this book, this example does not involve authorities urging or compelling local users to cut back on their exploitation of resources or to relocate because of a protected area designation. Instead, in this example, local users requested that the national government recognize traditional fishing use of the area and work with them in cooperative management of regulated use.

Efforts to integrate stewardship of human well-being and biodiversity have resulted in some inspiring accomplishments, yet in other cases the results have been disappointing. Histories of conquest, exploitation, violence, and environmental degradation due to domination make it difficult to establish trust between potential collaborators. In addition, stakeholders bring diverse histories, assumptions, belief systems, types of knowledge, and objectives to the table. Even under the best of circumstances, meaningful communication between people from different cultures requires a lot of commitment because of language barriers and contrasting world views.

One common thread that emerges is that top-down conservation programs are less likely to be effective than decentralized projects which involve collaborative consultation among affected parties and sharing of concerns, knowledge, and insights. Including stakeholders, educating affected parties, and sharing benefits are frequently cited as ways to create sustainable and ethical solutions to environmental problems. Conservation projects tend to be most successful where outside entities – such as international agencies, government departments, scientists, or NGOs – share their knowledge, expertise, and connections in a collaborative environment *(see Q12.1)*.

Mark Horstman and Glenn Wightman, two resource managers in northwestern Australia, suggest that the Aboriginal concept of *karparti* provides a useful way to think about such interactions. According to Horstman and Wightman, the *karparti* approach involves unhurried, respectful discussions and, at least in cultures influenced by English traditions, "a pannikin [cup] or two of tea." *Karparti* has been used to guide the development of policies regarding tourism in culturally sensitive places in northwestern Australia (Horstman and Wightman, 2001:103).

Regardless of how good the communication between stakeholders is, however, monitoring and critical evaluation are always essential.

12.8 Evaluating Conservation of Biocultural Diversity

> Not everything that can be faced can be changed, but nothing can be
> changed unless it is faced.
>
> *(Attributed to James Baldwin)*

Evaluating efforts to balance the needs of people and the nonhuman world is challenging. Funding for rigorous studies of effectiveness is often inadequate, and there are many actors with varied, and sometimes conflicting, objectives. Different players often have contrasting values by which they think conservation should be judged.

There is no consensus about what measures should be used to evaluate program success. Researchers generate unique lists of criteria they are interested in, making it difficult to compare different studies. When a group of ecologists and anthropologists at American universities evaluated 28 integrated-conservation-and-development projects, they found that most of them were inadequately monitored and assessed. Many projects had no quantitative assessment, and those that did measured a wide diversity of variables (Brooks et al., 2006).

When attempting to evaluate stewardship conservation, it is important to keep in mind that many projects include an optimistic array of objectives to satisfy funders (Section 12.5), perhaps including popular but poorly defined and hard-to-measure buzzwords. But setting out to be all things to all people creates unrealistic expectations and "vicious cycles of optimism and disenchantment" (McShane et al., 2011:967).

One way to avoid these pitfalls is to acknowledge tradeoffs at the outset. Because most projects cannot do everything that everyone might want them to do, it makes sense to be explicit and realistic about what can and cannot be accomplished rather than to hope everybody wins. For example, the production of plant biomass for a biofuel such as ethanol

Table 12.1 *Potential tradeoffs from biofuel production in a developing nation (After McShane, et al., 2011)*

	Reality of tradeoffs in a tropical forest context	
	Gains	Losses
Environmental conservation	Cleaner and more renewable fuel source	Loss of ecosystem services because of conversion of rainforests to biofuel plantations
Human well being	Biofuel jobs on plantations and in factories	Rise in food prices
	New economic development	Water scarcity
		Increased deforestation
		Increased conflicts over land tenure

(a type of alcohol) might seem to be a win-win technology that creates jobs and produces energy without burning fossil fuels. However, land that is used to produce biomass for fuel is not available to grow food or to maintain forests. Thus, social costs (rising food prices and food insecurity) and environmental costs (deforestation) are tied to biofuel production (Table 12.1). When the rosy expectations associated with win-win scenarios do not come to pass, disillusionment typically follows, and backlash may even result.

As the writer James Baldwin noted in the quote that begins this section, acknowledging problems does not mean they can be solved, but it may start the process. Admitting tradeoffs sets the stage for rigorous analysis, tough decisions, and earnest negotiations among actors. If combined with equitable opportunities for participation, explicit discussions of tradeoffs make it less likely that that unrealistic expectations will prevail with the negative consequences that they entail.

In other words, conservation inevitably involves hard choices and clear thinking.

Postscript

There is much to worry about in these pages. Many species are gone; our soil, air, and water are degraded; threats from habitat loss, overexploitation, pollution, invasive species, and climate change persist and many are getting worse. Some of these changes are irreversible. The possibility of crossing future tipping points looms.

There is also much to suggest a way forward. We have a more nuanced (and perhaps more honest) understanding of the ways that people influence the biosphere – for better or for worse. More players, with varied perspectives and skills, participate in the process of addressing our challenges. We recognize that ecosystems are complex, diverse, and potentially resilient. We are getting better at recognizing when our interventions promote desirable ecosystem states and when they are likely to have unwelcome consequences. We acknowledge the urgency of our situation.

A friend who received a cancer diagnosis once said to me: "with cancer you never have enough up-to-date information when you need it." But you have to make decisions anyway. (I am grateful that my friend recovered.)

We don't always have the answers to our questions when we need them. But we can admit what we don't know, frame clear questions, and set about trying to get answers. As new information becomes available, we can revise our understanding, update our information, come up with new questions, and continue. We can carefully evaluate evidence and arguments. We can examine our biases and try to avoid having them cloud our judgment. We can temper our hubris about our relationship to the natural world with humility (Hobbs et al., 2011).

There will be surprises. There will be disappointments and setbacks. There is no guarantee that we will get everything that we want.

But it is worth the effort.

Appendix: Types of Ecosystems

Widely separated regions around the globe that have similar climates tend to have similar types of vegetation (although fire, soil, topography, and interactions with animals and people can override the effects of climate (Section 2.2.2)). This appendix describes these basic ecosystem types, which are sometimes termed *biomes*.

The characteristics of the world's ecosystems have important implications for conservation because they can help us to predict how their plants and animals respond to human activities.

A.1 Terrestrial Ecosystems

Terrestrial ecosystems are generally defined by the major types of vegetation they support in the absence of widespread disturbance. Temperature and the availability of water throughout the growing season are major determinants of vegetation. In addition, a region's historical plant geography influences the composition of vegetation.

Anything that removes vegetation – fire, herbivory, agriculture, logging, etc. – might alter a dominant community type. In some contexts, this sets back succession, and a sequence of early successional stages follows (Section 2.2). In other contexts, a new stable state develops (Section 10.1.5.2).

A.1.1 Tundra

Tundra (from a Russian word referring to the marshy plains of northern Eurasia) occurs where there is too little heat to support the growth of trees. This occurs at high latitudes (the Arctic and Antarctic regions of the Northern and Southern Hemispheres, respectively) and at high altitudes (above tree line on mountain slopes, where it is termed *alpine tundra*). In the Northern Hemisphere, arctic tundra forms a circumpolar belt across North America and Eurasia. In the Southern Hemisphere, tundra occurs in parts of Antarctica, on a few islands, and at high elevations.

Grasses, sedges, mosses, lichens, and dwarf shrubs such as willows dominate tundra vegetation. These short plants do not provide a lot of biomass, so there is relatively little dead organic matter in tundra soils. What there is decays slowly in the cold climate. Thus, tundra soils are shallow and infertile.

Because of its poor soils and short growing season, tundra vegetation grows slowly and grows back slowly after it is disturbed. In addition, permafrost adds to the vulnerability of tundra. In inland arctic areas, where the climate is not moderated by proximity to an ocean, only the surface soil thaws during spring and summer. The soil beneath the surface remains permanently frozen in a state of permafrost. In addition, the annual cycle of freezing and thawing in the upper soil causes it to expand and contract repeatedly, a phenomenon termed *frost churning*. Tundra vegetation insulates the soil, so when frost churning removes this plant cover, the soil warms and thaws to greater depths. Frost churning increases, and more vegetation is uprooted. As a result, the limited biomass covering the soil is destroyed. This allows partially decomposed vegetation, which is widespread in tundra, to melt. When that happens, large amounts of carbon dioxide and the even more potent greenhouse gas methane are released (Section 10.3). Areas of permafrost that are further disturbed by activities related to oil exploration and pipeline construction remain bare for centuries.

Relatively few land animals are adapted to the tundra; however, tundra in the Northern Hemisphere supports large herds of caribou (reindeer) as well as smaller populations of wolves that prey on them. Reindeer herding is central to cultures in the far north. Arctic tundra is also of great importance to the millions of migratory waterfowl and *shorebirds* (sandpipers and their relatives) that breed there.

A.1.2 Boreal Forest

As we move toward the equator from the poles or, in the temperate zone, downslope from mountain peaks, we encounter coniferous forest. This is termed *subarctic* or *boreal forest* if it is adjacent to arctic tundra, or *subalpine forest* when it abuts alpine tundra. These communities are dominated by coniferous trees such as spruce and fir along with deciduous trees (primarily birch, aspen, and poplar).

Fire spreads easily through conifer foliage, which contains high levels of flammable oils and resins and decays slowly. Many plant species characteristic of coniferous forests have adaptations that allow them to survive fires or to reproduce after fires. But, because of the cold climate at high latitudes and altitudes, subarctic and subalpine forests regrow slowly after fire.

A.1.3 Temperate Forest

Temperate forests occur either downslope of subalpine forests or closer to the equator. In relatively moist climates, deciduous trees dominate. In the temperate deciduous forest of eastern North America, high-energy acorns, chestnuts, beechnuts, and fruits of other deciduous trees provided food for the passenger pigeon prior to its extinction.

In much of western North America, where the climate is drier, temperate coniferous forests prevail at moderate altitudes. Pines and junipers are often prevalent in this ecosystem.

Temperate evergreen forests in dry climates burn readily and contain some species that persist only where fires are frequent. These include lodgepole pine (Section 11.1.2.3) and jack pine (Box 7.4).

Wet temperate evergreen forests (*temperate rainforests*) occur along the west coasts of New Zealand, Chile, southern Australia, and North America.

The principal economic use of temperate forest is timber harvest. Many temperate forests are also cleared for agriculture or grazed. These uses do not necessarily destroy its capacity to regenerate. But overuse and poor management can exceed a forest's capacity to recover.

A.1.4 Chaparral

In temperate zones with a *Mediterranean climate* (mild, wet winters and warm, dry summers), plant communities dominated by shrubs with thick, evergreen leaves are common. This is known as *chaparral* in southwestern North America and *fynbos* in South Africa (Box 8.3). These communities have exceptionally high plant diversity.

The leaves of Mediterranean shrubs are high in flammable oils, and they decay slowly. These characteristics allow flammable litter to build up. As a result, fires are common during the dry summers that are typical of chaparral. Many chaparral shrubs sprout readily after a fire, and some have seeds that germinate only after being exposed to heat.

Because of decades of fire suppression in developed areas, highly flammable litter from chaparral vegetation has accumulated. Fires have become harder to control and more destructive because of this fuel buildup.

A.1.5 Steppe

Just as the distribution of trees at high latitudes and altitudes is limited by temperature, it is limited by available moisture at low elevations and latitudes. Regions with a pronounced dry season that are too dry for tree seedlings to become established except in microhabitats, and that are dominated by perennial grasses are a type of grassland known as *steppe*, or "prairie."

The best-known steppes are the prairies of North America and a vast belt of steppe that extends from Central Europe across Russia and Mongolia to China. In North America, steppe occurs on both sides of the Rocky Mountains. On the east side of the Rockies, steppe extends eastward across the midwestern USA and Canada to the deciduous forest. The climate of this steppe is driest near the mountains and becomes wetter to the east. As the climate becomes more moist, the dominant vegetation changes from *shortgrass prairie* to *tallgrass prairie*. At its eastern margin the tallgrass prairie has the potential to support forest if fire is absent (Box 8.5) (Daubenmire, 1978).

Steppes that occur in climates which are almost moist enough to support trees are sometimes accompanied by abundant wildflowers, a type of community termed *meadow steppe*.

The soils of steppes are deep, well drained, and high in organic matter because they form from the extensive roots of decaying perennial grasses. (Meadow steppe soils were the principal reason for designating Russia's Central Chernozem (Black Earth) Biosphere Reserve (Box 8.1)).

Agriculture and grazing are the principal uses of steppe. Some of the world's most productive croplands, including major wheat producing regions, are in steppe. The response of steppe vegetation to livestock is shaped by a region's evolutionary history of interactions with herbivores (Box 11.1).

Because of their annual dry season, steppe grasses and forbs dry out each year. However, the perennial grasses of steppes do not die when they dry out. They regrow when the rains return. During the seasonal dry period, fires – set either by lightning or by Indigenous people – were common before fire suppression policies were adopted.

A.1.6 Desert

Where the climate is too hot and dry during the daytime to support perennial grasses, shrubs dominate, and if grasses are present, they are annuals. This type of ecosystem is termed *desert*. (Some steppe communities are also popularly known as "desert.")

Because deserts have high daytime temperatures, water evaporates quickly and is not available for plants.

Desert productivity is low. Soils are low in organic matter and relatively infertile. Because evaporation rates are high, salts and some other chemicals dissolved in water accumulate in the soil instead of being carried downward.

Wikipedia describes desert as "hostile to plant and animal life." But desert plants and animals have adaptations to the temperature extremes, water scarcity, and low productivity of deserts. To a desert-adapted species, a more moderate environment might seem hostile. Indigenous people of desert cultures – such as the Bedouin of North Africa and the Middle East, Aboriginal Australians, and Native Americans of deserts – also have developed ways of using desert resources and coping with desert environments.

Because of their low productivity and limited water supplies, desert ecosystems are vulnerable to disturbance from anything that removes plant biomass or uses water. In addition, they are very susceptible to erosion because of their sparse plant cover.

Deserts are used for grazing livestock, for irrigated agriculture, and for hunting and gathering.

A.1.7 Savanna

The geographic region where lowland areas remain free of frost throughout the year is known as the *tropics*. This occurs at low latitudes, usually south of the Tropic of Cancer and north of the Tropic of Capricorn. Parts of the tropics where the climate is wetter than desert but not wet enough to support forest are termed *savanna*. Grassy plains with perennial grasses and scattered trees and shrubs characterize this type of vegetation. The so-called Kalahari "desert" of southern Africa is actually savanna (Box 12.4). Many contemporary savannas have been created and maintained by human-caused fires in regions that could otherwise support dry tropical forest.

Like steppes, savannas have a dry season when the aboveground parts of grasses die back. At this time, the vegetation burns easily (Section 11.2.1).

In the Serengeti-Mara Ecosystem, fire and herbivores affect the balance between trees and grasses, sometimes tipping the balance away from grasses and toward a community dominated by trees and shrubs (Section 10.1.5.2). Such communities are sometimes termed *woodland*. In woodland, as opposed to forest, the tree canopy does not cover most of the land surface.

Savannas are grazed by both livestock and wild herbivores. In the Serengeti, enormous herds of wildebeest and other ungulates undertake annual migrations throughout the ecosystem to take advantage of seasonal changes in forage quality (Box 9.2).

Open habitats such as steppe and savanna are widely regarded as having less conservation value than forests. This is reflected in many national and international policies as well as in private conservation programs that support *afforestation* (the planting of trees in areas where they did not historically occur) (Box 11.3). This results in policies that lead to declines in the biodiversity of plants and animals adapted to steppes and savannas (Veldman et al., 2015a).

A.1.8 Tropical Forest

Tropical forest extends on both sides of the equator in parts of South and Central America, Mexico, Africa, India, Southeast Asia, and Oceania including Australia. Tropical dry forests and tropical moist forests support some of the highest species diversity in the world. In these species-rich environments, many mutualistic interactions have evolved such as the relationship in which acacias provide shelter and food for ants and ants defend acacias against herbivores. Some mutualistic interactions that were important in the past have been interrupted by the extinction of one or more of the species that originally participated in the interaction (Box 11.2).

A.1.8.1 Dry Tropical Forest

Dry tropical forest has annual wet and dry seasons. The *monsoon forest* of southeastern Asia is a type of dry tropical forest because even though very heavy rains occur during the wet season there is also a dry season.

This type of forest once accounted for half of all tropical forest, but it has been reduced to a small fraction of its original area. Fire and conversion to pastures and farms are largely responsible for this decline in the extent of tropical dry forest (Box 11.2).

A.1.8.2 Moist Tropical Forest

Moist tropical forest or *tropical rainforest* is rainy all year long; there is no pronounced dry season. The Amazon tropical rainforest of South and Central America is well known, but this ecosystem also occurs in Africa, Asia, and Oceania.

Tropical rainforests have multi-layered deciduous tree canopies. Because the dense, tall, broad leaves in the top layer of the canopy intercept a lot of sunlight, rates of photosynthesis are high in tropical moist forest. Since photosynthesis uses carbon dioxide, high rates of photosynthesis mean that a lot of the heat-trapping gas carbon dioxide is removed from the air in moist tropical forest. This is termed *carbon capture*. Thus, moist tropical forest stores a great deal of carbon and has a big effect on the Earth's climate. These forests are in the news a lot because of this and because of the rapidity with which they are being cut down and their species are being lost.

Like dry tropical forests, tropical rainforests have some of the highest species diversity on Earth. Many species found in moist tropical forests have not even been described by scientists yet. Typically, large numbers of species coexist in a moist tropical forest, but they exist at low population densities and often they occupy only a small area. This combination of small population size and limited geographic range puts tropical moist forest species at high risk of extinction from deforestation, because the global populations of a species with limited geographic ranges can easily be wiped out or reduced to a perilous level when its habitat is degraded.

Most of the nutrients in tropical rainforest are above the ground rather than in the soil. For this reason, moist tropical forests recover slowly after disturbances such as large-scale logging or grazing that remove biomass frequently or from a large area. Even small-scale disturbances can be problematic too, if they occur often enough. On the other hand, recovery in gaps due to fallen trees, which create small forest openings but do not remove a lot of vegetation, can be fairly rapid (Section 11.1).

Many groups of Indigenous hunter-gatherers, as well as shifting cultivators and settlers, live in moist tropical forests. Representatives of multinational corporations and non-governmental organizations also have interests in rainforest resources. Conflict over resource use between these groups is intense.

Tropical rainforests produce many high-value products for export. In addition to timber and beef, coffee, chocolate, bananas, mangoes, palm oil, and other crops are grown in plantations and marketed mainly to countries in the Northern Hemisphere.

Some marketable forest products can be obtained from moist tropical forests without harvesting trees. Because these products can sometimes be obtained without killing the tree that produces them, there is interest in developing systems for extracting such nontimber forest products on a sustainable basis (Section 12.3.1).

A.2 Aquatic Ecosystems and Wetlands

In *aquatic* ecosystems, water is usually too deep for vegetation to be rooted on the bottom and extend above the surface into air. Many kinds of algae (popularly known as "seaweed") and a few kinds of plants such as water lilies, along with the food chains they support, inhabit aquatic zones.

A.2.1 Marine Aquatic Ecosystems

Marine aquatic ecosystems are not easy to delineate on a map for a couple of reasons. First, oceans and seas in different parts of the world are interconnected and mixed by currents, so organisms range widely. Second, because vertical stratification is important in oceans, their zones can't be shown on two-dimensional maps.

A.2.1.1 Photic Zone

The upper layer of the ocean, where sunlight penetrates and photosynthesis takes place, is termed the *photic zone* (*phot*, light). Almost all life in the Earth's oceans depends upon photosynthesis carried out in the photic zone by some kinds of plankton. Hence, the productivity of the Earth's oceans is limited by the amount of photosynthesis that can take place in the photic zone. Animals below the photic zone consume organic matter such as feces and dead organisms that are produced in the photic zone and sink beneath it.

The polar regions of the oceans are very productive because in those regions upward movements of water known as *upwellings* bring nutrients from deep water toward the surface. These cold, nutrient-rich waters support the growth of *krill*, small crustaceans that feed on the plankton. Squid, fish, penguins, seals, and baleen whales (Section 1.3.2.4) feed on krill, forming the basis of a critical marine food chain.

Photosynthetic plankton capture a great deal of heat-trapping carbon dioxide. Other photosynthetic marine organisms – including kelp, seagrass, and coral – also influence the Earth's atmosphere through carbon capture.

Coral reefs: Corals are tiny marine animals related to jellyfish. For most of their life, corals live together in colonies attached to the sea floor (although they also live as free-swimming larvae that resemble little jellyfish for part of their life cycle). Each individual coral secretes a hard outer skeleton. Collectively these skeletons form complex, species-specific structures that form coral reefs. A type of red algae that contains hard deposits in its tissues also contributes to reef structure.

Reef-building corals live in the photic zone of tropical oceans. They depend upon mutualistic, photosynthetic algae that live within their tissues. Because corals depend on these photosynthetic mutualists, coral reefs can grow only in the photic zone and cannot tolerate suspended material that impedes the penetration of light.

Coral reefs are the most biodiverse shallow water marine ecosystems in the world.

Corals are adapted to frequent natural disturbances from storms and exceptionally low tides. After a disturbance kills corals on a reef's surface, it is normally repopulated by free-swimming coral larvae. However, hazards associated with human activities threaten reef ecosystems. Furthermore, fluctuations in the warm-water environment of coral reefs are associated with a phenomenon known as *bleaching*, in which corals expel their algal mutualists.

Kelp forests: *Kelp* are large, brown algae that dominate shallow oceans in some places. Kelp form forest-like stands, where they are considered keystone species. Because of its large biomass, kelp stores large amounts of carbon.

Seagrass meadows: Seagrasses are not grasses (although they have long, narrow leaves that superficially resemble grass), and they are not seaweed. They are related to land plants (not algae). Seagrasses form dense underwater meadows in the photic zone of temperate and tropical waters where they photosynthesize. This productive and diverse environment provides habitat for many invertebrates, seahorses and other fishes, birds, and marine turtles.

A.2.1.2 Benthic Zone

The *benthic zone* is the zone at the bottom of a body of water. Benthic organisms on the ocean floor depend on organic material from the photic zone. This zone also includes unusual communities in which the only producers are heat-tolerant microorganisms (Box A.1). In these communities, bacteria and archaea (Section 2.3.2.1) obtain energy from chemosynthesis, a process that is similar to photosynthesis but does not depend on sunlight. The energy for chemosynthesis comes from the chemical bonds of some inorganic compounds that are emitted by underwater hot springs.

Benthic ecosystems face threats from exploration and exploitation. Commercial operations that catch bottom-dwelling fish or shellfish drag large nets or other gear along the sea floor. This buries, crushes, or exposes benthic structures and animals (Watling and Norse, 1998). Ocean communities recover slowly from this type of disturbance.

A.2.2 Freshwater Aquatic Ecosystems

Rivers, streams, lakes, and ponds are freshwater aquatic ecosystems. The waters of different watersheds are not connected. Due to this isolation by land barriers, freshwater

---

Box A.1
"Astounding Discoveries" on the Sea Floor

In 1977, scientists on a research vessel near the Galápagos Islands found a *hydrothermal vent* (*hydro*, water; *thermal*, hot), a crack in the ocean floor where rising *magma* (melted rock) heats seawater to temperatures as high as 750° F (400° C), forming a hot, chemical-laden soup below the photic zone. A deep-sea submersible research apparatus named Alvin allowed the researchers to obtain data at great depths. (The Alvin was named for its creator, Allyn Vine, but coincidentally Alvin was the name of a popular cartoon chipmunk at the time.)

The team had been looking for the vents, so finding them was no surprise. But when Alvin returned to the surface, the scientists were surprised indeed. Alvin brought up thirteen photos of live clams and mussels. "We all started jumping up and down. We were dancing off the walls," a geochemist on board wrote later. No one had expected to find life in this hot, dark environment.

They were not prepared for this. No biologists were on board, and no one had brought preservatives for the specimens they collected. Forced to improvise, the team used what they had on hand: Russian vodka.

Follow-up expeditions, this time with biologists included, discovered unique communities in the vents (Figure A.1):

On each of its dives, *Alvin's* front basket and cameras captured a remarkable variety of animals that never had been seen before: unknown mussels, anemones, whelks, limpets, featherduster worms, snails, lobsters, brittle stars, and blind white crabs. One crustacean seemed to have teeth on the end of eyestalks, which scientists speculated were used to scrape food off rocks. A new species of giant white clams with blood-red flesh ....

[A] delicate, orange, dandelion-looking creature ... turned out to be ... [related to jellyfish].

[R]ed-tipped tubeworms that were an astonishing 8-feet tall. Aboard ship, they found that the tubeworms ... had no mouth to take in food and no guts to digest food.

(Woods Hole Oceanographic Institution (No date) www.whoi.edu/feature/
history-hydrothermal-vents/discovery/1979.html)

Figure A.1   Deepwater community in Galápagos hydrothermal vent. White animals with tentacles in center of the photo are anemones; dark animals are mussels. Animals with white, rope-like stalks and reddish, cylindrical tips are tube worms (NOAA, 2011).

drainages tend to develop genetically distinct populations and subspecies and to have many endemic species (Allan and Flecker, 1993). However, many of these endemics have been displaced by non-native species.

In addition to exotic species, the major threats to freshwater ecosystems are habitat alteration (from dams, water diversion, and channelization) and pollution.

A.3 Wetlands: The Terrestrial-Aquatic Interface

Wetlands occur where aquatic and terrestrial environments meet. Organisms that inhabit wetlands have adaptations that allow them to tolerate intermediate and/or fluctuating conditions.

A.3.1 Intertidal Zones

Intertidal zones occur along shorelines in places that are exposed to rising and falling tides on a daily basis. They may be rocky, sandy, or muddy. Salt marshes, mangrove forests, tide pools, mudflats, and sandy shorelines (beaches) are located in intertidal zones.

A.3.1.1 Tide Pools, Mudflats, and Sandy Shorelines

Organisms that are exposed to advancing and receding tides each day must be adapted to fluctuating environmental conditions. They have to withstand changes in temperature and salinity, as well as wave action and alternating exposure to air and submersion in water.

Vegetation is very sparse on rocky shores and mudflats.

Clams, worms, and other invertebrates that inhabit sandy shores and mudflats survive low tides by burrowing in the soft substrate. Sandpipers and other shorebirds use their long bills to reach these buried invertebrates. Where such areas occur along migration routes, they provide a vital resource for birds that need to stop and replenish their energy stores (Weidensaul, 2021).

On rocky shorelines, burrowing to escape low tide is not an option. Animals such as starfish and snails must get through low tides in rock crevices or pools of water or be able to withstand exposure (in some cases by withdrawing into their shells). Relatively few organisms are adapted to these conditions, so species diversity in rocky intertidal environments is not impressive.

A.3.1.2 Mangroves

The term *mangrove* or *mangrove forest* refers both to salt-tolerant trees or shrubs that are rooted in the intertidal zone and to the community dominated by those plants. Mangrove vegetation grows along the tropical and southern parts of temperate coastlines of Asia, Australia, and the Americas.

Mangroves are ecologically and economically important. The roots of mangrove trees stabilize coastlines in the face of wave action during storms. Decomposition of the mangrove leaves adds nutrients to the marine community. The mud in which mangroves are rooted provides habitat for many kinds of organisms, and the roots, trunks, branches, and leaves of mangroves provide habitat for spiders, insects, and many vertebrates. (Recall that Simberloff and Wilson did their classic study of extinction and colonization rates on mangrove islands in Florida Bay (Box 6.12)).

A.3.1.3 Estuaries

The environment where a river meets salt water is termed an *estuary*. Estuaries are characterized by *brackish* water, which is intermediate in salinity between ocean water and fresh water. Estuaries typically contain salt marsh vegetation (Box 8.6) consisting of non-woody plants that are rooted in wet soil and are able to tolerate fluctuating tides.

Like mangroves, estuaries are very productive. Rivers bring nutrients washed from terrestrial ecosystems into estuaries, where tides circulate them. These conditions allow estuaries to produce large amounts of biomass and to support high densities of animals. Salt marshes provide critical habitat for many fishes and crustaceans that spend part of their life in salt water and part in fresh water (Woodwell et al., 1973). Estuaries also serve as stopover areas for migrating birds and provide nesting habitat for resident breeders.

A.3.2 Freshwater Wetlands

Marshes, swamps, bogs, and seasonally flooded meadows are examples of freshwater wetlands. They may be dry at times, but there are periods when the soil is saturated. Wetland plants have adaptations that allow them to survive those times of low oxygen availability.

Wetlands provide ecosystem services such as storing floodwaters, filtering pollutants, and reducing erosion. They also provide habitat for a variety of breeding, wintering, and migrating species of fish and wildlife, and plant and animal resources that have sustained human societies for millennia (Mitsch and Gosselink, 2015).

Riparian zones are likewise productive and diverse. They are especially important in arid or semiarid climates such as steppes and deserts, where trees and shrubs along streams provide habitat for a variety of birds and small mammals that might not survive and reproduce as well in the surrounding habitat.

Throughout history people have settled in and near wetlands. However, the same habitats are in demand for urban, industrial, agricultural, and recreational uses.

A.3.3 Saline Lakes

Inland wetlands usually contain fresh water, but there are exceptions in arid environments, where salinity is high because evaporation is high. This occurs in the Great Salt Lake and the Dead Sea (which is actually a lake) and in numerous small lakes that lack outlets to an ocean. Few plants, animals, and microorganisms are adapted to these aquatic environments, but in some cases, they provide food for substantial populations of migratory waterfowl and shorebirds.

A.4 Ecosystem Productivity

Plants, algae, and those microorganisms that photosynthesize are the producers in all ecosystems except hydrothermal vents. They use carbon dioxide and water to synthesize sugar and other organic compounds.

Ecosystems differ markedly in their productivity (Table A.1).

Biomass production is influenced both by climate, especially moisture and temperature, and by nutrient availability. In terrestrial ecosystems, temperature and moisture are critical determinants of primary productivity. Tropical forests, estuaries, coral reefs, beds of algae, swamps, and marshes have high primary productivity, comparable to or exceeding

Table A.1 *Estimated primary production of major ecosystems.* Based on Whittaker and Likens (2012) and Leith (2012).

Ecosystem type	Net primary production (dry matter)	
	Normal range (g/m^2/yr)	Mean (g/m^2/yr)
Terrestrial		
Tundra (arctic and alpine)	10–400	140
Boreal forest (subarctic and subalpine)	400–2,000	800
Temperate forest		

Table A.1 (*cont.*)

Ecosystem type	Net primary production (dry matter)	
	Normal range (g/m^2/yr)	Mean (g/m^2/yr)
Evergreen temperate forest	600–2,500	1,300
Deciduous temperate forest	600–2,500	1,200
Chaparral	250–1,500	800
Grassland		
Temperate grassland	100–1,500	500
Tropical grassland	200–2,000	700
Tropical forest		
Moist tropical forest	1,000–3,500	2,200
Dry tropical forest	1,000–2,500	1,600
Woodland and shrubland	250–1,200	700
Savanna	200–2,000	900
Desert and semidesert scrub	10–250	90
Extreme desert – rock, sand, ice	0–10	3
Cultivated land	100–4,000	650
Aquatic and wetland		
Swamp and marsh	800–6,000	3,000
Lake and stream	100–1,500	400
Open ocean	2–400	125
Upwelling zones	400–1,000	500
Continental shelf	200–600	360
Algae beds and reefs	500–4,000	2,500
Estuaries (excluding marsh)	200–4,000	1,500
Cultivated land	100–4,000	650

croplands. In these ecosystems, high levels of nutrients are available under climatic conditions that allow them to be utilized by photosynthetic organisms. In the ocean, productivity is high in warm, shallow, sunlit water and in places where currents bring nutrients to the surface, but low in cold, dark, deep water.

Primary production is low where water is scarce (deserts), where the growing season is short (tundra), and where nutrient availability is low (open ocean). Despite their low productivity, these environments are important for a variety of reasons. They may support organisms that form the basis of food chains of ecological, economic, and cultural importance; supply resources essential in the life cycles of migrating species; play critical roles in moderating climate; and contain concentrations of rare and endemic organisms.

Bibliography

Abreu, R. C. R., W. A. Hoffmann, H. L. Vasconcelos, *et al.* (2017). The biodiversity cost of carbon sequestration in tropical savanna. *Science Advances* **3**:e1701284.

Acheson, J. M. (1987). The lobster fiefs revisited: Economic and ecological effects of territoriality in Maine lobster fishing. In B. J. McCay and J. M. Acheson, eds., *The Question of the Commons: The Culture and Ecology of Communal Resources*. Tucson, AZ: University of Arizona Press, pp. 37–65.

Adewale, C. (2018). CornBot—Everyday Farmers [sic] Virtual Assitant [sic] for combating fall armyworm in Africa. www.youtube.com/watch?v=GlbrO_zHnuE.

Alaska Eskimo Whaling Commission. (No date). www.aewc-alaska.org/about-us.html.

Agarwal, B. (1992). The gender and environment debate: Lessons from India. *Feminist Studies* **18**:119–58.

Agee, J. K. (1988). Successional dynamics in forest riparian zones. In K. J. Raedeke, ed., *Streamside Management: Riparian Wildlife and Forestry Interactions*. Seattle, WA: University of Washington Press, pp. 31–43.

Agrawal, A. and K. Redford. (2009). Conservation and displacement: An overview. *Conservation and Society* **7**:1–10.

Ainsworth, T. D., S. F. Heron, J. C. Ortiz, *et al.* (2016). Climate change disables coral bleaching protection on the Great Barrier Reef. *Science* **352**(6283):338–42.

Alison, R. M. (1981). The earliest traces of a conservation conscience. *Natural History* **90**(5):72–77.

Allan, J. D. and A. S. Flecker. (1993). Biodiversity conservation in running waters. *BioScience* **43**:32–43.

Allen, B. L., L. R. Allen, H. Andrén, *et al.* (2017). Can we save large carnivores without losing large carnivore science? *Food Webs* **12**:64–75.

Allen, D. L. (1962). *Our Wildlife Legacy*. New York, NY: Funk and Wagnalls.

Allen, W. H. (1988). Biocultural restoration of a tropical forest. *BioScience* **38**:156–61.

Allin, C. W. (1990). *International Handbook of National Parks and Nature Reserves*. New York, NY: Greenwood Press.

Altermatt, F. (2010). Climatic warming increases voltinism in European butterflies and moths. *Proceedings of the Royal Society B* **277**:1281–87.

Anderson, M. K. (2005). *Tending the Wild: Native American Knowledge and the Management of California's Natural Resources*. Berkeley, CA: University of California Press.

Andrewartha, H. G. and L. C. Birch. (1954). *The Distribution and Abundance of Animals*. Chicago, IL: University of Chicago Press.

Anon. (2005–2016). Edward Wilson of the Antarctic. www.edwardawilson.com/biography-the-grouse-disease-inquiry-1905-1910.

Anon. (2012). Hiding in plain sight, a new frog species with a "weird" croak is identified in New York City – *ScienceDaily*, March 14, 2012. www.sciencedaily.com/releases/2012/03/120314124016.htm.

Archer, M. (1974). New information about the quaternary distribution of the thylacine (Marsupialia, Thylacinidae) in Australia. *Journal of the Royal Society of Western Australia* **57** (Part 2):43–50.

Armstrong, D. P. and J. L. Craig. (1995). Effects of familiarity on the outcome of translocations, I. A test using saddlebacks *Philesturnus carunculatus rufusater*. *Biological Conservation* **71**:133–41.

Atkinson, C. T. and D. A. LaPointe. (2009). Introduced avian diseases, climate change, and the future of Hawaiian honeycreepers. *Journal of Avian Medicine and Surgery* **23**:53–63.

Bailey, T. A. (1935). The North Pacific Sealing Convention of 1911. *Pacific Historical Review* **4**:1–14.

Baker, R. C., F. Wilke, and C. H. Baltzo. (1970). *The Northern Fur Seal*. Circular 336. Washington, DC: US Department of the Interior, US Fish and Wildlife Service, Bureau of Commercial Fisheries.

Balser, D. S., H. H. Dill, and H. K. Nelson. (1968). Effect of predator reduction on waterfowl nesting success. *Journal of Wildlife Management* **32**:669–82.

Barkham, P. (2022). Butterflies: out of the blue. *The Guardian*, July 18, 2010. www.theguardian.com/environment/2010/jul/18/large-blue-butterflies-conservation.

Barnett, A. (2001). In Africa the Hoodia cactus keeps men alive: Now its secret is "stolen" to make us thin. *The Guardian*, June 17, 2001. www.theguardian.com/world/2001/jun/17/internationaleducationnews.businessofresearch.

Barnett, L. (1954). The world we live in: The woods of home. *Life* **37**(November 8):78–100.

Barnett, L. R. (2010). Michigan's war against birds. *Michigan Historical Review* **36**:95–124.

Beck, D. R. M. (1995). Return to *Namä'o Uskíwämît*: The importance of sturgeon in Menominee Indian history. *Wisconsin Magazine of History* **79**(1):32–48.

Begon, M., M. Mortimer, and D. J. Thompson. (1996). *Population Ecology: A Unified Study of Animals and Plants*, 3rd edn., Malden, MA: Blackwell.

Beiler, K. J., D. M. Durall, S. W. Simard, *et al.* (2010). Architecture of the wood-wide web: *Rhizopogon* spp. genets link multiple Douglas-fir cohorts. *New Phytologist* **185**:543–53.

Beletsky, L. D. and G. H. Orians. (1989). Familiar neighbors enhance breeding success in birds. *Proceedings of the National Academy of Sciences* **86**:7933–36.

Bell, S. S., E. D. McCoy, and H. R. Mushinsky. (2012). *Habitat Structure: The Physical Arrangement of Objects in Space*, vol. 8. Dordrecht: Springer.

Berkes, F., J. Colding, and C. Folke. (2000). Rediscovery of traditional ecological knowledge as adaptive management. *Ecological Applications* **10**:1251–62.

Bevill, R. L., S. M. Louda, and L. M. Stanforth. (1999). Protection from natural enemies in managing rare plant species. *Conservation Biology* **13**:1323–31.

Bhatt, C. P. (1990). The Chipko Andolan: Forest conservation based on people's power. *Environment and Urbanization* **2**:7–18.

Bindoff, N. L., J. Willebrand, V. Artale, *et al.* (2007). Observations: Oceanic climate change and sea level. In S. Solomon, D. Qin, M. Manning, *et al.*, eds., *Climate Change 2007: The Physical Science Basis. Contribution of Working Group I to the Fourth Assessment Report of the Intergovernmental Panel on Climate Change*. Cambridge, UK: Cambridge University Press, pp. 385–432.

BirdLife International. (2020a). *Gymnogyps californianus*, California condor. IUCN Red List of Threatened Species 2020. https://dx.doi.org/10.2305/IUCN.UK.2020-3.RLTS.T22697636A181151405.en.

BirdLife International. (2020b). *Milvus milvus*. IUCN Red List of Threatened Species 2020. https://dx.doi.org/10.2305/IUCN.UK.2020-3.RLTS.T22695072A181651010en.

Blackburn, T. C. and K. Anderson, eds. (1993). *Before the Wilderness: Environmental Management by Native Californians*. Menlo Park, CA: Malki-Ballena Press.

Blanchan, N. (1913). *How to Attract the Birds: And Other Talks About Bird Neighbors*. Garden City, NY: Doubleday.

Bland, A. (2012). Can brown bears survive in the Pyrenees? *Smithsonian*, June 12, 2012. www.smithsonianmag.com/travel/can-brown-bears-survive-in-the-pyrenees-118565664/.

Blockstein, D. E. and H. B. Tordoff. (1985). Gone forever: A contemporary look at the extinction of the passenger pigeon. *American Birds* **39**:845–51.

Blumm, M. and J. Brunberg. (2006). "Not much less necessary than the atmosphere they breathed": Salmon, Indian treaties, and the Supreme Court – a centennial remembrance of United States v. Winans and its enduring significance. *Natural Resources Journal* **46**:489–546.

Boeken, B. and M. Shachak. (1994). Desert plant communities in human-made patches – implications for management. *Ecological Applications* **4**:702–16.

Bonnell, M. L. and R. K. Selander. (1974). Elephant seals: Genetic variation and near extinction. *Science* **184**:908–9.

Botkin, D. B. (1990). *Discordant Harmonies: A New Ecology for the Twenty-first Century*. New York, NY: Oxford University Press.

Botkin, D. B. (2012). *The Moon in the Nautilus Shell: Discordant Harmonies Reconsidered*. Oxford, UK: Oxford University Press.

Brayard, A., L. J. Krumenacker, J. P. Botting, *et al.* (2017). Unexpected Early Triassic marine ecosystem and the rise of the Modern evolutionary fauna. *Science Advances* **3**:e1602159.

Brittingham, M. C. and S. A. Temple. (1983). Have cowbirds caused forest songbirds to decline? *BioScience* **33**:31–35.

Brockington, D. (2002). *Fortress Conservation: The Preservation of the Mkomazi Game Reserve, Tanzania*. Bloomington, IN: Indiana University Press.

Brooks, J. S., M. A. Franzen, C. M. Holmes, *et al.* (2006). Testing hypotheses for the success of different conservation strategies. *Conservation Biology* **20**:1528–38.

Brown, J. H. (1971). Mammals on mountaintops: Nonequilibrium insular biogeography. *American Naturalist* **105**:467–78.

Brown, J. H. (1978). The theory of insular biogeography and the distribution of boreal birds and mammals. *Great Basin Naturalist Memoirs* **2**:209–27.

Brown, J. H. and A. Kodric-Brown. (1977). Turnover rates in insular biogeography: Effect of immigration on extinction. *Ecology* **58**:445–49.

Bryant, R. L. (1997). Beyond the impasse: The power of political ecology in Third World environmental research. *Area* **29**:5–19.

Bucher, E. H. (1992). The causes of extinction of the passenger pigeon. *Current Ornithology* **9**:1–36.

Budyko, M. I. (1967). On the causes of the extinction of some animals at the end of the Pleistocene. *Soviet Geography* **8**:783–93.

Buell, M. (1954). Fire in the history of Mettler's Woods. *Torreya* **81**:252–55.

Bullard, R. D., P. Mohai, R. Saha, and B. Wright. (2008). Toxic wastes and race at twenty: Why race still matters after all of these years. *Environmental Law* **38**:371–411.

Burrows, B. (2005). Preface. In B. Burrows, ed., *The Catch: Perspectives in Benefit Sharing*. Edmonds, WA: The Edmonds Institute, pp. i–v.

Butchart, S. H. M., M. Walpole, B. Collen, *et al.* (2010). Global biodiversity: Indicators of recent declines. *Science* **328**:1164–68.

Butler, J. (1907). *The History, Work, and Aims of the Michigan Audubon Society*. Detroit, MI: Michigan Audubon Society.

Cade, T. J. and S. A. Temple. (1995). Management of threatened bird species: Evaluation of the hands-on approach. *Ibis* **137**:S161–S172.

California Academy of Sciences. (No date). Darwin's hawkmoth. www.calacademy.org/learn-explore.

Carson, R. (1962). *Silent Spring*. Greenwich, CT: Fawcett Publications.

Carstens, M. (2020). Supreme Court (Sweden) recognizes Sami group's exclusive right to confer hunting and fishing rights in traditional area [Billet]. *NORDEUROPAforum.blog*, September 8, 2020. https://nofoblog.hypotheses.org/669.

Cartmill, M. (1983). "Four legs good, two legs bad": Man's place (if any) in nature. *Natural History* **92**(11):64–79.

Cassidy, K. M., C. E. Grue, M. R. Smith, *et al.* (2001). Using current protection status to assess conservation priorities. *Biological Conservation* **97**:1–20.

Caughley, G. (1970). Eruption of ungulate populations, with emphasis on Himalayan thar in New Zealand. *Ecology* **51**:53–72.

Caughley, G. (1985). Harvesting of wildlife: Past, present, and future. In S. L. Beasom and S. F. Roberson, eds., *Game Harvest Management*. Kingsville, TX: Caesar Kleberg Wildlife Research Institute, pp. 3–14.

Caughley, G. and A. Gunn. (1996). *Conservation Biology in Theory and Practice*. Cambridge, MA: Blackwell Science.

Chapin, F. S., III, M. E. Power, S. T. A. Picklett, *et al.* (2011). Earth stewardship: Science for action to sustain the human-earth system. *Ecosphere* **2**:art89:1–20.

Chapin, M. (1998). Defending Kuna Yala: PEMASKY. In A. Gary, A. Parellada, and H. Newman, eds., *From Principles to Practice: Indigenous Peoples and Biodiversity Conservation in Latin America. Proceeding of the Pucallpa Conference*. Copenhagen: IWGIA, pp. 240–80.

Chatty, D. and M. Colchester. (2002). Introduction. In D. Chatty and M. Colchester, eds., *Conservation and Mobile Indigenous Peoples: Displacement, Forced Settlement, and Sustainable Development*. New York, NY: Berghahn Books, pp. 1–20.

Chazdon, R. L. (2003). Tropical forest recovery: Legacies of human impact and natural disturbances. *Perspectives in Plant Ecology, Evolution and Systematics* **6**:51–71.

Chazdon, R. L. and J. P. Arroyo. (2013) Tropical forests as complex adaptive systems. In C. Messier, K. J. Puettmann, and K. D. Coates, eds., *Managing Forests as Complex Adaptive Systems: Building Resilience to the Challenge of Global Change*. London: Routledge, pp. 49–73.

Cheke, A. S. and R. Bour. (2014). Unequal struggle – how humans displaced the dominance of tortoises in island ecosystems. In J. Gerlach, ed., *Western Indian Tortoises: Ecology, Diversity, Evolution, Conservation, Paleontology*. Manchester, UK: Siri Scientific Press, pp. 31–120.

Cheng, G. and T. Wu. (2007). Responses of permafrost to climate change and their environmental significance, Qinghai-Tibet Plateau. *Journal of Geophysical Research: Earth Surface* **112**:F02S03.

Chivas, B. J., Jr. (1987). *Toxic Wastes and Race in the United States: A National Report on the Racial and Socio-economic Characteristics of Communities with Hazardous Waste Sites*. New York, NY: United Church of Christ, Commission for Racial Justice.

Christensen, N. L., J. K. Agee, P. F. Brussard, *et al.* (1989). Interpreting the Yellowstone Fires of 1988. *BioScience* **39**:678–85.

Christensen, N. L., J. K. Agee, P. F. Brussard, *et al.* (No date). Ecological consequences of the 1988 fires in the Greater Yellowstone Area (Greater Yellowstone Coordinating

Committee). Final report, the Greater Yellowstone Postfire Ecological Assessment Workshop. http://npshistory.com.

Chun, Y. W. (2009). *The Red Pine: Korean's Tree of Life*. Seoul, South Korea: Book's Hill.

Church, J. A. and N. J. White. (2011). Sea-level rise from the late 19th to the early 21st century. *Surveys in Geophysics* **32**:585–602.

Clements, F. E. (1916). *Plant Succession: An Analysis of the Development of Vegetation*. Washington, DC: Carnegie Institution of Washington.

Clements, F. E. (1936). Nature and structure of the climax. *Journal of Ecology* **24**:252–84.

Coale, A. J. (1970). Man and his environment. *Science* **170**(3954):132–36.

Coates, P. (2007). *American Perceptions of Immigrant and Invasive Species: Strangers on the Land*. Berkeley, CA: University of California Press.

Collett, D. (1987). Pastoralists and wildlife: The Maasai. In D. Anderson and R. Grove, eds., *Conservation in Africa: People, Policies, and Practice*. Cambridge, UK: Cambridge University Press, pp. 129–48.

Committee of Inquiry on Grouse Disease. (1911). *The Grouse in Health and in Disease*, vol. 1, London: Smith, Elder, and Company.

Commoner, B. (1966). *Science and Survival*, 3rd edn, New York, NY: Viking.

Commoner, B. (1971). *The Closing Circle*, 1st edn, New York, NY: Alfred A. Knopf.

Commoner, B., M. Corr, and P. J. Stamler. (1971). The causes of pollution. *Environment* **13**(3):2–19.

Congressional Globe. (1864). *Debates and proceedings, 1833–1873*. Thirty-eighth Congress, First Session. Washington, DC: John C. Rives.

Congressional Record (March 10, 1874). 43rd Congress, 1st Session. GPO-CREC-1874-pt 3-v2-5-2, p. 2107. www.govinfo.gov/app/collection/crecb/_crecb/Volume%20002%20(1874).

Connell, J. H. (1978). Diversity in tropical rain forests and coral reefs. *Science* **199**:1302–10.

Connell, J. H. (1980). Diversity and the coevolution of competitors, or the ghost of competition past. *Oikos* **35**:131–38.

Coope SoliDar, Conservación y derechos humanos. (No date). http://coopesolidar.org/.

Cooper, W. S. (1926). The fundamentals of vegetational change. *Ecology* **7**:391–413.

Cornelius, S. E., R. Arauz, J. Fretey, *et al.* (2007). Effect of land-based harvest of *Lepidochelys*. In P. T. Plotkin, ed., *Biology and Conservation of Ridley Sea Turtles*. Baltimore, MD: Johns Hopkins University Press, pp. 231–52.

Cowles, H. C. (1899). The ecological relations of the vegetation on the sand dunes of Lake Michigan. *Botanical Gazette* **27**:95–117, 167–202, 281–308, 361–91.

Cowles, H.C. (1911). The causes of vegetative cycles. *Botanical Gazette* **51**:161–83.

Cox, J. C. (1905). *The Royal Forests of England*. London: Methuen.

Cranworth, Lord. (1912). *A Colony in the Making. Or, Sport and Profit in British East Africa*. London: Macmillan.

Croat, T. B. (1972). The role of overpopulation and agricultural methods in the destruction of tropical ecosystems. *BioScience* **22**:465–67.

Cronon, W. (1983). *Changes in the Land: Indians, Colonists, and the Ecology of New England*. New York, NY: Hill and Wang.

Cronon, W. (1996). The trouble with wilderness. In W. Cronon, ed., *Uncommon Ground: Rethinking the Human Place in Nature*, 2nd edn. New York, NY: Norton, pp. 60–90.

Crosby, A. W. (1972). *The Great Columbian Exchange: Biological and Cultural Consequences of 1492*. Westport, CT: Greenwood Press.

Crosby, A. W. (1986). *Ecological Imperialism: The Biological Expansion of Europe, 900–1900*. Cambridge, UK: Cambridge University Press.

Culler, R. C. (1970). A progress report: Water conservation by removal of phreatophytes. *Eos, Transactions American Geophysical Union* **51**:684–89.

Curtis, J. T. and M. L. Partch. (1948). Effect of fire on the competition between blue grass and certain prairie plants. *American Midland Naturalist* **39**:437–43.

Darwin, C. (1871). *The Descent of Man and Selection in Relation to Sex.* Princeton, NJ: Princeton University Press.

Darwin, C. (1958). *The Origin of Species*, 7th printing. New York, NY: New American Library.

Daubenmire, R. (1978). *Plant Geography: With Special Reference to North America,* 1st edn. New York, NY: Academic Press.

Davidson, N. C. (2014). How much wetland has the world lost? Long-term and recent trends in global wetland area. *Marine and Freshwater Research* **65**:934–41.

Davis, M. B. (1981). Outbreaks of forest pathogens in Quaternary history. *Proceedings of the 4th International Palynology Conference B*:216–17. Lucknow, India.

Davis, P. E. and I. Newton. (1981). Population and breeding of red kites in Wales over a 30-year period. *Journal of Animal Ecology* **50**:759–72.

Davis, W., K. D. Harrison, and C. H. Howell. (2007). *Book of Peoples of the World: A Guide to Cultures.* Washington, DC: National Geographic Books.

Dawson, W. L. (1903). *The Birds of Ohio: A Complete Scientific and Popular Description of the 320 Species of Birds Found in the State.* Columbus, OH: Wheaton Publishing.

De'ath, G., J. Lough, and K. Fabricius. (2009). Declining coral calcification on the Great Barrier Reef. *Science* **323**(5910):116–19.

Deichmann, U. (2016). Epigenetics: The origins and evolution of a fashionable topic. *Developmental Biology* **416**:249–54.

Derocher, A. E. and I. Stirling. (1995). Estimation of polar bear population size and survival in western Hudson Bay. *Journal of Wildlife Management* **59**:215–21.

Diamond, J. M. (1969). Avifaunal equilibria and species turnover rates on Channel Islands of California. *Proceedings of the National Academy of Sciences, U.S.A.* **64**:57–63.

Diamond, J. M. (1975). The island dilemma: Lessons of modern biogeographical studies for the design of nature preserves. *Biological Conservation* **7**:129–53.

Diehm, C. (2013). Wolves, Wisconsin, and Aldo Leopold. *Minding Nature.* https://humansandnature.org/wolves-wisconsin-and-aldo-leopold/.

Donlan, J., H. W. Greene, J. Berger, *et al.* (2005). Re-wilding North America. *Nature* **436**:913–14.

Donnelly, C. A., R. Woodroffe, D. R. Cox, *et al.* (2003). Impact of localized badger culling on tuberculosis incidence in British cattle. *Nature* **426**:834–37.

Dove, M. R. (1995). Political versus techno-economic factors in the development of non-timber forest products: Lessons from a comparison of natural and cultivated rubbers in Southeast Asia (and South America). *Society and Natural Resources* **8**:193–208. https://doi.org/10.1080/08941929509380914.

Dowie, M. (2005). Conservation refugees: When protecting nature means kicking people out. *Orion* **24**:16–27, https://orionmagazine.org/article/conservation-refugees.

Dowie, M. (2009). *Conservation Refugees: The Hundred-Year Conflict between Global Conservation and Native Peoples,* Cambridge, MA: Massachusetts Institute of Technology.

Drummond, W. H. (1875). *Rough Notes on the Large Game and Natural History of South and South-east Africa: From the Journals of the Hon. W. H. Drummond.* Edinburgh: Edmonston and Douglas.

Druzin, H. (2020). Latest shot fired in lead ammo debate. Boise State Public Radio. August 18, 2020. www.boisestatepublicradio.org.

Dublin, H. T., A. R. E. Sinclair, and J. McGlade. (1990). Elephants and fire as causes of multiple stable states in the Serengeti-Mara woodlands. *Journal of Animal Ecology* **59**:1147–64.

Dudley, N. and J. Parish. (2006). *Closing the Gap: Creating Ecologically Representative Protected Area Systems.* CBD Technical Series 24. Montreal, Canada: Secretariat of the Convention on Biological Diversity.

Durham, W. H. (1979). *Scarcity and Survival in Central America: Ecological Origins of the Soccer War.* Stanford, CA: Stanford University Press.

Echo-Hawk, W. R. (2013). *In the Light of Justice: The Rise of Human Rights in Native America and the UN Declaration on the Rights of Indigenous Peoples,* Golden, CO: Fulcrum Publishing.

Ehrlich, P. (1968). *The Population Bomb.* New York, NY: Ballantine Books.

Ehrlich, P. (1974). *The End of Affluence.* New York, NY: Ballantine Books.

Ehrlich, P. and A. Ehrlich. (1985). *Extinction: The Causes and Consequences of the Disappearance of Species,* 2nd Ballantine printing. New York, NY: Ballantine Books.

Ehrlich, P. R. and D. D. Murphy. (1987). Conservation lessons from long-term studies of checkerspot butterflies. *Conservation Biology* **1**:122–31.

Ehrlich, P. R., A. H. Ehrlich, and J. P. Holden. (1977). *Ecoscience: Population, Resources, Environment.* San Francisco, CA: W. H. Freeman.

Ellis, D. H., J. C. Lewis, G. F. Gee, and D. G. Smith. (1992). Population recovery of the whooping crane with emphasis on reintroduction efforts: Past and future. *Proceedings North American Crane Workshop* **6**:142–150.

Elton, C. S. (1958). *The Ecology of Invasions by Animals and Plants.* London: Methuen.

Eng, K. F. (2016). A newly drawn tree of life reminds us to question what we know about the history of Earth. Interview with Hélène Morlon. https://fellowsblog.ted.com/a-newly-drawn-tree-of-life-reminds-us-to-question-what-we-know-about-the-history-of-earth-8f8f1dee0cd5.

Enright, S. R., R. Meneses-Orellana, and I. Keith. (2021). The Eastern Tropical Pacific Marine Corridor (CMAR): The emergence of a voluntary regional cooperation mechanism for the conservation and sustainable use of marine biodiversity within a fragmented regional ocean governance landscape. *Frontiers in Marine Science* **8**:674825.

Errington, P. L. (1946). Predation and vertebrate populations. *Quarterly Review of Biology* **21**:144–77, 221–45.

Evans, I. M., R. H. Dennis, D. C. Orr-Ewing, *et al.* (1997). The re-establishment of red kite breeding populations in Scotland and England. *British Birds* **90**:123–38.

Evans, I. M., J. A. Love, C. A. Galbraith, and M. W. Pienkowski. (1994). Propagation and range restoration of threatened raptors in the United Kingdom. In B.-U. Meyberg and R. D. Chancellor, eds., *Raptor Conservation Today, Proceedings of the IV World Conference on Birds of Prey and Owls.* Berlin: Pica Press, pp. 447–57.

FAO. (2001). *Genetically Modified Organisms, Consumers, Food Safety and the Environment, FAO Ethics Series 2.* Rome: Food and Agriculture Organization of the United Nations.

Farinotti, D., L. Longuevergne, G. Moholdt, *et al.* (2015). Substantial glacier mass loss in the Tien Shan over the past 50 years. *Nature Geoscience* **8**:716–22.

Farnsworth, N. R. (1988). Screening plants for new medicines. In E. O. Wilson, ed., *Biodiversity.* Washington, DC: National Academy Press, pp. 83–97.

Fedrowitz, K., J. Koricheva, S. C. Baker, *et al.* (2014). Review: Can retention forestry help conserve biodiversity? A meta-analysis. *Journal of Applied Ecology* **51**:1669–79.

Fiedler, P. L., P. S. White, and R. L. Leidy. (1997). The paradigm shift in ecology and its implications for conservation. In S. T. A. Pickett, R. S. Ostfeld, M. Shachak, and G. E. Likens, eds., *The Ecological Basis of Conservation: Heterogeneity, Ecosystems, and Biodiversity.* New York, NY: Chapman and Hall, pp. 83–92.

Foreman, D. (2004). *Rewilding North America: A Vision for Conservation in the 21st Century.* Washington, DC: Island Press.

Fox, J., D. M. Truong, A. T. Rambo, *et al.* (2000). Shifting cultivation: A new old paradigm for managing tropical forests. *BioScience* **50**:521–28.

Frankham, R., J. D. Ballou, K. Ralls, *et al.* (2017). Managing gene flow among isolated population fragments. II. Management based on kinship. In R. Frankham, J. D. Ballou, K. Ralls, *et al.*, eds., *Genetic Management of Fragmented Animal and Plant Populations.* Oxford, UK: Oxford University Press, pp. 266–90.

Frankham, R. and K. Ralls. (1998). Inbreeding leads to extinction. *Nature* **392**(6675): 441–42.

Franklin, J. F. (1990). Biological legacies: A critical management concept from Mount St. Helens. *Transactions of the 55th North American Wildlife and Natural Resources Conference*: 216–19.

Franklin, J. F., K. Cromack, Jr., W. Denison, *et al.* (1981). Ecological characteristics of old-growth Douglas-fir forests. General Technical Report, PNW-GTR-118. Portland, OR: US Department of Agriculture Forest Service, Pacific Northwest Forest and Range Experiment Station.

Frazer, N. (1992). Sea turtle conservation and halfway technology. *Conservation Biology* **6**:179–84.

Freuling, C. M., K. Hampson, T. Selhorst, *et al.* (2013). The elimination of fox rabies from Europe: Determinants of success and lessons for the future. *Philosophical Transactions of the Royal Society B: Biological Sciences* **368**:20120142.

Gadgil, M., F. Berkes, and C. Folke. (1993). Indigenous knowledge for biodiversity conservation. *Ambio* **22**:151–56.

Gadgil, M. and R. Guha. (1995). *Ecology and Equity: The Use and Abuse of Nature in Contemporary India.* New Delhi: Penguin Books India.

Gadgil, M. and V. D. Vartak. (1975). Sacred groves of India: A plea for continued conservation. *Journal of the Bombay Natural History Society* **72**:313–26.

Galbraith, J. K. (1958a). *The Affluent Society.* Boston, MA: Houghton Mifflin.

Galbraith, J. K. (1958b). How much should a country consume? In H. Jarrett, ed., *Perspectives on Conservation: Essays on America's Natural Resources.* Baltimore, MD: Johns Hopkins Press, pp. 89–99.

Galli, A., C. F. Leck, and R. T. T. Forman. (1976). Avian distribution patterns in forest islands of different sizes in central New Jersey. *Auk* **93**:356–64.

Geldmann, J., A. Manica, N. D. Burgess, *et al.* (2019). A global-level assessment of the effectiveness of protected areas at resisting anthropogenic pressures. *PNAS* **116**:23209–15.

Gentry, R. L. (1998). *Behavior and Ecology of the Northern Fur Seal.* Princeton, NJ: Princeton University Press.

Gershon, D. (1992). If biological diversity has a price, who sets it and who should benefit? *Nature* **359**:565.

Gese, E. M. and M. Bekoff. (2004). Coyote (*Canis latrans*). In C. Sillero-Zubiri, M. Hoffman, and D. W. Macdonald, eds., *Canids: Foxes, Wolves, Jackals and Dogs, Status Survey and Conservation Action Plan.* Gland, Switzerland: IUCN – The World Conservation Union. SSC Canid Specialist Group, pp. 81–86.

Gesner, J., M. Chebanov, and J. Freyhof. (2010). *Huso huso.* IUCN Red List of Threatened Species 2010. http://dx.doi.org/10.2305/IUCN.UK.2010-1.RLTS.T10269A3187455.en.

Ghoddousi, S., P. Pintassilgo, J. Mendes, *et al.* (2018). Tourism and nature conservation: A case study in Golestan National Park, Iran. *Tourism Management Perspectives* **26**:20–7.

Gibson, C. C. (2001). Forest resources: Institutions for local governance in Guatemala. In J. Burger, E. Ostrom, R. B. Norgaard, *et al.*, eds., *Protecting the Commons: A Framework for Resource Management in the Americas.* Washington, DC: Island Press, pp. 71–89.

Gillespie, G. (2007). The Empire's Eden: British hunters, travel writing, and imperialism in nineteenth-century Canada. In J. L. Manore and D. G. Miner, eds., *The Culture of Hunting in Canada*. Vancouver, Canada: University of British Columbia Press, pp. 42–55.

Gleason, H. A. (1917). The structure and development of the plant association. *Bulletin of the Torrey Botanical Club* **44**:463–81.

Gleason, H. A. (1926). The individualistic concept of the plant association. *Bulletin of the Torrey Botanical Club* **53**:7–26.

Glen, A. S. and J. Short. (2000). The control of dingoes in New South Wales in the period 1883–1930 and its likely impact on their distribution and abundance. *Australian Zoologist* **31**:432–42.

Global Alliance for the Rights of Nature. (No date). www.garn.org/universal-declaration/.

Glover, J. D., S. W. Culman, S. T. DuPont, *et al.* (2010). Harvested perennial grasslands provide ecological benchmarks for agricultural sustainability. *Agriculture, Ecosystems & Environment* **137**:3–12.

Goettsch, B., A. P. Durán, and K. J. Gaston. (2019). Global gap analysis of cactus species and priority sites for their conservation. *Conservation Biology* **33**:369–76.

Gould, S. J. (1977). Why we should not name human races – a biological view. In S. J. Gould, ed., *Ever Since Darwin: Reflections in Natural History*. New York, NY: Norton, pp. 231–36.

Gould, S. J. (1996) *The Mismeasure of Man. (Revised and Expanded)*. New York, NY: Norton.

Gould, S. J. (2011). *I Have Landed: The End of a Beginning in Natural History*. Cambridge, MA: Belknap Press.

Grayson, K. L., N. J. Mitchell, J. M. Monks, *et al.* (2014). Sex ratio bias and extinction risk in an isolated population of tuatara (*Sphenodon punctatus*). *PLoS One* **9**:e94214.

Greeley, W. B. (1999). "Piute Forestry" or the fallacy of light burning. *Forest History Today* 1999(Spring):33–37. Reprinted from *The Timberman*, March, 1920.

Gregory, T. R. (2009). Understanding natural selection: Essential concepts and common misconceptions. *Evolution* **2**:156–75.

Griffiths, C. J., C. G. Jones, D. M. Hansen, *et al.* (2010). The use of extant non-indigenous tortoises as a restoration tool to replace extinct ecosystem engineers. *Restoration Ecology* **18**:1–7.

Groshong, K. (2007). The noisy reception of Silent Spring. In H. Chang and C. Jackson, eds., *An Element of Controversy: The Life of Chlorine in Science, Medicine, Technology and War*. British Society for the History of Science, BSHS Monographs, vol. 13. Cambridge, UK: Cambridge University Press, pp. 360–82.

Gross, J. E. (1969). Optimum yield in deer and elk populations. *Transactions of the North American Wildlife Conference* **34**:372–86.

Grove, R. H. (1990). Colonial conservation, ecological hegemony, and popular resistance: Towards a global synthesis. In J. M. MacKenzie, ed., *Imperialism and the Natural World*. Manchester, UK: Manchester University Press, pp. 15–50.

Grove, R. H. (1992). Origins of western environmentalism. *Scientific American* **267**(July):42–47.

Guha, R. (1989a). Radical American environmentalism: A third world critique. *Environmental Ethics* **11**:71–83.

Guha, R. (1989b). *The Unquiet Woods: Ecological Change and Peasant Resistance in the Himalaya*. Oxford, India: Oxford University Press.

Guimarães, P. R., M. Galetti, and P. Jordano. (2008). Seed dispersal anachronisms: Rethinking the fruits extinct megafauna ate. *PloS One* **3**:e1745.

Gustafsson, L., M. Hannerz, M. Koivula, *et al.* (2020). Research on retention forestry in northern Europe. *Ecological Processes* **9**:3.

Haag-Wackernagel, D. (1995). Regulation of the street pigeon in Basel. *Wildlife Society Bulletin* **23**:256–60.

Haltiner, J., J. Zedler, K. Boyer, *et al.* (1997). Influence of physical processes on the design, functioning and evolution of restored tidal wetlands in California (USA). *Wetlands Ecology and Management* **4**:73–91.

Hardin, G. (1968). The tragedy of the commons. *Science* **162A**:1243–48.

Hardin, G. (1974). Living on a lifeboat. *BioScience* **24**:561–68.

Harris, L. D. (1984). *The Fragmented Forest: Island Biogeography Theory and the Design of Nature Reserves*. Chicago, IL: University of Chicago Press.

Harwell, M. A. (1997). Ecosystem management of South Florida. *BioScience* **47**:499–512.

Haugan, P. M. and H. Drange. (1996). Effects of CO_2 on the ocean environment. *Energy Conversion and Management* **37**:1019–22.

Hays, S. P. (1959). *Conservation and the Gospel of Efficiency: The Progressive Conservation Movement, 1890–1920*. Cambridge, MA: Harvard University Press.

Helms, D. (1990). Conserving the plains: The Soil Conservation Service in the Great Plains. *Agricultural History* **64**:58–73.

Hess, G. R. (1994). Conservation corridors and contagious disease: A cautionary note. *Conservation Biology* **8**:256–62.

Hickling, R., D. B. Roy, J. K. Hill, *et al.* (2006). The distributions of a wide range of taxonomic groups are expanding polewards. *Global Change Biology* **12**:450–55.

Higgs, E. S. (1997). What is good ecological restoration? *Conservation Biology* **11**:338–48.

Hilderbrand, G. V., T. A. Hanley, C. T. Robbins, and C. C. Schwartz. (1999). Role of brown bears (*Ursus arctos*) in the flow of marine nitrogen into a terrestrial ecosystem. *Oecologia* **121**:546–50.

Hinton, J. W., M. J. Chamberlain, and D. R. Rabon. (2013). Red wolf (*Canis rufus*) recovery: A review with suggestions for future research. *Animals* **3**:722–44.

Hobbs, R. J., L. M. Hallett., P. R Ehrlich, and H. A. Mooney. (2011). Intervention ecology: Applying ecological science in the twenty-first century. *Bioscience* **61**:442–50.

Hoddle, M. S. (2004). Restoring balance: Using exotic species to control invasive exotic species. *Conservation Biology* **18**:38–49.

Hodgkins, G. A., I. C. James, and T. G. Huntington. (2002). Historical changes in lake ice-out dates as indicators of climate change in New England, 1850–2000. *International Journal of Climatology: A Journal of the Royal Meteorological Society* **22**:1819–27.

Homewood, K., E. F. Lambin, E. Coast, *et al.* (2001). Long-term changes in Serengeti-Mara wildebeest and land cover: Pastoralism, population, or policies? *Proceedings of the National Academy of Sciences* **98**:12544–49.

Homewood, K. and W. A. Rodgers. (1987). Pastoralism, conservation and the overgrazing controversy. In D. Anderson and R. H. Grove, eds., *Conservation in Africa: People, Policies and Practice*. Cambridge, UK: Cambridge, pp. 111–28.

Homewood, K. M., P. C. Trench, and D. Brockington. (2012). Pastoralist livelihoods and wildlife revenues in East Africa: A case for coexistence? *Pastoralism: Research, Policy and Practice* **2**:1–23.

Hornaday, W. T. (1887). *The Extermination of the American Bison: A Sketch of its Discovery and Life History*. Washington, DC: U.S. Government Printing Office.

Hornaday, W. T. (1913). *Our Vanishing Wildlife: Its Extermination and Preservation*. New York, NY: Scribner's.

Horstman, M. and G. Wightman. (2001). *Karparti* ecology: Recognition of Aboriginal ecological knowledge and its application to management in north-western Australia. *Ecological Management & Restoration* **2**:99–109.

Hug, L. A., B. J. Baker, K. Anantharaman, *et al.* (2016). A new view of the tree of life. *Nature Microbiology* **1**:1–6.

Hunn, E. S. with James Selam. (1990). *Nch'i-wana, "The Big River": Mid-Columbia Indians and Their Land.* Seattle, WA: University of Washington Press.

Hunt, E. G. and A. I. Bischoff. (1960). Initial effects on wildlife of periodic DDD applications to Clear Lake. *California Fish and Game* **46**:91–106.

Hunter, M., Jr. (1996a). Benchmarks for defining ecosystems: Are human activities natural? *Conservation Biology* **10**:695–97.

Hunter, M. L., Jr. (1996b). *Fundamentals of Conservation Biology.* Cambridge, MA: Blackwell Science.

Hunter, M. L. Jr., G. L. Jacobson, Jr., and T. Webb, III. (1988). Paleoecology and the coarse-filter approach to maintaining biological diversity. *Conservation Biology* **2**:375–85.

Hurt, R. D. (1987). *Indian Agriculture in America: Prehistory to the Present.* Lawrence, KS: University Press of Kansas.

Illius, A. W. and T. G. O'Connor. (1999). On the relevance of nonequilibrium concepts to arid and semiarid grazing systems. *Ecological Applications* **9**:798–813.

Ingrouille, M. (1995). *Historical Ecology of the British Flora.* London: Chapman and Hall.

International Whaling Commission (1982). Aboriginal/subsistence whaling (with special reference to the Alaska and Greenland fisheries). Reports of the International Whaling Commission, Special Issue 4.

International Whaling Commission. (No date). Description of the USA Aboriginal Subsistence Hunt: Alaska. https://iwc.int/management-and-conservation/whaling/aboriginal/usa/alaska.

Ise, J. (1961). *Our National Park Policy: A Critical History.* Baltimore, MD: Johns Hopkins University Press.

Jacoby, K. (2014). *Crimes Against Nature: Squatters, Poachers, Thieves, and the Hidden History of American Conservation.* Berkeley, CA: University of California Press.

Janzen, D. H. (1981). *Enterolobium cyclocarpum* seed passage rate and survival in horses, Costa Rican Pleistocene seed dispersal agents. *Ecology* **62**:593–601.

Janzen, D. H. (1983). No park is an island: Increase in interference from outside as park size decreases. *Oikos* **41**:403–10.

Janzen, D. H. and P. S. Martin. (1982). Neotropical anachronisms: The fruits the gomphotheres ate. *Science* **215**:19–27.

Jensen, M. N. (2000). Common sense and common pool resources. *BioScience* **50**:638–44.

Johnson, N. K. (1975). Controls on number of bird species on montane islands in the Great basin. *Evolution* **29**:545–74.

Johnston, R. (1997). *Introduction to Microbiotic Crusts.* Washington, DC: United States Department of Agriculture, Natural Resources Conservation Service, Soil Quality Institute, Grazing Lands Technology Institute.

Johnstone, J. F., C. D. Allen, J. F. Franklin, *et al.* (2016). Changing disturbance regimes, ecological memory, and forest resilience. *Frontiers in Ecology and the Environment* **14**:369–78.

Joppa, L. N. and A. Pfaff. (2009). High and far: Biases in the location of protected areas. *PLoS One* **4**:e8273.

Jordan, W. R. III. (1988). Ecological restoration: Reflections on a half-century of experience at the University of Wisconsin-Madison Arboretum. In E. O. Wilson, ed., *Biodiversity.* Washington, DC: National Academy Press, pp. 311–16.

Kareiva, P., R. Lalasz, and M. Marvier. (2012). Conservation in the Anthropocene: Beyond solitude and fragility. *Breakthrough Journal* **2**(Fall):29–37.

Kareiva, P. and M. Marvier. (2012). What is conservation science? *BioScience* **62**:962–69.

Karr, J. R. (1982). Avian extinction on Barro Colorado island, Panama: A reassessment. *American Naturalist* **119**:220–39.

Kellaway, K. (2010). How the Observer brought the WWF into being. *The Guardian*, November 6, 2010. www.theguardian.com/environment/2010/nov/07/wwf-world-wildlife-fund-huxley.

Kelly, S. T. and M. E. DeCapita. (1982). Cowbird control and its effect on Kirtland's warbler reproductive success. *Wilson Bulletin* **94**:363–65.

King, F. H. (1907). *A Textbook of the Physics of Agriculture*. Madison, WI: Published by the author.

King, W. (1685). Of the bogs, and loughs of Ireland by Mr. William King, Fellow of the Dublin Society, as it was presented to that Society. *Philosophical Transactions of the Royal Society of London* **15**:948–60.

Kirkpatrick, J. F. and J. W. Turner. (1991). Compensatory reproduction in feral horses. *Journal of Wildlife Management* **55**:649–52.

Klein, D. R. (1968). The introduction, increase, and crash of reindeer on St. Matthew Island. *Journal of Wildlife Management* **32**:350–67.

Knowlton, F. F. (1972). Preliminary interpretations of coyote population mechanics with some management implications. *Journal of Wildlife Management* **36**:369–82.

Korstian, C. F. and W. Maughan. (1935). *The Duke forest, a demonstration and research laboratory*. Forestry Bulletin 1. Durham, NC: Duke University.

Kuhn, T. S. (1970). *The Structure of Scientific Revolutions*, 2nd ed. Chicago, IL: University of Chicago Press.

La Sorte, F. A. and W. Jetz. (2010). Avian distributions under climate change: Towards improved projections. *Journal of Experimental Biology* **213**:862–69.

Landres, P. B., J. Verner, and J. W. Thomas. (1988). Ecological use of vertebrate indicator species: A critique. *Conservation Biology* **2**:316–28.

Lasgorceix, A. and A. Kothari. (2009). Displacement and relocation of protected areas: A synthesis and analysis of case studies. *Economic and Political Weekly* **44**:37–47.

Laurance, W. F. (2007). Ecosystem decay of Amazonian forest fragments: Implications for conservation. In T. Tscharntke, C. Leuschner, M. Zeller, *et al.*, eds., *Stability of Tropical Rainforest Margins: Linking Ecological, Economic, and Social Constraints of Land Use and Conservation*. Berlin: Springer, pp. 9–35.

Lay, D. W. (1938). How valuable are woodland clearings to birdlife? *Wilson Bulletin* **50**:254–56.

Laycock, W. A. (1991). Stable states and thresholds of range condition on North American rangelands: A viewpoint. *Journal of Range Management* **44**:427–33.

Lear, L. J. (1997). *Rachel Carson: Witness for Nature*. New York, NY: Henry Holt.

Leith, H. (2012). Primary production of the major vegetation units of the world. In H. Lieth and R. H. Whittaker, eds., *Primary Productivity of the Biosphere*. New York, NY: Springer-Verlag, pp. 203–15.

Leopold, A. (1920). "Piute Forestry" vs. forest fire prevention. *Southwestern Magazine* **2**:12–13.

Leopold, A. (1933). *Game Management*. New York, NY: Scribner's.

Leopold, A. (1936a) Deer and Dauerwald in Germany. I. History. *Journal of Forestry* **34**:366–75.

Leopold, A. (1936b). Deer and Dauerwald in Germany. II. Ecology and Policy. *Journal of Forestry* **34**:460–66.

Leopold, A. (1943). Deer irruptions. *Wisconsin Conservation Bulletin* **8**(8):3–11. https://babel.hathitrust.org/cgi/pt?id=umn.31951t002451202&view=1up&seq=280.

Leopold, A. (1966). *A Sand County Almanac*, 7th edn, New York, NY: Ballantine Books.

Leopold, A. S. (1955). Too many deer. *Scientific American* **193**(110):101–8.

Likens, G. E., R. F. Wright, J. N. Galloway, and T. J. Butler. (1979). Acid rain. *Scientific American* **241**(4):43–51.

Likens, G. E., and F. H. Bormann. (1974). Acid rain: A serious regional environmental problem. *Science* **184**:1176–79.

Limerick, P. N. (2010). Foreword to W. R. Echo-Hawk, *In the Courts of the Conqueror: The Ten Worst Indian Law Cases Ever*. Golden, CO: Fulcrum.

Ljung, P. E., S. J. Riley, T. A. Heberlein, and G. Ericsson. (2012). Eat prey and love: Game-meat consumption and attitudes toward hunting. *Wildlife Society Bulletin* **36**:669–75.

Locher, F. (2013). Cold War pastures: Garrett Hardin and the "Tragedy of the Commons." *Revue d'Histoire Moderne et Contemporaine* **60–1**(1): 7–36.

Loh, J. and D. Harmon. (2005). A global index of biocultural diversity. *Ecological Indicators* **5**:231–41.

Loladze, I. (2014). Hidden shift of the ionome of plants exposed to elevated CO_2 depletes minerals at the base of human nutrition. *ELife* **3**:e02245.

Lothian, W. F. (1987). A brief history of Canada's national parks. QS-9056-000-EE-A1 Environment Canada, Parks. www.parkscanadahistory.com/.

Louda, S. M. and P. Stiling. (2004). The double-edged sword of biological control in conservation and restoration. *Conservation Biology* **18**:50–53.

Louda, S. M., D. Kendall, J. Connor, and D. Simberloff. (1997). Ecological effects of an insect introduced for the biological control of weeds. *Science* **277**:1088–90.

Lovejoy, T. E. and C. Nobre. (2018). Amazon tipping point. *Science Advances* **4**:eaat2340.

Lowry, L. (2015). *Neomonachus tropicalis*. The Caribbean monk seal, IUCN Red List of Threatened Species 2015. https://dx.doi.org/10.2305/IUCN.UK.2015-2.RLTS .T13655A45228171.en.

MacArthur, R. H. (1972). *Geographical Ecology: Patterns in the Distribution of Species*. New York, NY: Harper and Row.

MacArthur, R. H. and E. O. Wilson. (1967). *The Theory of Island Biogeography*. Princeton, NJ: Princeton University Press.

Mack, R. N. (1981). Invasion of *Bromus tectorum* L. into western North America: An ecological chronicle. *AgroEcosystems* **7**:145–65.

Mack, R. N. and J. N. Thompson. (1982). Evolution in steppe with few large hooved mammals. *American Naturalist* **119**:757–73.

MacKenzie, J. M. (1987). Chivalry, social Darwinism, and ritualised killing: The hunting ethos in Central Africa up to 1914. In D. Anderson and R. Grove, eds., *Conservation in Africa: People, Policies and Practice*. Cambridge, UK: Cambridge University Press, pp. 41–81.

MacMillan, D. C. and J. Han. (2011). Cetacean by-catch in the Korean Peninsula – by chance or by design? *Human Ecology* **39**:757–68.

Maffi, L. and E. Woodley. (2012) *Biocultural Diversity Conservation: A Global Sourcebook*. London: Earthscan.

Main, A. R. (1987). Evolution and radiation of the terrestrial fauna. In G. R. Dyne and D. W. Walton, eds., *Fauna of Australia: Volume 1A, General Articles*. Canberra: Bureau of Flora and Fauna, Australian Government Printing Service, pp. 136–55.

Malthus, T. R. (1798). *An Essay on the Principle of Population*, 7th edn, Fairfield, NJ: Augustus M. Kelley.

Manore, J. L. (2007) Contested terrains of space and place: Hunting and the landscape known as Algonquin Park, 1890–1950. In J. L. Manore and D. G. Miner, eds., *The Culture of Hunting in Canada*. Vancouver, Canada: University of British Columbia Press, pp. 121–47.

Margules, C. R. and R. L. Pressey. (2000). Systematic conservation planning. *Nature* **405**:243–53.

Marris, E. (2011). *Rambunctious Garden: Saving Nature in a Post-Wild World*. New York, NY: Bloomsbury.

Marsh, G. P. (1874). *The Earth as Modified by Human Action*. New York, NY: Scribner, Armstrong, and Company.

Marsh, G. P. (1965). *Man and Nature: Or, Physical Geography as Modified by Human Action*. D. Lowenthal, ed., Cambridge, MA: Harvard University Press.

Martin, P. S. (1973). The discovery of America. *Science* **179**(4077): 969–74.

Martin, P. S. and H. E. Wright. (1967). *Pleistocene Extinctions: The Search for a Cause*. New Haven, CT: Yale University Press.

Mascia, M. B. and C. A. Claus. (2009). A property rights approach to understanding human displacement from protected areas: The case of marine protected areas. *Conservation Biology* **23**:16–23.

Maycock, P. F. (1967). Josef Paczoski: Founder of the science of phytosociology. *Ecology* **48**:1031–34.

McCay, B. J. and J. M. Acheson, eds. (1987). *The Question of the Commons: The Culture and Ecology of Communal Resources*. Tucson, AZ: University of Arizona Press.

McIntosh, R.P. (1983). Excerpts from the work of L. G. Ramensky. *Bulletin of the Ecological Society of America* **64**:7–12.

McShane, T. O., P. D. Hirsch, T. C. Trung, *et al.* (2011). Hard choices: Making trade-offs between biodiversity conservation and human well-being. *Biological Conservation* **144**:966–72.

Meadows, D. H., D. L. Meadows, J. Randers, *et al.* (1974). *The Limits to Growth*, 2nd edn, New York, NY: Signet.

Mech, D. L. (2012). Is science in danger of sanctifying the wolf? *Biological Conservation* **150**:143–49.

Merunková, K., Z. Preislerová, and M. Chytrý. (2012). White Carpathian grasslands: Can local ecological factors explain their extraordinary species richness? *Preslia* **84**:311–25.

Michalcová, D., M. Chytrý, V. Pechanec, *et al.* (2014). High plant diversity of grasslands in a landscape context: A comparison of contrasting regions in Central Europe. *Folia Geobotanica* **49**:117–35.

Miller, B., M. E. Soulé, and J. Terborgh. (2014). "New conservation" or surrender to development?: *Animal Conservation* **17**:509–15.

Milner-Gulland, E. J. and E. L. Bennett. (2003). Wild meat: The bigger picture. *Trends in Ecology and Evolution* **18**:351–57.

Mitchell, P. W. (2018). The fault in his seeds: Lost notes to the case of bias in Samuel George Morton's cranial race science. *PLOS Biology* **16**:e2007008.

Mitsch, W. J. and J. G. Gosselink. (2015) *Wetlands*. 5th edition. Hoboken, NJ: Wiley.

Molenaar, F. M., J. E. Jaffe, I. Carter, *et al.* (2017). Poisoning of reintroduced red kites (*Milvus milvus*) in England. *European Journal of Wildlife Research* **63**:94.

Mori, A. S. and R. Kitagawa. (2014). Retention forestry as a major paradigm for safeguarding forest biodiversity in productive landscapes: A global meta-analysis. *Biological Conservation* **175**:65–73.

Mosimann, J. E. and P. S. Martin. (1975). Simulating overkill by Paleoindians: Did man hunt the giant mammals of the New World to extinction? *American Scientist* **63**:304–13.

Muir, J. (1911). *My First Summer in the Sierra*. Boston, MA: Houghton Mifflin.

Muir, J. (1912). *The Yosemite*. New York, NY: Century Company.

Murkin, H. R., R. M. Kaminski, and R. D. Titman. (1982). Responses by dabbling ducks and aquatic invertebrates to an experimentally manipulated cattail marsh. *Canadian Journal of Zoology* **60**:2324–32.

Murphree, M. W. (2005). Congruent objectives, competing interests, and strategic compromise: Concept and process in the evolution of Zimbabwe's CAMPFIRE, 1984–1996. In J. P. Brosius, A. L. Tsing, C., Zenner, *et al.*, eds., *Communities and Conservation: Histories and Politics of Community-Based Natural Resource Management*. Lanham, MD: AltaMira Press, pp. 105–47.

Murphy, D. D. and S. B. Weiss. (1992). Effects of climate change on biological diversity in North America. In R. L. Peters and T. E. Lovejoy, eds., *Global Warming and Biological Diversity*. New Haven, CT: Yale University Press, pp. 355–68.

Myers, N. (1979). *The Sinking Ark*. Oxford, UK: Pergamon Press.

Myers, N. and R. Tucker. (1987). Deforestation in Central America: Spanish legacy and North American consumers. *Environmental Review* **11**:55–71.

Naess, A. (1973). The shallow and the deep, long-range ecology movement: A summary. *Inquiry* **16**:95–100.

Naff, J. M., Jr. (1972). Reflections on the dissent of Douglas J., in Sierra Club v. Morton. *American Bar Association Journal*. www.abajournal.com/news/article/earth_day_reflections_from_1972_aba_journal_archives.

Naiman, R. J., J. J. Latterell, N. E. Pettit, and J. D. Olden. (2008). Flow variability and the biophysical vitality of river systems. *Comptes Rendus Geoscience* **340**:629–43.

National Oceanic and Atmospheric Administration (NOAA). (2011). Riftia tube worm colony Galapagos 2011. Okeanos Explorer Program, Galapagos Rift Expedition. https://commons.wikimedia.org/wiki/File:Riftia_tube_worm_colony_Galapagos_2011.jpg.

National Oceanic and Atmospheric Administration (NOAA). (2021). NOAA Fisheries. National Oceanic and Atmospheric Administration, US Department of Commerce. Northern Fur Seal, www.fisheries.noaa.gov/species/northern-fur-seal#management.

Nature Conservancy. (1982). *Natural Heritage Program Operations Manual*. Arlington, VA: The Nature Conservancy.

Newman, J. R. (1979). Hunting and hunter education in Czechoslovakia. *Wildlife Society Bulletin (1973–2006)* **7**:155–61.

Newsome, A., I. Parer, and P. Catling. (1989). Prolonged prey suppression by carnivores – predator-removal experiments. *Oecologia* **78**:458–67.

Nietschmann, B. (1994). Defending the Miskito reefs with maps and GPS: Mapping with sail, scuba, and satellite. *Cultural Survival Quarterly* **18**(4):34–37. www.culturalsurvival.org/publications/cultural-survival-quarterly/defending-miskito-reefs-maps-and-gps-mapping-sail-scuba.

Noss, R. F. (1987). Corridors in real landscapes: A reply to Simberloff and Cox. *Conservation Biology* **1**:159–64.

Noss, R. F. and A. Y. Cooperrider. (1994). *Saving Nature's Legacy*. Washington, DC: Island Press.

Noss, R., R. Nash, P. Paquet, *et al.* (2013). Humanity's domination of nature is part of the problem: A response to Kareiva and Marvier. *BioScience* **63**:241–42.

Novellino, D. (2010). From indigenous customary practices to policy interventions: The ecological and sociocultural underpinnings of the non-timber forest trade on Palawan Island, the Philippines. In S. A. Laird, R. J. McLain, and R. P. Wynberg, eds., *Wild Product Governance*. London: Earthscan, pp. 183–97.

Nowak, R. M. (2002). The original status of wolves in eastern North America. *Southeastern Naturalist* **1**:95–130.

Nowak, R. M. and J. L. Paradiso. (1983). *Walker's Mammals of the World*, vol. 2. Baltimore, MD: Johns Hopkins University Press.

O'Brien, S. J. and E. Mayr. (1991). Bureaucratic mischief: Recognizing endangered species and subspecies. *Science* **251**:1187–88.

Odén, S. (1976). The acidity problem – an outline of concepts. *Water, Air, and Soil Pollution* **6**:137–66.

Olson, D. M., E. Dinerstein, E. D. Wikramanayake, *et al.* (2001). Terrestrial ecoregions of the world: A new map of life on earth. *Bioscience* **51**: 933–38.

Olson, S. H. (1984). The robe of the ancestors: Forests in the history of Madagascar. *Journal of Forest History* **28**:174–86.

Oosting, H. J. (1942). An ecological analysis of the plant communities of Piedmont, North Carolina. *American Midland Naturalist* **28**:1–126.

Ordway, S. H., Jr. (1953). *Resources and the American Dream: Including a Theory of the Limit of Growth*. New York, NY: Ronald Press Company.

Ordway, S. H., Jr. (1956). Possible limits of raw material consumption. In W. L. Thomas, Jr., ed., *Man's Role in Changing the Face of the Earth*, vol. 2. Chicago, IL: University of Chicago Press, pp. 987–1009.

Orta-Martínez, M. and M. Finer. (2010). Oil frontiers and Indigenous resistance in the Peruvian Amazon. *Ecological Economics* **70**:207–18.

Ostfeld, R. S., S. T. A. Pickett, M. Shachak, and G. E. Likens. (1997). Defining the scientific issues. In S. T. A. Pickett, R. S. Ostfeld, M. Shachak, and G. E. Likens, eds., *The Ecological Basis of Conservation: Heterogeneity, Ecosystems, and Biodiversity*. New York, NY: Chapman and Hall, pp. 3–10.

Ostrom, E. (1990). *Governing the Commons: The Evolution of Institutions for Collective Action*. New York, NY: Cambridge University Press.

Palazón, S., A. Batet, I. Afonso, *et al.* (2012). Space use patterns and genetic contribution of a reintroduced male brown bear (*Ursus arctos*) in the Pyrenees between 1997 and 2011: The risk of genetic dominance of few males in reintroduced populations. *Galemys* **24**:93–6.

Palmer T. S. (1899). A review of economic ornithology in the United States. In: *United States Department of Agriculture, Yearbook, 1899*. Paper 1195. Lincoln, NE: Publications from USDA-ARS/UNL Faculty, pp. 259–92.

Parker, I. and M. Amin. (1983). *Ivory Crisis*. London: Chatto & Windus.

Parmesan, C. (1996). Climate and species' range. *Nature* **382** (6594):765–66.

Parmesan, C. and G. Yohe. (2003). A globally coherent fingerprint of climate change impacts across natural systems. *Nature* **421**(6918):37–42.

Parmesan, C., N. Ryrholm, C. Stephanescu, *et al.* (1999). Poleward shifts in geographical ranges of butterfly species associated with regional warming. *Nature* **399**(7636):579–83.

Parsons, M., C. A. McLoughlin, M. W. Rountree, and K. H. Rogers. (2006). The biotic and abiotic legacy of a large infrequent flood disturbance in the Sabie River, South Africa. *River Research and Applications* **22**:187–201.

Patterson, B. D. (1984). Mammalian extinction and biogeography in the southern Rocky Mountains. In M. H. Nitecki, ed., *Extinctions*. Chicago, IL: University of Chicago Press, pp. 247–93.

Paul, D. B. (2003). Darwin, social Darwinism, and eugenics. In J. Hodge and G. Radick, eds., *The Cambridge Companion to Darwin*. Cambridge, UK: Cambridge University Press, pp. 219–44.

Pearson, R. M., J. P. van de Merwe, C. J. Limpus, and R. M. Connolly. (2017). Realignment of sea turtle isotope studies needed to match conservation priorities. *Marine Ecology Progress Series* **583**:259–71.

Peluso, N. L. (1992). *Rich Forests, Poor People: Resource Control and Resistance in Java*. Berkeley, CA: University of California Press.

Peluso, N. L. and P. Vandergeest. (2001). Genealogies of the political forest and customary rights in Indonesia, Malaysia, and Thailand. *Journal of Asian Studies* **60**:761–812.

Peñuelas, J., I. Filella, and P. Comas. (2002). Changed plant and animal life cycles from 1952 to 2000 in the Mediterranean Region. *Global Change Biology* **8**:531–44.

Peres, C. A. (1994). Indigenous reserves and nature conservation in Amazonian forests. *Conservation Biology* **8**:586–88.

Peters, C. M., A. H. Gentry, and R. O. Mendelsohn. (1989). Valuation of an Amazonian rainforest. *Nature* **339** (6227):655–56.

Pfeiffer, J. and Y. Uril. (2003). The role of indigenous parataxonomists in botanical inventory: From Herbarium Amboinense to Herbarium Floresense. *Telopea* **10**:61–72.

Pfeiffer, J. M., S. Dun, K. J. Rice, and B. Mulawarman. (2006). Biocultural diversity in traditional rice-based agroecosystems: Indigenous research and conservation of *mavo* (*Oryza sativa* L.) upland rice landraces of eastern Indonesia. *Environment, Development, and Sustainability* **8**:609–25.

Philbrick, N. (2001). *In the Heart of the Sea: The Tragedy of the Whaleship Essex.* New York, NY: Penguin.

Phippen, J. W. (2016). Busting cactus smugglers in the American West; how undercover agents infiltrated the global black market for cacti. *The Atlantic*, February 22, 2016. www.theatlantic.com/science/archive/2016/02/cactus-thieves.

Picardi, A. C. and W. W. Seifert. (1977). A tragedy of the commons in the Sahel. *Ekistics* **43**:297–304.

Pickett, S. T. A. and R. S. Ostfeld. (1995). The shifting paradigm in ecology. In R. L. Knight and S. F. Bates, eds., *A New Century for Natural Resources Management.* Washington, DC: Island Press, pp. 261–79.

Pickett, S. T. A., V. T. Parker, and P. L. Fiedler. (1992). The new paradigm in ecology: Implications for conservation biology above the species level. In P. L. Fiedler and S. K. Jain, eds., *Conservation Biology: The Theory and Practice of Nature Conservation Preservation and Management.* New York, NY: Chapman & Hall, pp. 65–88.

Pickford, G. D. and E. H. Reid. (1943). Competition of elk and domestic livestock for summer range forage. *Journal of Wildlife Management* **7**:328–32.

Pimentel, D., ed. (1993). *World Soil Erosion and Conservation.* Cambridge, UK: Cambridge University Press.

Pinchot, G. (1947). *Breaking New Ground.* Seattle, WA: University of Washington Press.

Pisani, D. J. (1985). Forests and conservation, 1865–1890. *Journal of American History* **72**:340–59.

Pi-Sunyer, O. (1982). The cultural costs of tourism. *Cultural Survival Quarterly* **6**(3)7–10. www.culturalsurvival.org/publications/cultural-survival-quarterly/.

Poff, N. L., J. D. Allan, M. B. Bain, *et al.* (1997). The natural flow regime. *BioScience* **47**:769–84.

Poff, N. L., B. D. Richter, A. H. Arthington, *et al.*, 2010. The ecological limits of hydrologic alteration (ELOHA): A new framework for developing regional environmental flow standards. *Freshwater Biology* **55**:147–70.

Political Database of the Americas. (2011). Republic of Ecuador, Constitution of 2008, Article 71. English translation from Spanish. Washington, DC: Georgetown University, https://pdba.georgetown.edu/Constitutions/Ecuador/english08.html.

Pounds, J. A., M. R Bustamante, L. A. Coloma, *et al.* (2006). Widespread amphibian extinctions from epidemic disease driven by global warming. *Nature* **439**(7073):161–67.

Powledge, T. M. (2011). Behavioral epigenetics: How nurture shapes nature. *BioScience* **61**: 588–592.

Proctor, J. D. (1996). Whose nature? The contested moral terrain of ancient forests. In W. Cronon, ed., *Uncommon Ground: Rethinking the Human Place in Nature*, 2nd edn. New York, NY: Norton, pp. 269–97.

Prudham, W. S. (1998). Timber and town: Post-war federal forest policy, industrial organization, and rural change in Oregon's Illinois Valley. *Antipode* **30**:177–96.

Pulliam, H. R. (1988). Sources, sinks, and population regulation. *American Naturalist* **132**:652–61.

Pyne, S. J. (2017). *Fire in America: A Cultural History of Wildland and Rural Fire*. Seattle: WA: University of Washington Press.

Rasmussen, D. I. (1941). Biotic communities of Kaibab Plateau, Arizona. *Ecological Monograph* **11**:229–75.

Rattner, B. A., G. M. Haramis, D. S. Chu, *et al.* (1987). Growth and physiological condition of black ducks reared on acidified wetlands. *Canadian Journal of Zoology* **65**:2953–58.

Redford, K. H. (1991). The ecologically noble savage. *Cultural Survival Quarterly* **15**(1):46–48. www.culturalsurvival.org/publications/cultural-survival-quarterly/ecologically-noble-savage.

Redford, K. H. and A. M. Stearman. (1993). Forest-dwelling native Amazonians and the conservation of biodiversity: Interests in common or in collision? *Conservation Biology* **7**:248–55.

Reed, D. H. (2010). Albatrosses, eagles and newts, Oh My!: Exceptions to the prevailing paradigm concerning genetic diversity and population viability? *Animal Conservation* **13**:448–57.

Reid, W. V., S. A. Laird, R. Gomez, *et al.* (1993). A new lease on life. In W. V. Reid, S. A. Laird, and C. A. Meyer, eds., *Biodiversity Prospecting: Using Genetic Resources for Sustainable Development*. Washington, DC: World Resources Institute, pp. 1–52.

Reining, C. and R. Heinzman. (1992). Nontimber forest products in the Petén, Guatemala: Why extractive reserves are critical for both conservation and development. In M. Plotkin and L. Famolare, eds., *Sustainable Harvest and Marketing of Rain Forest Products*. Washington, DC: Island Press, pp. 110–17.

Richardson, B. J. (2009). The ties that bind: Indigenous peoples and environmental governance. In B. J. Richardson, S. Imai, and K. McNeil, eds., *Indigenous Peoples and the Law: Comparative and Critical Perspectives*. Portland, OR: Hart Publishing, pp. 337–70.

Robbins, C. S. (1973). Introduction, spread, and present abundance of the house sparrow in North America. *Ornithological Monographs*, No. **14**:3–9.

Rogers, K. H. (1999). Operationalizing ecology under a new paradigm: An African perspective. In S. T. A. Pickett, R. S. Ostfeld, M. Shachak, and G. E. Likens, eds., *The Ecological Basis of Conservation: Heterogeneity, Ecosystems, and Biodiversity*. New York, NY: Chapman & Hall, pp. 60–77.

Rogers, P. (1996). Disturbance ecology and forest management: A review of the literature. General Technical Report, INT-GTR-336. Ogden, UT: US Department of Agriculture Forest Service, Intermountain Research Station.

Rolland, J., F. L. Condamine, C. R. Beeravolu, *et al.* (2015). Dispersal is a major driver of the latitudinal diversity gradient of Carnivora. *Global Ecology and Biogeography* **24**:1059–71.

Romme, W. H. and D. G Despain. (1989). Historical perspective on the Yellowstone Fires of 1988. *BioScience* **39**:695–99.

Roosevelt, T. (1910). *African Game Trails: An Account of the African Wanderings of an American Hunter-naturalist*. London: John Murray.

Rosen, G. E. and K. F. Smith. (2010). Summarizing the evidence on the international trade in illegal wildlife. *EcoHealth* **7**:24–32.

Rotem, G., A. Bouskila, and A. Rothschild. (2014). Ecological effects of afforestation in the northern Negev. Society for the Protection of Nature in Israel, Tel Aviv, Israel. Translated from Hebrew by Zev Labinger. www.researchgate.net/publication/272161151.

Rothé, J. R. (1968). Fill a lake: Start an earthquake. *Ekistics* **26**:432–35.

Rozzi, R. (1999). The reciprocal links between evolutionary-ecological sciences and environmental ethics. *BioScience* **49**:11–21.

Rozzi, R. (2015). Implications of the biocultural ethic for earth stewardship. In R. Rozzi, F. S. Chapin, III, J. B. Callicott, *et al.*, eds., *Earth Stewardship: Linking Ecology and Ethics in Theory and Practice*. Cham, Switzerland: Springer, pp. 113–36.

Rubenstein, D. R., D. I. Rubenstein, P. W. Sherman, and T. A. Gavin. (2006). Pleistocene Park: Does re-wilding North America represent sound conservation for the 21st century? *Biological Conservation* **132**:232–38.

Saccheri, I., M. Kuussaari, Ma Kankare, *et al.* (1998). Inbreeding and extinction in a butterfly metapopulation. *Nature* **392**(6675):491–94.

Saini, A. (2019). *Superior: The Return of Race Science*. Boston, MA: Beacon Press.

Salafsky, N., B. L. Dugelby, and J. W. Terborgh. (1993). Can extractive reserves save the rain forest? An ecological and socioeconomic comparison of nontimber forest product extraction systems in Petén, Guatemala, and West Kalimantan, Indonesia. *Conservation Biology* **7**:39–52.

Sampson, A. W. (1926). Grazing periods and forage production on the national forests. Department Bulletin No. 1405, US Department of Agriculture.

Santos, F. L., J. Nogueira, R. A. F. de Souza, *et al.* (2021). Prescribed burning reduces large, high-intensity wildfires and emissions in the Brazilian savanna. *Fire* **4**, 56. https://doi.org/10.3390/fire4030056.

Santos-Díaz, M. S., E. Pérez-Molphe, R. Ramírez-Malagón, *et al.* (2011). Mexican threatened cacti: Current status and strategies for their conservation. In G. E. Tepper, ed., *Species Diversity and Extinction*. New York, NY: Nova Science, pp. 1–60.

Sanz, J. J. (2002). Climate change and breeding parameters of great and blue tits throughout the Western Palaearctic. *Global Change Biology* **8**:409–22.

Scatena, F. N. and M. C. Larsen. (1991). Physical aspects of Hurricane Hugo in Puerto Rico. *Biotropica* **23**:317–23.

Scatena, F. N., S. Moya, C. Estrada, and J. D. Chinea. (1996). The first five years in the reorganization of aboveground biomass and nutrient use following Hurricane Hugo in the Bisley experimental watersheds, Luquillo Experimental Forest, Puerto Rico. *Biotropica* **28**:424–40.

Schumm, S. A. and R. W. Lichty. (1965). Time, space, and causality in geomorphology. *American Journal of Science* **263**:110–99.

Schwartz, M. D. and B. E. Reiter. (2000). Changes in North American spring. *International Journal of Climatology* **20**:929–32.

Scott, J. M., F. Davis, B. Csuti, *et al.* (1993). Gap analysis: A geographic approach to protection of biological diversity. *Wildlife Monographs* **123**:1–41.

Scott, J. M., S. Mountainspring, F. L. Ramsey, and C. B. Kepler. (1986). Forest bird communities of the Hawaiian Islands: Their dynamics, ecology, and conservation. Studies in Avian Biology No. 9. Lawrence, KS: Allen Press.

Shachak, M. and S. T. A. Pickett. (1997). Linking ecological understanding and application: Patchiness in a dryland system. In S. T. A. Pickett, R. S. Ostfeld, M. Shachak, and G. E. Likens, eds., *The Ecological Basis of Conservation: Heterogeneity, Ecosystems, and Biodiversity*. New York, NY: Chapman & Hall, pp. 108–19.

Shachak, M., M. Sachs, and I. Moshe. (1998). Ecosystem management of desertified shrublands in Israel. *Ecosystems* **1**:475–83.

Sheldrick, D. (1973). *The Tsavo Story*. London: Collins and Harvill Press.

Shiva, V. (2016). *Biopiracy: The Plunder of nature and knowledge*. Berkeley, CA: North Atlantic Books.

Shtilmark, F. (2003). *History of the Russian zapovedniks, 1895–1995*. Edinburgh, UK: Russian Nature Press.

Simard, S. W., W. J. Roach, J. Beauregard, *et al.* (2021). Partial retention of legacy trees protect mycorrhizal inoculum potential, biodiversity, and soil resources while promoting natural regeneration of interior Douglas-fir. *Frontiers in Forests and Global Change* **3**:620436.

Simberloff, D. S. (1976). Species turnover and equilibrium island biogeography. *Science* **194**:572–78.

Simberloff, D. S. (1997). Biogeographic approaches and the new conservation biology. In S. T. A. Pickett, R. S. Ostfeld, M. Shachak, and G. E. Likens, eds., *The Ecological Basis of Conservation: Heterogeneity, Ecosystems, and Biodiversity.* New York, NY: Chapman & Hall, pp. 274–84.

Simberloff, D. (1998). Flagships, umbrellas, and keystones: Is single-species management passé in the landscape era? *Biological Conservation* **83**:247–57.

Simberloff, D. (2015). Non-native invasive species and novel ecosystems. *F1000Prime Reports* **7**:47.

Simberloff, D. S. and L. G. Abele. (1976a). Island biogeography theory and conservation practice. *Science* **191**:285–86.

Simberloff, D. S. and L. G. Abele. (1976b). Response: Island biogeography and conservation: Strategy and limitations. *Science* **193**:1032.

Simberloff, D. S. and J. Cox. (1987). Consequences and costs of conservation corridors. *Conservation Biology* **1**:63–71.

Sinclair, A. R. E. (1977). *The African Buffalo: A Study of Resource Limitation of Populations.* Chicago, IL: University of Chicago Press.

Sinclair, A. R. E. (1995). Equilibria in plant-herbivore interactions. In A. R. E. Sinclair and P. Arcese, eds., *Serengeti II: Dynamics, Management, and Conservation of an Ecosystem.* Chicago, IL: University of Chicago Press, pp. 91–113.

Sinclair, A. R. E. and J. M. Fryxell. (1985). The Sahel of Africa: Ecology of a disaster. *Canadian Journal of Zoology* **63**:987–94.

Sluyter, A. (1996). The ecological origins and consequences of cattle ranching in sixteenth-century New Spain. *Geographical Review* **86**:161–77.

Smith, R. and G. Smith. (1949). Supervised control of insects: Utilizes parasites and predators and makes chemical control more efficient. *California Agriculture* **3**:3,12.

Solís Rivera, V. and P. Madrigal Cordero. (2013). *¿Qué es lo que el InBio devuelve al Estado?* [What does InBio give back to the State?] *La Nación*, April 9, 2013. www.nacion.com/opinion/foros/que-es-lo-que-el-inbio-devuelve-al-estado/.

Solís Rivera, V., P. Madrigal Cordero, D. C. Rojas, and B. O'Riordan. (2017). Institutions and collective action in a Costa Rican small-scale fisheries cooperative: The case of CoopeTárcoles R.L. *Maritime Studies* **16**:1–19.

Solís Rivera, V., A. M. Rivera, and M. F. Borrás. (2015). Integrating traditional and scientific knowledge for the management of small scale fisheries: An example from Costa Rica. In J. Fischer, J. Jorgensen, H. Josupeit, *et al.*, eds., *Fishers' Knowledge and the Ecosystem Approach to Fisheries: Applications, Experiences and Lessons from Latin America.* Food and Agriculture Organization of the United Nations, FAO Fisheries and Agriculture Technical Paper No. 591. Rome: Food and Agriculture Organization of the United Nations, pp. 179–90.

Soma, T. (2013). Ethnoarchaeology of ancient falconry in East Asia. Asian Conference on Cultural Studies 2013, *Official Conference Proceedings*, 81–102. https://papers.iafor.org/proceedings_category/accs-official-conference-proceedings/.

Soma, T. and B. Sukhee. (2014). Altai Kazakh falconry as heritage tourism: The golden eagle festivals of Western Mongolia. *International Journal of Intangible Heritage* **9**:135–47.

Soulé, M. and R. Noss. (1998). Rewilding and biodiversity: Complementary goals for continental conservation. *Wild Earth* (Fall):18–28.

Soulé, M. E. and B. A. Wilcox, eds. (1980). *Conservation Biology: An Evolutionary Ecological Perspective*. Sunderland, MA: Sinauer Associates.

Spaeder, J. J. (2005). Co-management in a landscape of resistance: The political ecology of wildlife management in Western Alaska. *Anthropologica* **47**:165–78.

Spalding, M. D., H. E. Fox, G. R. Allen, *et al.* (2007). Marine ecoregions of the world: A bioregionalization of coastal and shelf areas. *Bioscience* **57**:573–83.

Spellerberg, I. F. (2005). *Monitoring Ecological Change*. Cambridge, NY: Cambridge University Press.

Spence, M. D. (1999). *Dispossessing the Wilderness: Indian Removal and the Making of the National Parks*. New York, NY: Oxford University Press.

Spizzirri, M. (2019). *Annual Report for 2019*. Revelstoke, Canada. Revelstoke Bear Aware Society. http://revelstokebearaware.org.

Sprugel, D. G. (1991). Disturbance, equilibrium, and environmental variability: What is "natural" vegetation in a changing environment? *Biological Conservation* **58**:1–18.

Stare, F. J. (1963). Some comments on "Silent Spring." *The Sanitarian's Journal of Environmental Health* **25**:242–44, 246.

Stavi, I. and E. Argaman. (2016). Soil quality and aggregation in runoff water harvesting forestry systems in the semi-arid Israeli Negev. *Catena* **146**:88–93.

Steneck, R. S., T. P. Hughes, J. E. Cinner, *et al.* (2011). Creation of a gilded trap by the high economic value of the Maine lobster fishery. *Conservation Biology* **25**:904–12.

Stern, V. M. R. F., R. Smith, R. van den Bosch, and K. S. Hagen. (1959). The integration of chemical and biological control of the spotted alfalfa aphid: The integrated control concept. *Hilgardia* **29**: 81–101.

Stewart, B. S., P. K. Yochem, H. R. Huber, *et al.* (1994). History and present status of the northern elephant seal population. In B. J. Le Boeuf and R. M. Laws, eds., *Elephant Seals*. Berkeley, CA: University of California Press, pp. 29–48.

Stone, C. D. (1972). Should trees have standing?–toward legal rights for natural objects. *Southern California Law Review* **45**:450–501.

Stuart, A. J. (2015). Late Quaternary megafaunal extinctions on the continents: A short review. *Geological Journal* **50**:338–63.

Stuessy, T. F., U. Swenson, D. J. Crawford, and G. Anderson. (1998). Plant conservation in the Juan Fernández archipelago, Chile. *Aliso: A Journal of Systematic and Evolutionary Botany* **16**:89–101.

Sustainable Human. (2014). How wolves change rivers. Narrated by George Monbiot, Produced by Chris Agnos and Dawn Agnos. www.youtube.com/.

Swaisgood, R., D. Wang, and F. Wei. 2016. *Ailuropoda melanoleuca* (errata version published in 2017). IUCN Red List of Threatened Species 2016. https://dx.doi.org/10.2305/IUCN.UK.2016-2.RLTS.T712A45033386.en.

Swetnam, T. W., C. D. Allen, and J. L. Betancourt (1999). Applied historical ecology: Using the past to manage for the future. *Ecological Applications* **9**:1189–206.

Taberlet, P. and J. Bouvet. (1994). Mitochondrial DNA polymorphism, phylogeography, and conservation genetics of the brown bear *Ursus arctos* in Europe. *Proceedings of the Royal Society of London B* **255**:195–200.

Taggart, D. A., D. Schultz, C. White, *et al.* (2005). Cross-fostering, growth, and reproductive studies in the brush-tailed rock-wallaby, *Petrogale penicillata* (Marsupialia:Macropodidae): Efforts to accelerate breeding in a threatened marsupial species. *Australian Journal of Zoology* **53**:313–23.

Tanasescu, M. (2017).When a river is a person: From Ecuador to New Zealand, nature gets its day in court. *The Conversation*, June 19, 2017. https://theconversation.com/when-a-river-is-a-person-from-ecuador-to-new-zealand-nature-gets-its-day-in-court-79278.

Tansley, A. G. (1920). The classification of vegetation and the concept of development. *Journal of Ecology* **8**:118–49.

Tansley, A. G. (1935). The use and abuse of vegetational concepts and terms. *Ecology* **16**:284–307.

Tausch, R. J. (1996). Past changes, present and future impacts, and the assessment of community or ecosystem condition. In J. R. Barrow, E. D. McArthur, R. E. Sosebee, and R. J. Tausch, compilers, *Proceedings: Shrubland Ecosystem Dynamics in a Changing Environment*. General Technical Report, INT-GTR-338. Ogden, UT: US Department of Agriculture Forest Service, Rocky Mountain Research Station, pp. 97–101.

Taylor, N. P. (1997). Cactaceae. In S. Oldfield, comp., *Cactus and succulent plants: Status survey and conservation action plan*. Gland, Switzerland: IUCN/SCC Cactus and Succulent Specialist Group. International Union for Conservation of Nature and Natural Resources, pp. 17–20.

Templeton, A. R. (1998). Human races: A genetic and evolutionary perspective. *American Anthropologist* **100**:632–50.

ten Kate, K. and S. A. Laird. (2000). Biodiversity and business: Coming to terms with the "grand bargain." *International Affairs* **76**:241–64.

Thompson, J. N. (1982). *Interaction and Coevolution*. New York, NY: Wiley.

Tomback, D. F. (1982). Dispersal of whitebark pine seeds by Clark's nutcracker: A mutualism hypothesis. *Journal of Animal Ecology* **51**:451–67.

Trefethen, J. B. (1975). *An American Crusade for Wildlife*. New York, NY: Winchester Press and the Boone and Crockett Club.

Treves, A., M. Krofel, and J. McManus. (2016). Predator control should not be a shot in the dark. *Frontiers in Ecology and the Environment* **1**:380–88.

Turner II, B. I. and K. W. Butzer. (1992). The Columbian encounter and land-use change. *Environment* **34**:16.

Turner, M. G., V. H. Dale, and E. I. Everham (1997). Fires, hurricanes, and volcanoes: Comparing large disturbances. *Bioscience* **47**:758–68.

Udall, S. L. (1963). *The Quiet Crisis*. New York, NY: Holt, Rinehart, and Winston.

Ullrey, D. E., W. G Youatt, H. E. Johnson, *et al.* (1967). Protein requirement of white-tailed deer fawns. *Journal of Wildlife Management* **31**:679–85.

UNESCO. (No date). Biosphere Reserves. https://en.unesco.org/biosphere/.

van Vliet, N., O. Mertz, A. Heinimann, *et al.* (2012).Trends, drivers and impacts of changes in swidden cultivation in tropical forest-agriculture frontiers: A global assessment. *Global Environmental Change* **22**:418–29.

Veblen, T. T. and D. C. Lorenz. (1988). Recent vegetation changes along the forest/steppe ecotone of northern Patagonia. *Annals of the Association of American Geographers* **78**:93–111.

Veldman, J. W., G. E. Overbeck, D. Negreiros, *et al.* (2015a). Tyranny of trees in grassy biomes. *Science* **347**(6221):484–85.

Veldman, J. W., G. E. Overbeck, D. Negreiros, *et al.* (2015b). Where tree planting and forest expansion are bad for biodiversity and ecosystem services. *BioScience* **65**:1011–18.

Velicogna, I. (2009). Increasing rates of ice mass loss from the Greenland and Antarctic ice sheets revealed by GRACE. *Geophysical Research Letters* **36**:19503.

Ventocilla, J. with Olaidi. (1995). Submarine "deforestation." In J. Ventocilla, H. Heraclio, and V. Núñez, eds., *Plants and Animals in the Life of the Kuna*. Translated by E. King. Austin: University of Texas Press, pp. 54–66.

Vitz, A. C. and A. D. Rodewald. (2006). Can regenerating clearcuts benefit mature-forest songbirds? An examination of post-breeding ecology. *Biological Conservation* **127**:477–86.

von Heland, J. and C. Folke. (2014). A social contract with the ancestors – culture and ecosystem services in southern Madagascar. *Global Environmental Change* 24:251–64.

Walker, K. F. (1979). Regulated streams in Australia: The Murray-Darling river system. In J. Ward, ed., *The Ecology of Regulated Streams*. New York, NY: Plenum Press, pp. 143–63.

Walker, P. A. (2005). Political ecology: Where Is the ecology? *Progress in Human Geography* 29:73–82.

Walkinshaw, L. H. (1983). *Kirtland's Warbler: The Natural History of an Endangered Species*. Bloomfield Hills, MI: Cranbrook Institute of Science.

Watling, L. and E. A. Norse. (1998). Disturbance of the seabed by mobile fishing gear: A comparison to forest clearcutting. *Conservation Biology* 12:1180–97.

Wayne, R. K. and S. M. Jenks. (1991). Mitochondrial DNA analysis implying extensive hybridization of the endangered red wolf, *Canis rufus*. *Nature* 351:565–83.

Webb, W. J. (1960). Forest wildlife management in Germany. *Journal of Wildlife Management* 24:147–161.

Weber, D. S., B. S. Stewart, J. C. Garza, and N. Lehman. (2000). An empirical genetic assessment of the severity of the northern elephant seal population bottleneck. *Current Biology* 10:1287–90.

Weddell, B., L. Carpenter-Boggs, and S. Higgins. (2012). *Global climate change fact sheet FS069E*. Pullman, WA: Washington State University Extension.

Wehi, P. M. (2009). Indigenous ancestral sayings contribute to modern conservation partnerships: Examples using *Phormium tenax*. *Ecological Applications* 19:267–75.

Weidensaul, S. (2021). *A World on the Wing: The Global Odyssey of Migratory Birds*. New York, NY: Norton.

Weisberg, M. (2014). Remeasuring man. *Evolution & Development* 16:166–78.

Westerling, A. L., H. G. Hidalgo, D. R. Cayan, and T. W. Swetnam. 2006. Warming and earlier spring increase in western U.S. forest wildfire activity. *Science* 313(5789):940–43.

Western, D. and R. M. Wright. (1994). *Natural Connections: Perspectives in Community-Based Conservation*. Washington, DC: Island Press.

Westman, W. H. (1977). How much are nature's services worth? *Science* 197(4307):960–964.

Whitcomb, R. F., J. F. Lynch, M. K. Klimkiewicz, *et al.* (1981). Effects of forest fragmentation on avifauna of the eastern deciduous forest. In R. L. Burgess and D. M. Sharpe, eds., *Forest Island Dynamics in Man-dominated Landscapes*. New York, NY: Springer-Verlag, pp. 125–205.

White, L. (1967). The historical roots of our ecologic crisis. *Science* 155(3767):1203–07.

White, R. (1991). *"It's Your Misfortune and None of My Own": A New History of the American West*. Norman, OK: University of Oklahoma Press.

White, R. (1996). "Are you an environmentalist or do you work for a living?": Work and nature. In W. Cronon, ed., *Uncommon Ground: Rethinking the Human Place in Nature*, 2nd edn. New York, NY: Norton, pp. 171–85.

White, R. (1999). Environmentalism and Indian peoples. In J. K. Conway, K. Keniston, and L. Marx, eds., *Earth, Air, Fire, Water: Humanistic Studies of the Environment*. Amherst, MA: University of Massachusetts Press, pp. 125–44.

Whittaker, R. H. and G. E. Likens (2012). The biosphere and man. In H. Lieth and R. H. Whittaker, eds., *Primary Productivity of the Biosphere*. New York, NY: Springer-Verlag, pp. 305–28.

Wilkinson, C. F. (2005). *Blood Struggle: The Rise of Modern Indian Nations*. New York, NY: Norton.

Williams, C. K., I. Parer, B. J. Coman, *et al.* (1995). *Managing Vertebrate Pests: Rabbits. Bureau of Resource Sciences/CSIRO Division of Wildlife and Ecology*. Canberra, Australia: Australian Government Publishing Service.

Williams, G. R. (1984). Has island biogeography any relevance to the design of biological reserves in New Zealand? *Journal of the Royal Society of New Zealand* **14**:7–10.

Williams, M. (2003). *Deforesting the Earth: From Prehistory to Global Crisis*. Chicago, IL: University of Chicago Press.

Winkler, M. G. and C. B. DeWitt. (1985). Environmental impacts of peat mining in the United States: Documentation for wetland conservation. *Environmental Conservation* **12**:317–30.

Winkler, W. G. and K. Bögel (1992). Control of rabies in wildlife. *Scientific American* **266**(6):86–92.

Wohl, E. (2013). Landscape-scale carbon storage associated with beaver dams. *Geophysical Research Letters* **40**:3631–36.

Wohl, E. (2019). *Saving the Dammed: Why We Need Beaver Modified Ecosystems*. Oxford, UK: Oxford University Press.

Wohl, E., S. N. Lane, and A. C. Wilcox. (2015). The science and practice of river restoration. *Water Resources Research* **51**:5974–97.

Woinarski, J. and A. A. Burbidge. (2016). *Petrogale penicillata*. IUCN Red List of Threatened Species 2016. http://dx.doi.org/10.2305/IUCN.UK.2016-1.RLTS.T16746A21955754.en.

Wolfe, D. W., M. D. Schwartz, A. N. Lakso, *et al.* (2005). Climate change and shifts in spring phenology of three horticultural woody perennials in northeastern USA. *International Journal of Biometeorology* **49**:303–09.

Wolfheim, J. H. (1976). The perils of primates. *Natural History* **85**(8):90–99.

Woodroffe, R., C. A. Donnelly, D. R. Cox, *et al.* (2006). Effects of culling on badger *Meles meles* spatial organization: Implications for the control of bovine tuberculosis. *Journal of Applied Ecology* **43**:1–10.

Woods Hole Oceanographic Institution. (No date). www.whoi.edu/feature/history-hydrothermal-vents/discovery/1979.html.

Woodwell, G. M. (1967). Toxic substances and ecological cycles. *Scientific American* **213**(3):24–31.

Woodwell, G. M., P. H. Rich, and C. A. S. Hall. (1973). Carbon in estuaries. In G. M. Woodwell and E. V. Pecan, eds., *Carbon and the Biosphere*. Springfield, VA: Technical Information Center, Office of Information Services, US Atomic Energy Commission, pp. 221–39.

Woody Guthrie. (No date). Grand Coulee Dam. www.woodyguthrie.org/Lyrics/Grand_Coulee_Dam.htm.

Worster, D. (2008). *A Passion for Nature: The Life of John Muir*. New York, NY: Oxford University Press.

Wright, R. G. (1992). *Wildlife Research and Management in the National Parks*. Urbana, IL: University of Illinois Press.

Wu, B., C. Zhu, D. Li, *et al.* (2006). Setting biodiversity conservation priorities in the forests of the Upper Yangtze Ecoregion based on ecoregion conservation methodology. *Biodiversity Science* **14**: 87–97.

Wynberg, R. and R. Chennells. (2009). Green diamonds of the south: An overview of the San-*Hoodia* Case. In R. Wynberg, D. Schroeder, and R. Chennells, eds., *Indigenous Peoples, Consent and Benefit Sharing: Lessons from the San-Hoodia Case*. Dordrecht: Springer Netherlands, pp. 89–124.

Yellowstone National Park Protection Act of 1872, www.nps.gov/yell/learn/management/yellowstoneprotectionact1872.htm.

Yosemite Valley Grant Act of 1864, www.nps.gov/parkhistory/online_books/anps/anps_1a.htm.

Zedler, J. B. (1988). Restoring diversity in salt marshes: Can we do it? In E. O. Wilson, ed., *Biodiversity*. Washington, DC: National Academy Press, pp. 317–25.

Zedler, J. B. (2000). Progress in wetland restoration ecology. *Trends in Ecology & Evolution* **15**:402–07.

Zhang, S. (2018). No one knows exactly what would happen if mosquitoes were to disappear. *The Atlantic*, September 24, 2018. www.theatlantic.com/science/archive/2018/09/mosquito-target-malaria/570937/.

Zimmerman, C. and K. Cuddington. (2007). Ambiguous, circular and polysemous: Students' definitions of the "balance of nature" metaphor. *Public Understanding of Science* **16**:393–406.

Zimmerman, D. R. (1975). *To Save a Bird in Peril*. New York, NY: Coward, McCann, & Geoghegan.

Index

Page numbers in *italics* indicate pages where a keyword is defined. Page numbers in **bold** indicate a keyword appears in a figure.